GRANITES

Granites

Petrology, Structure, Geological Setting, and Metallogeny

Anne Nédélec and Jean-Luc Bouchez

translated and updated with the participation of
Peter Bowden

OXFORD
UNIVERSITY PRESS

Great Clarendon Street, Oxford, OX2 6DP,
United Kingdom

Oxford University Press is a department of the University of Oxford.
It furthers the University's objective of excellence in research, scholarship,
and education by publishing worldwide. Oxford is a registered trade mark of
Oxford University Press in the UK and in certain other countries

Original Edition: *Pétrologie des granites – Structure – Cadre géologique*
© Editions Vuibert – Paris 2011
English Translation © Oxford University Press 2015

The moral rights of the authors have been asserted

Impression: 1

All rights reserved. No part of this publication may be reproduced, stored in
a retrieval system, or transmitted, in any form or by any means, without the
prior permission in writing of Oxford University Press, or as expressly permitted
by law, by licence or under terms agreed with the appropriate reprographics
rights organization. Enquiries concerning reproduction outside the scope of the
above should be sent to the Rights Department, Oxford University Press, at the
address above

You must not circulate this work in any other form
and you must impose this same condition on any acquirer

Published in the United States of America by Oxford University Press
198 Madison Avenue, New York, NY 10016, United States of America

British Library Cataloguing in Publication Data

Data available

Library of Congress Control Number: 2014942245

ISBN 978–0–19–870561–1

Printed and bound by
CPI Group (UK) Ltd, Croydon, CR0 4YY

Links to third party websites are provided by Oxford in good faith and
for information only. Oxford disclaims any responsibility for the materials
contained in any third party website referenced in this work.

Acknowledgements

Many people have contributed to the successful completion of this book.
They include:

- Christiane Cavaré, a *Paganini* in drawings and layouts, who patiently and kindly helped the authors to complete every page, never complaining about the successive changes made to diagrams and text, up to manuscript completion.
- Our laboratory (*Laboratoire des Mécanismes et Transferts en Géologie*, that became *Géosciences Environnement Toulouse*) and our university where we had all facilities to aid our teaching and research during more than twenty years.
- Our students from France, Brazil, Burkina Faso, Madagascar, Cameroon, Iran, Belgium, Portugal, Spain, UK . . . , who gave us so much and helped us to progress.
- Our colleagues, from the Petrophysics team, in particular Gérard Gleizes for his boundless energy and Philippe Olivier for his picky kindness.
- Our first reviewers, Professor Adolphe Nicolas (Montpellier University) and Professor Hervé Martin (Clermont-Ferrand University), for their expertise, enthusiasm and long-standing friendship.
- Jean-François Moyen and Patrick Thommen for their friendly support.
- Our former publisher, who confidently gave us credit for the French edition *Pétrologie des granites: structure, cadre géologique*, under the heading of Patrick de Wever, Professor at the *Museum National d'Histoire Naturelle* of Paris.
- Peter Bowden, for the highly valuable assistance brought to this book, with so many suggestions in both Science and the English language through systematic changes toward eliminating our gallicisms. Your culture and energy helped us to be accepted by a prestigious publisher, Oxford University Press.
- Sonke Adlung, our commissioning editor at OUP, for his encouragements.

Toulouse, March 2014
Anne Nédélec
Jean-Luc Bouchez

Contents

1 **What is a granite?**   1
  - 1.1 Classification of granites   1
    - *Box 1.1 Granite and granit*   4
  - 1.2 Petrography   6
    - 1.2.1 Texture   6
    - 1.2.2 Mineralogy   6
  - 1.3 Granites: rocks representative of the continental crust   8
    - 1.3.1 Structure of the continental crust   8
    - 1.3.2 Composition of the continental crust   9
  - 1.4 Granites and related rocks   10
    - 1.4.1 Concept of magmatic series   10
    - 1.4.2 Granite types and magmatic series   11
  - 1.5 Conclusions   11

2 **Origin of granitic magmas**   12
  - 2.1 Field data: migmatites and granites   12
    - *Box 2.1 From neptunism to modern magmatism*   12
  - 2.2 Experimental data   15
    - 2.2.1 Early experiments   15
    - 2.2.2 The importance of water   18
    - 2.2.3 Melting experiments with hydrous phases   20
  - 2.3 Fertility of crustal protoliths and melt compositions   25
  - 2.4 Tracing the sources with isotopes   26
    - 2.4.1 The Rb–Sr pair and the $^{87}Sr/^{86}Sr$ isotopic tracer   26
    - 2.4.2 The Sm–Nd pair and the $^{143}Nd/^{144}Nd$ isotopic tracer   28
    - 2.4.3 Coupling Sr and Nd tracers   30
    - 2.4.4 The $^{18}O/^{16}O$ ratio   31
    - 2.4.5 Hf isotopes in zircon   32
    - *Box 2.2 Oxygen isotope palaeothermometry*   33
  - 2.5 Crustal vs. mantle contributions   34
  - 2.6 Conclusions   39

3 **Segregation of granitic melts**   40
  - 3.1 Viscosity of granitic melts   40
  - 3.2 Melt behaviour during partial melting   41
    - *Box 3.1 Viscosity as a function of composition, temperature and water content*   42

3.3 Progressive partial melting experiments — 43
    3.3.1 Melt location in a static aggregate — 44
    3.3.2 Dynamic melting of a granite — 45
    3.3.3 Dynamic partial melting of an amphibolite — 46
3.4 Melt distribution in migmatites — 47
    3.4.1 Melt segregation rate — 48
    3.4.2 Role of the melting reaction — 51
    3.4.3 Role of gravity-driven compaction — 51
    3.4.4 Fracturing assisted by melt pressure — 51
    Box 3.2 Filter-press mechanism: a dynamic compaction mode — 52
3.5 Migmatites in the field — 53
    3.5.1 Layering in migmatites — 54
    3.5.2 Other migmatitic structures — 55
3.6 Conclusions — 57

4 Genesis of hybrid granitoids: mingling and mixing — 58

4.1 Field observations — 58
4.2 Influence of viscosity on magma mingling or mixing — 59
    Box 4.1 Viscosity of a suspension: Einstein's model — 60
4.3 Characteristics of hybrid rocks — 62
    4.3.1 Mineralogy — 62
    4.3.2 Major and trace element chemistry — 63
    4.3.3 Isotopic signatures — 65
4.4 Processes of magma mingling and mixing — 66
    4.4.1 Element interdiffusion in silicate liquids — 66
    4.4.2 Transfer of granitic liquid into mafic magma enclaves — 67
    4.4.3 Dispersion of mafic magma into a granitic magma host — 67
4.5 Where do hybrid magmas form? — 69
4.6 Conclusions — 72

5 Transport of granitic magma — 73

5.1 Diapirism — 74
5.2 Transport by dykes — 77
    5.2.1 Analysis and orders of magnitude — 77
    5.2.2 Transfer from the source — 79
    5.2.3 Role of magma pressure — 79
    5.2.4 Fracturing the crust — 80
    5.2.5 Magma ascent rate — 83
5.3 Field data — 84
5.4 Conclusions — 86

6 Emplacement and shape of granite plutons — 87

6.1 Cessation of magma ascent — 87
    Box 6.1 Emplacement in an extensional context: a simple exercise of mechanics — 88

	6.2	Meeting dilatant sites	90
		6.2.1 Fracturing in extensional environments	90
		6.2.2 Fracturing in compressive environments	94
	6.3	Granite emplacement in the ductile crust	98
		6.3.1 Sills and sheet-like granites	98
		6.3.2 Migmatitic domes	100
	6.4	Emplacement depth	101
	6.5	Three-dimensional shapes of plutons	102
		6.5.1 Bouguer anomaly map	102
		6.5.2 Residual anomaly and density contrast	104
		6.5.3 Modelling pluton depth and floor shape	104
		6.5.4 Pluton shapes	105
	6.6	Passive versus forced emplacement	107
	6.7	Conclusions	108
7	**Thermomechanical aspects in the country rocks around granite plutons**		**109**
	7.1	Conductive heat transfer	109
		Box 7.1 Conductive heat transfer: the virtue of non-dimensional variables	110
	7.2	Convective heat transfer	113
	7.3	Diachronic metamorphism and rheological changes at the contact	114
	7.4	Parageneses of contact metamorphism	116
		7.4.1 Pelitic rocks	116
		7.4.2 Carbonate rocks	118
	7.5	Thermomechanical aspects at the crustal scale	120
		7.5.1 Regional contact metamorphism	120
		7.5.2 Crustal decoupling and HT-LP metamorphism	124
	7.6	Conclusions	126
8	**Crystallization of granitic magmas**		**127**
	8.1	General considerations on nucleation and crystal growth	127
		8.1.1 Nucleation	127
		8.1.2 Crystal growth	128
		8.1.3 Crystal size distributions (CSDs)	131
		8.1.4 Evolution of crystallinity with time and its rheological consequences	134
	8.2	Order of crystallization of minerals	137
		8.2.1 Textural observations	137
		8.2.2 Influence of magma composition on the order of crystallization	138
	8.3	Fractional crystallization and magmatic differentiation	141
		8.3.1 History of magmatic differentiation	141
		8.3.2 Evidence of fractional crystallization	142
		Box 8.1 Compatible and incompatible elements	145

		8.3.3 Mechanisms of fractional crystallization	148
		8.3.4 Assimilation and fractional crystallization	150
	8.4	Late-magmatic processes	152
		8.4.1 Composition of residual granitic melts	152
		8.4.2 Magma water saturation and its consequences	153
		8.4.3 Pegmatites and aplites	155
	8.5	Subsolidus mineral transformations	156
		8.5.1 Hydrothermal alteration	156
		8.5.2 Exsolution	157
	8.6	Conclusions	158

9 Microstructures and fabrics of granites — 159

- 9.1 Granite microstructures — 159
  - *Box 9.1 A tale of dislocation* — 161
  - 9.1.1 Magmatic microstructures — 162
  - *Box 9.2 From microstructures to nanostructures* — 163
  - 9.1.2 'Submagmatic' microfractures — 165
  - 9.1.3 Solid-state deformation microstructures — 166
- 9.2 Fabrics in granites — 169
  - 9.2.1 Foliation and lineation — 169
  - 9.2.2 Origin of magmatic fabrics — 171
  - *Box 9.3 Pure shear? Simple shear?* — 172
  - 9.2.3 The non-cyclicity of magmatic fabrics — 174
  - 9.2.4 Magmatic lineation and finite extension direction — 176
- 9.3 Fabric of the Brâme–St Sylvestre–St Goussaud complex — 178
- 9.4 Fabrics and digital imagery — 179
  - 9.4.1 The intercept method — 180
  - 9.4.2 2D wavelet analysis technique — 181
- 9.5 Conclusions — 184

10 Magnetic fabrics in granites — 185

- 10.1 Magnetic properties of materials — 185
  - 10.1.1 Basic concepts — 185
  - 10.1.2 Types of magnetic behaviour — 186
  - *Box 10.1 The ferromagnetic behaviour of magnetite* — 188
- 10.2 Magnetic susceptibility of granitic rocks — 190
  - 10.2.1 Paramagnetic granite and ferromagnetic granite — 191
  - 10.2.2 The AMS ellipsoid — 194
- 10.3 Magnetic fabrics in granites — 195
  - 10.3.1 Fabrics in paramagnetic granites — 195
  - 10.3.2 Fabrics in ferromagnetic granites — 196
  - 10.3.3 Fabrics with mixed magnetic mineralogy — 197
  - 10.3.4 AMS in practice — 199
  - 10.3.5 AAR fabrics — 201

			Contents	xi

	10.4	The conspicuous structural homogeneity of granites: examples and implications	203
		10.4.1 Sidobre granite (Massif Central, France)	204
		10.4.2 Tesnou granitic complex (Hoggar, Algeria)	206
		10.4.3 Mono Creek granite pluton (California)	206
		10.4.4 Cauterets–Panticosa granite complex (Pyrenees)	209
		10.4.5 Bassiès pluton (Pyrenees)	210
	10.5	Conclusions	211
11	**Zoning in granite plutons**		212
	11.1	Examples of zoned plutons	212
		11.1.1 Concentric zoning	212
		11.1.2 Vertical zoning	212
		11.1.3 Complex zoning	216
		11.1.4 Rhythmic zoning or layering	218
		Box 11.1 *What is a magma chamber?*	218
	11.2	Origin of zoning	219
		11.2.1 In situ magmatic differentiation	219
		11.2.2 Successive magma intrusions	225
	11.3	Conclusions	226
12	**Granites and plate tectonics**		227
	12.1	Granites and oceanic accretion	227
	12.2	Granites and hot spots or continental rifting	227
		12.2.1 The Bushveld granite	229
		12.2.2 Jurassic granites of Nigeria	230
	12.3	Granites and oceanic subduction	232
		12.3.1 Thermotectonic setting	232
		12.3.2 Island arc granitoids	235
		12.3.3 Granitoids in active continental margins	237
	12.4	Granites and continental collision	238
		12.4.1 Thermotectonic setting	238
		12.4.2 Himalayan leucogranites	241
		12.4.3 Hercynian granites of western Europe	241
	12.5	Granites and continental extension	246
		12.5.1 Thermotectonic setting	246
		12.5.2 Granites related to Hercynian late-orogenic extension	248
		12.5.3 Granites related to the Oslo rift in Norway	249
		Box 12.1 *What are the links between geochemistry and tectonics?*	250
	12.6	Conclusions	251
13	**Precambrian granitic rocks**		252
	13.1	Age determination of granitic rocks	252

		13.1.1	Crystallization ages	252
		13.1.2	Cooling ages	253
		Box 13.1 The mystery of the Hadean crust		255
	13.2	Archaean granitoids: tonalites, trondhjemites and granodiorites (TTGs)		256
		13.2.1	Lithology and structure	256
		13.2.2	Geochemistry	260
		13.2.3	Petrogenesis: geochemical and experimental constraints	260
		13.2.4	Evidence from the Barberton migmatitic amphibolites	262
		13.2.5	Geodynamic context	264
	13.3	Late-Archaean granitoids		267
		13.3.1	Temporal evolution of Archaean TTGs	267
		13.3.2	Potassic granitoids of continental origin	268
		13.3.3	Sanukitoids: granitoids of the Archaean–Proterozoic transition	268
	13.4	Proterozoic granitoids		269
		13.4.1	Eburnian/Transamazonian granitoids (*ca.* 2 Ga)	269
		13.4.2	AMCG (anorthosite–mangerite–charnockite–granite) associations	271
	13.5	Impact granites		276
	13.6	Conclusions		279
14	**Granite metallogeny**			**280**
	14.1	Development of ore mineral concentrations		280
		14.1.1	Importance of magmatic source and granite type	281
		14.1.2	Orthomagmatic processes	284
		14.1.3	The pegmatitic stage	284
		Box 14.1 The periodic classification of chemical elements		285
		14.1.4	Concentration processes in the aqueous fluid phase	287
		14.1.5	Transfer of metals and deposition of ore minerals	290
	14.2	Examples of granite-related mineral deposits		293
		14.2.1	Tin and kaolin in the Cornubian batholith, Cornwall	293
		14.2.2	Uranium from Limousin province (Massif Central, France)	295
		Box 14.2 Kaolin and china porcelain		297
		14.2.3	Tantalum from Tanco (Canada)	300
		14.2.4	Porphyry copper deposits in Chile	301
		14.2.5	Tin and fluorine in Bushveld granites	304
		14.2.6	Uranium at Olympic Dam, Australia	305
	14.3	Conclusions		307
Glossary				**309**
References				**313**
Index				**333**

1
What is a granite?

A granite is a massive crystalline rock exposed at the Earth's surface by weathering; it derives from the cooling of granitic magma at depth a long time ago. It has millimetre- to centimetre-sized grains, which are usually white to grey in colour reflecting its chemical composition, i.e. rich in silica and poor in iron. Its mineral composition is principally made of quartz, alkali feldspar, plagioclase and sometimes white mica for light-coloured minerals, as well as biotite and occasionally amphibole, clinopyroxene or orthopyroxene for the dark-coloured minerals. Other minerals occur in minor amounts, hence are called 'accessories'. These may include tourmaline, garnet, apatite, zircon, monazite, ilmenite, magnetite, topaz and occasionally rare earths and metal ore minerals.

1.1 Classification of granites

The QAP Streckeisen modal diagram (Fig. 1.1) provides a simple nomenclature based on the percentage of light-coloured minerals. Granites in their broadest sense (*granitoïdes* in French) are composed of 20–60% quartz (Q) as well as variable proportions of alkali feldspar (A) and plagioclase (P). In the QAP diagram, the A end-member naturally refers to potassic feldspar (orthoclase or microcline) but also to sodic feldspar (perthitic alkali feldspar and albite). The relative proportions of A and P help to define four distinct granite domains. From left to right (A to P in Fig. 1.1) are: (1) alkali-feldspar granite, in which K-Na feldspar is dominant; (2) granite *sensu stricto*; (3) granodiorite; and finally (4) tonalite in which plagioclase is dominant. The largest granite domain may be subdivided into syenogranite (the alkali side of the diagram) and monzogranite. From A to P the rocks darken in colour, reflecting the increasing content in dark minerals (biotite, amphibole, pyroxene) along with increasing concentrations of Fe + Mg (Fig. 1.2). At the same time, plagioclase becomes more calcic, namely closer to the anorthite end-member (as its Ca/Ca + Na ratio increases).

This classification, based on quartz–alkali feldspar and plagioclase proportions, ignores the petrographic and geochemical subtleties which will be revealed later. To use this classification we must know the modal composition of the rock. It can be precisely determined from thin sections observed under the petrological microscope using

2 What is a granite?

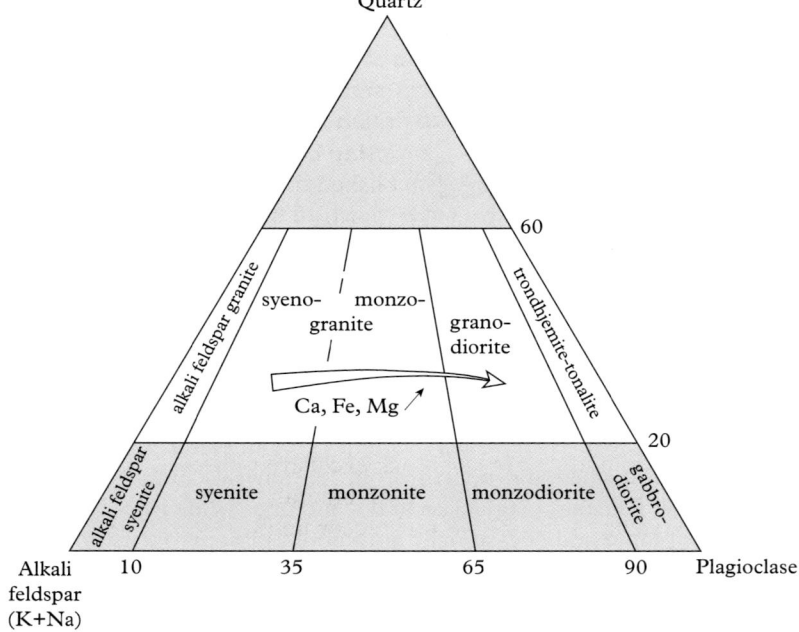

Figure 1.1 *Modal classification of plutonic rocks, after Streckeisen (1976). In white: the granitoid or granite domain,* sensu lato.

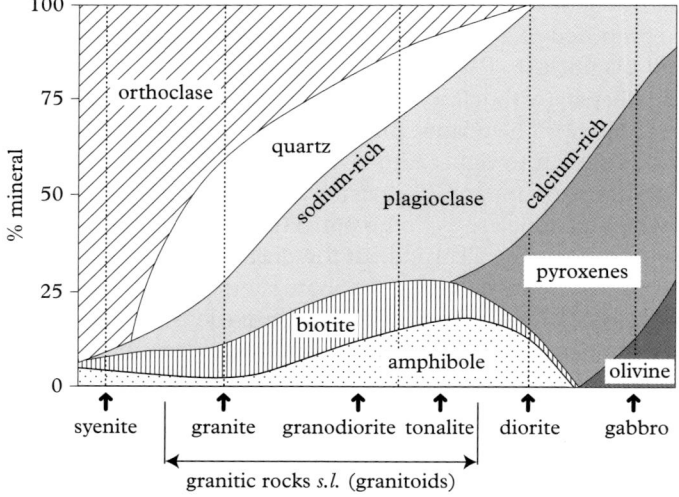

Figure 1.2 *Mineralogy of granitic rocks.*

polarized light. Traditionally, the mode is performed using a point-counting device by attributing Q, A or P at each node of a square grid. A modern way of determining the mode consists of using an analysis technique of digitized thin-section images (for example by scanning) with the help of computer software. However, since in many laboratories a chemical analysis may be easier to obtain (at least for the major elements) than modal estimations of thin sections, the relationship between the mode of a granite and its percentage of oxides (weight %) can be established through calculating the virtual mineralogical composition of the rock, called a 'norm'. The corresponding normative data may then be plotted into the Streckeisen and Lemaître normative diagram (1979, see Fig. 1.3), which presents the same subdivisions as in the Streckeisen's modal QAP plot.

In order to better understand the geochemical diversity of granites, the simplest and most fruitful classification is based on the saturation in alumina, through calculation of the A/CNK (Al_2O_3/CaO + Na_2O + K_2O) and A/NK (Al_2O_3/Na_2O + K_2O) molar ratios from weight percentages in oxides obtained by chemical analyses (Shand, 1943). Be careful: the A of the A/CNK is different from the A of the QAP diagram. A/CNK equals 1 for the haplogranites, a (virtual) rock exclusively composed of quartz and feldspar (orthoclase and/or albite). Introduction of minerals having different Al contents will modify the A/CNK value of the whole rock accordingly. For example, the addition of hornblende (A/CNK = 0.7) will decrease this ratio (A/CNK < 1), while the introduction of garnet (A/CNK = ∞) or other Al-rich minerals (for example, muscovite) will increase it (A/CNK > 1).

On such a basis (Fig. 1.4), peraluminous granites are characterized by A/CNK > 1. For the metaluminous granites (A/CNK < 1), an additional distinction is made, according to the A/NK value, between metaluminous granites *sensu stricto*, also called calc-alkaline granites (A/NK > 1), and peralkaline granites (A/NK < 1). The latter approach provides no more than a classification of granites as a function of their principal minerals,

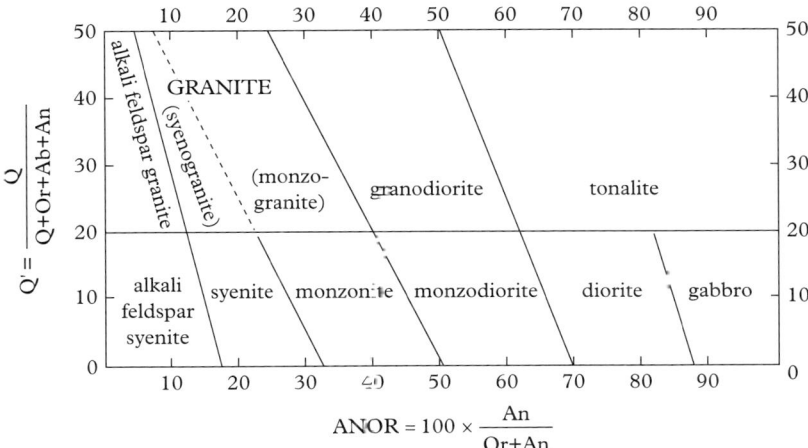

Figure 1.3 *Normative classification after Streckeisen and Le Maître (1979).*

4 *What is a granite?*

Box 1.1 Granite and granit

Granite has a Latin origin, *granum* for grain. A granite is a magmatic rock which contains mineral grains. Although all geologists spell the word 'granite' with an 'e' at the end, a common practice in the French literature is to write *granit* (without an 'e'). For example, Victor Hugo wrote in *Contemplations*: 'the Earth is of granit, rivers of marble …'. The word *granit* has been used since the seventeenth century. Buffon, in his first volume of *Histoire Naturelle des Minéraux* (1783), noticed that 'at present granits cover most parts of the globe', a fact difficult to argue against. In France, *granit* is improperly used by rock dealers who propose granite counter tops made of 'black granit' (*sic*), actually of gabbro. In fact marble workers call 'granit' any rock surface whose grains are distinct when polished and 'marble' a rock whose grains are indiscernible. The word 'marble' is restricted to metamorphosed limestones by geologists.

---

or major chemical element composition. No genetic implication concerning the source (origin), or the extraction mechanism, can be derived from these classifications, even though, for example, granites derived from the melting of pelites are well-known to be peraluminous (see Chapter 2).

A genetic meaning has been given by the letters S and I first introduced by Chappell and White (1974) to signify the granitic source or protolith. S stands for 'supracrustal', indicating partial melting of metasediments, and I for 'infracrustal' or 'igneous', suggesting partial melting of an igneous protolith. Chemically, S-type granites are approximately equivalent to peraluminous granites of Shand, and I-type granites are equivalent to metaluminous and Al-poor granites, A/CNK = 1.1 being the limit between these two groups (Fig. 1.4). This alphabetical classification was further complemented by Loiselle and Wones (1979) who defined another group of granites, known as A-type, namely 'alkaline' or 'anorogenic'. The A-type granite group is larger than the peralkaline group

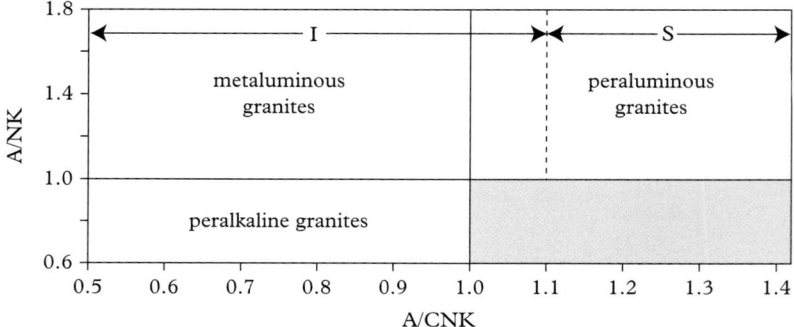

Figure 1.4 *Shand's (1943) classification with superimposed I- and S-type domains.*

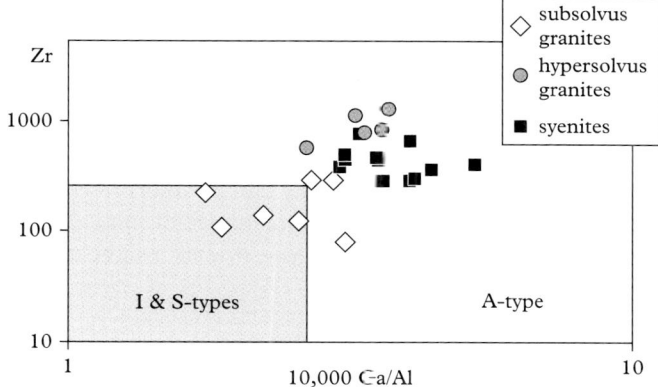

Figure 1.5 *Classification diagram of Whalen et al. (1987) applied to the stratoid granites of Madagascar (after Nédélec et al., 1995).*

of Shand. A-type granites are always rich in silica, have high Fe/Mg ratios as well as high contents of high field strength elements (HFSE), i.e. elements having high charge/ionic radius ratios, such as Zr, Ce, Nb, Hf, Ta, etc. This is why Whalen et al. (1987) use the contents of Zr, Ce and Nb (normalized to Ga/Al) in order to distinguish the A-type granites from the S- and I-type granites (Fig. 1.5). Table 1.1 gives the average chemical compositions of I-, S- and A-type granites.

Table 1.1 *Representative compositions of the main granite types, after Cox et al. (1979) and Whalen et al. (1987).*

Type	I			S	A
Oxides (wt %)	Tonalite	Granodiorite	Granite I	Granite S	Granite A
SiO_2	61.52	66.09	67.89	69.08	73.81
TiO_2	0.73	0.54	0.45	0.55	0.26
Al_2O_3	16.48	15.73	14.49	14.30	12.40
FeO (total Fe)	5.47	3.97	3.72	3.89	2.43
MnO	0.08	0.08	0.08	0.06	0.06
MgO	2.80	1.74	1.75	1.82	0.20
CaO	5.42	3.83	3.78	2.49	0.75
Na_2O	3.63	3.75	2.95	2.20	4.07
K_2O	2.07	2.73	3.05	3.63	4.65
P_2O_5	0.25	0.18	0.11	0.13	0.04

Whoever would like to unravel the multiple classifications proposed so far will find the well-documented synthesis by Barbarin (1999) a useful reference. Barbarin does not refrain from proposing another classification aimed at best integrating this multi-parameter problem.

In this chapter and elsewhere we shall refer to the I, S, A granite classification, which has the advantage of simplicity. However, we shall try to avoid any systematic deductions about the nature of the source which essentially can be interpreted from isotopic data. Similarly, any deductions concerning the geodynamic environment, which should arise mainly from discussions of structural and tectonic studies, will be avoided initially.

1.2 Petrography

1.2.1 Texture

The texture (sometimes also called 'structure') describes the shapes and sizes of the grains that compose the granite as well as the relationships between them. Such a description provides useful information about the order of crystallization and, eventually, about the conditions of crystallization of the constituent minerals and their subsequent history such as deformation or disequilibrium reactions. If a texture is rather easy to describe, its interpretation with confidence is rather difficult. In contrast to most volcanic rocks, granitic rocks are entirely crystallized, and the grains are always perceptible to the naked eye. The grain size is characterized as follows: fine-grained (< 1 mm), also qualified as 'aplitic' if the rock is light-coloured, hence dominated by quartz and feldspars; medium-grained (1–5 mm); coarse-grained (5–20 mm) and pegmatitic (> 20 mm). Then, one can examine whether the grains have all the same size (equigranular texture) or not (heterogranular texture). In the latter case, the texture is seriate if a full range of sizes is present, or porphyritic if a given mineral species, with generally idiomorphic to sub-idiomorphic shapes, largely surpasses the size of the matrix minerals.

The relationships between grains are examined through the eventual shape or lattice preferred orientation (also called fabric, see Chapter 9). Grain intergrowths are also examined, such as graphic texture in which the sharp-edged interpenetration of quartz and alkali feldspar resembles Sumerian cuneiform writing. In granophyric texture, homogeneous phenocrysts of quartz and feldspars can be observed inside the graphic assemblage. These phenocrysts were more-or-less isolated in the melt at the beginning of magma crystallization, and this was followed at the end of crystallization by rapid co-precipitation of quartz and feldspar. Less common textures are also examined, such as overgrowths of plagioclase around alkali feldspars, observed in rapakivi granites (*rapakivi* is a Finish word, because rapakivi granites are common in southern Finland: see Chapter 13).

1.2.2 Mineralogy

A granite commonly contains quartz, feldspars and biotite. Such a simple mineralogy does not help in identifying to which category (I, S or A) the granite belongs.

The presence of additional aluminous mineral species should be mentioned since they participate in defining the chemical signature of the rock. For example, two-mica granites, containing biotite and muscovite, or granites containing only muscovite, and/or cordierite, garnet, tourmaline or aluminosilicates (andalusite or sillimanite), belong to the S-type granites (Clarke, 1981). By contrast, the I-type granites are characterized by the presence of calcium-bearing ferromagnesian minerals such as hornblende, clinopyroxene (augite or diopside) or epidote (provided it is magmatic, i.e. grown from a melt). Finally, sodic ferromagnesian minerals, such as Na-bearing clinopyroxene (aegirine, or aegirine-augite) or Na-bearing amphibole (arfvedsonite or riebeckite), are typical in A-types, and more precisely in peralkaline A-type granites. If these sodic ferromagnesian minerals have crystallized after quartz and feldspars, and hence occupy interstitial sites with xenomorphic shapes, an agpaitic texture can be concluded.

A leucocratic granite, light- to very light-coloured due to its paucity in ferromagnesian minerals, is also called a 'leucogranite'. It is rich in silica and alkaline minerals, hence is highly differentiated, but one cannot conclude that its composition is constant in every element. I-, A- and S-type leucogranites do exist, contrary to the common thought that a leucogranite always belongs to the S-type. I- and S-type leucogranites have convergent characters which make them difficult to distinguish. Presence or absence of primary (magmatic) muscovite is a distinctive feature in identifying the respective S- or I-type nature, as well as the contents in some minor and trace elements (Fig. 1.6).

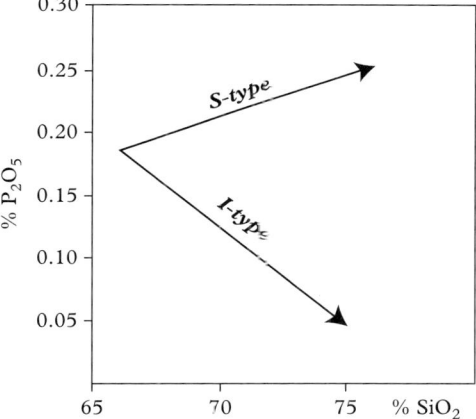

Figure 1.6 *Discrimination between S- and I-type granites as a function of their content in phosphorus. Apatite (calcium phosphate), a common accessory mineral, is more soluble in peraluminous magmas (S-type) than in metaluminous magmas (I-type). The first ones become richer in P with increasing differentiation indicating that saturation in apatite is not reached, hence apatite does not crystallize (Chappell, 1999).*

8 What is a granite?

Iron oxides (ilmenite and magnetite) are accessories revealing the oxygen fugacity that characterizes the magma. It has been recognized for a long time that two categories of granites exist: one the ilmenite-bearing and the other the magnetite-bearing granites (see Fig. 12.5) with few representatives in between. The S-types always belong to the first category, reflecting their formation from pelitic metasediments more-or-less associated with organic matter, thus explaining the reduced conditions during partial melting. I- and A-type granites may belong to either categories.

1.3 Granites: rocks representative of the continental crust

1.3.1 Structure of the continental crust

Except in a few places, such as in the Ivrea and Strona-Ceneri zones of the Italian Alps (Fig. 1.7a), it is not possible to observe a section of the whole continental crust. Due to the Alpine orogeny, the continental crust has been tilted by up to about 90° and offers

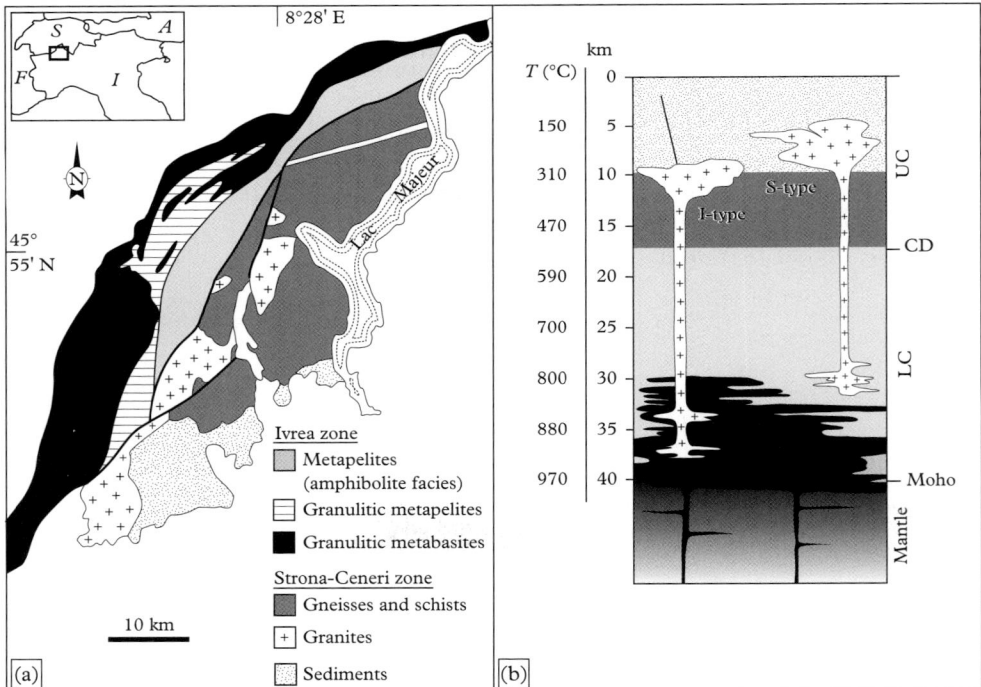

Figure 1.7 *(a) The Ivrea zone, a model of lower continental crust metamorphosed under amphibolite to granulite facies due to voluminous gabbroic intrusions; associated with the Strona-Cineri zone, the Ivrea zone offers a complete section of the continental crust. (b) Interpretative vertical section showing that, whatever their sources, the granitic intrusions occupy mostly the upper part of the crust. After Wedepohl (1991).*

a complete section typical of the crust (Fig. 1.7b) down to the Moho that is the base of the crust. In this section, some granitic rocks, although derived from anatexis (i.e. partial melting) of the lower part of the crust, are observed to have crystallized in the upper crust. The lower crust is highly metamorphic (it has reached granulite facies conditions) and is rich in mafic components derived from partial melting of the mantle. This structure was formed as a consequence of extensional tectonics characterized by both crustal and lithospheric thinning at the end of the Hercynian (also called 'Variscan') orogeny (see Chapter 12), but may not be typical of continental crust worldwide.

Wedepohl (1995) used the European GeoTraverse (EGT seismic profile) to build a representative section of the European continental crust. The 3000 km-long north-south profile, running from the Baltic Shield to the Alpine chain, cuts across about 60% of Precambrian basement rocks and 40% of more recent formations (Hercynian belt, Alpine belt and respective associated basins). Integration of the data over the profile together with the addition of geological information, help to locate granitic rocks within the upper part of the crust where seismic velocities range between 6 and 6.5 km/s (Fig. 1.8). These granitic rocks represent in volume 50% of the upper crust and 26.5% of the whole continental crust (Fig. 1.9).

1.3.2 Composition of the continental crust

The approach of Wedepohl (1995), essentially based on seismic profiles, leads one to define a tonalitic composition for the bulk continental crust (62 wt % of silica). Using a different approach based on the composition of sediments from North America, Taylor and McLellan (1985) concluded that the continental crust has an average dioritic composition (57 wt % of silica). Both approaches, however, define a granodioritic composition for the middle part of the continental crust.

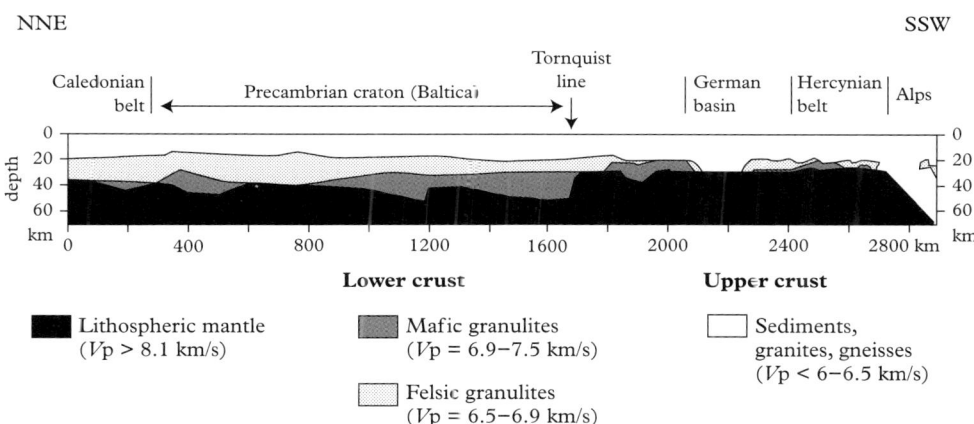

Figure 1.8 *Interpretative seismic section of Middle Europe (European GeoTraverse—EGT profile). After Wedepohl (1995).*

10 What is a granite?

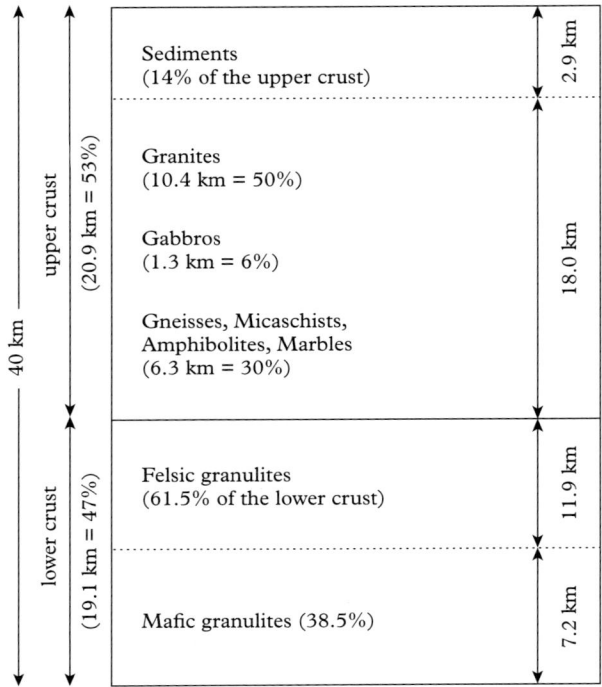

Figure 1.9 *Average composition of the European crust, reconstructed after the EGT seismic profile. After Wedepohl (1995).*

1.4 Granites and related rocks

1.4.1 Concept of magmatic series

As for volcanic rocks, plutonic rocks often form associations that appear as contemporaneous and cogenetic in the field. A few simple diagrams help to identify the main types of associations, also called 'magmatic series'. They are interpreted as resulting from the evolution of a parental magma subjected to increasing degrees of magmatic differentiation. The TAS diagram (total alkalis versus silica: $Na_2O + K_2O = f(SiO_2)$) devised by MacDonald and Katsura (1964) has been used to distinguish between alkaline and subalkaline series, the whole series being then subdivided into three series: alkaline, calc-alkaline and tholeiitic (Fig. 1.10). The calc-alkaline series are themselves subdivided according to the Na_2O/K_2O ratio. In the diagram representing K_2O as a function of SiO_2 (see Fig. 12.6) one can make the distinction between calc-alkaline series poor in potassium (in fact of tholeiitic character), from series with intermediate contents in potassium defining the classical calc-alkaline series, and then from high-K calc-alkaline series.

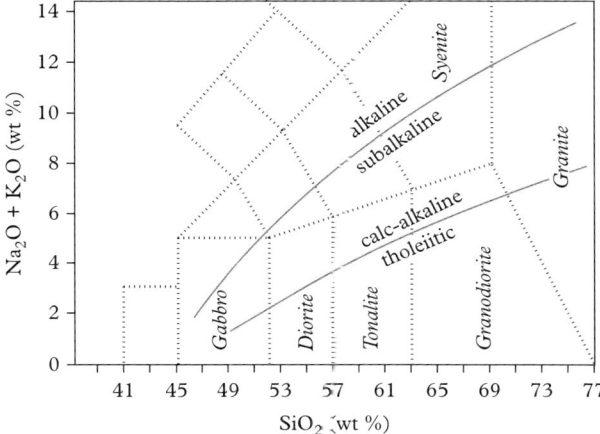

Figure 1.10 *TAS diagram: total alkalis ($Na_2O + K_2O$) versus silica (SiO_2) in wt %.*

1.4.2 Granite types and magmatic series

The most evolved plutons (i.e. the richest in silica) belonging to the principal magmatic series, correspond mainly to granites *sensu stricto* (see Fig. 12.19). Potassium-poor calc-alcaline series, or tholeiitic series, constitute an exception since their magmatic differentiation leads to trondhjemites or plagiogranites, which are equivalent to leucocratic tonalites, as rich in silica as granites but with a high Na/K ratio preventing K-feldspar from crystallizing. The majority of I-type granites are classified as monzo- or syeno-granites resulting from the differentiation of either classical calc-alkaline series or high K calc-alkaline series (sometimes called monzonitic series). In the first case, I-type granites are associated with tonalites and granodiorites, while in the second case they are associated with monzonites. A-type granites, either syenogranites or alkali feldspar granites, are often associated with syenites. Finally, S-type granites form a distinct group; they do not correspond to a real magmatic series although such granites span compositions from granodiorite to granite (see Fig. 12.19).

1.5 Conclusions

The very first observations collected in the field about the mineralogy and the nature of the rocks associated with granite provide precious information about the type of granitic rock under study. A structural study must be undertaken before considering geochemistry. Only then will it be time to propose a model of magma genesis, the origin of granite and to define the tectonic context of its emplacement into the crust.

2
Origin of granitic magmas

The origin of granites has been a matter of controversy for more than a century. The main steps of this debate are presented briefly in Box 2.1. First, let's go to into the field!

2.1 Field data: migmatites and granites

The word 'migmatite' (from the Greek *migma*, mixture) was coined by Sederholm (1907), a Finnish geologist who first used it in 1907 to describe ice-polished outcrops along the coast of the Gulf of Finland. A migmatite is a rock that was partially molten and that is presently made of a gneissic part and a granitic part.

In the field, migmatites appear as the end of prograde metamorphism and often make the transition towards a so-called anatectic granite ('anatexis' is a word created by Sederholm to define the process of regional melting).

> **Box 2.1** From neptunism to modern magmatism
>
> Towards the end of the eighteenth century, Abraham Gottlob Werner (1750–1817) founded and popularized the idea of Neptunism, meaning that the geological strata were deposited or precipitated at the bottom of a primordial ocean covering all the surface of the Earth. The same century, this theory was contradicted by James Hutton, the father of modern geology. Two points were debated: are granites magmatic rocks and do they have any link with basaltic rocks? Hutton (1794) asserted the magmatic origin of granite after his own observations in Scotland. Lyell (1830), another Scotsman, continued to popularize Hutton's ideas that granite originates at great depths by melting due to high temperature and pressure. These conditions are responsible for the formation of both metamorphic and granitic rocks, together called 'plutonic' rocks (an adjective now used only for magmatic rocks). At the same time, French authors, influenced by their observations in the French Massif central, emphasized the differences between basaltic and granitic rocks. Durocher (1857) suggested the existence of two different magmas: one granitic, formed at relatively shallower levels, and the other basaltic of deeper origin. Unfortunately, this proposal received little attention. Until 1950, most
>
> *continued*

Box 2.1 *continued*

people opposed volcanic rocks, as being undoubtedly magmatic, mainly mafic in composition and formed in non-orogenic settings, to plutonic rocks, mainly granitic (or granodioritic) and formed in orogenic settings by poorly understood and still debated processes. Read (1948) summarized this controversy in a successful book. At that time, supporters of the magmatic origin of granite were still struggling with proponents of solid-state origin by a transformation called 'granitization'. In the magmatist faction, two opinions were discussed: is the granitic magma primary or secondary, i.e. derived from the fractional crystallization of basaltic magma? A few years later, the results of the first melting experiments by Tuttle and Bowen (1958) quietly terminated the idea of 'granitization' and crustal melting was then accepted as the main origin for the formation of granites.

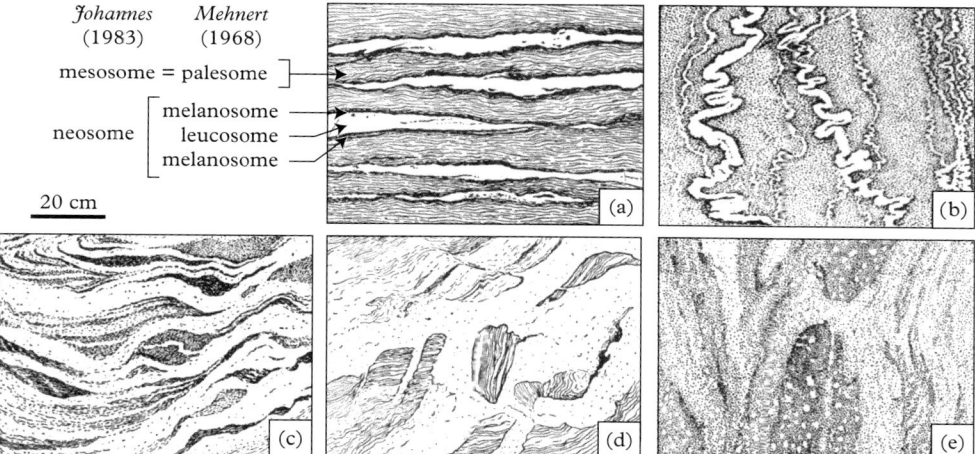

Figure 2.1 *Different types of migmatites after Mehnert (1968). (a) Layered or stromatic metatexite (from the Greek stroma, carpet). (b) Folded or ptygmatitic metatexite. (c) Diatexite with schlieren (elongate and frayed concentration of dark minerals). (d) Rafts of gneisses in diatexite. (e) Nebulitic diatexite.*

Mehnert (1968) proposes a detailed classification of migmatites, using specific terminology corresponding to the increased degree of partial melting: metatexites, followed by diatexites. Other specific terminology refers to characteristic structures (Fig. 2.1). Mehnert's classical description of migmatites calls 'leucosome' (from the Greek *leucos*: white) the quartzo-feldspathic domains, that resemble granite, and paleosome the other gneissic and foliated domains. The leucosome always displays a larger grain size than the paleosome (see Fig. 2.2b). This feature, together with an irregular shape and a centimetre-scale rather than a millimetre-scale thickness, helps to distinguish leucosomes due to *in situ* partial melting from quartz-feldspathic layers due to metamorphic solid-state mineral segregation in gneissic rocks.

14 Origin of granitic magmas

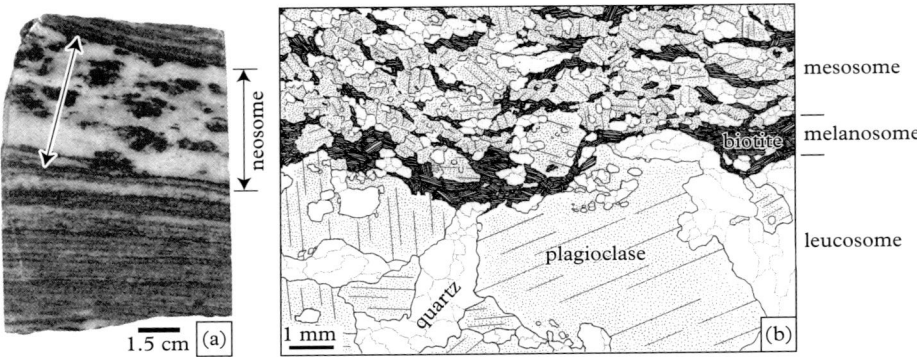

Figure 2.2 *Migmatite structures. (a) Migmatite from Eseka (Cameroon) with a spotted neosome characterized by new hornblende megacrysts, after Nédélec et al. (1993). (b) Detail of another migmatite after Mehnert (1968).*

In layered metatexites, the leucosomes may be edged by dark melanosomes (from the Greek *melanos*: black). A melanosome is often biotite-rich and called 'restite', referring to its refractory (not molten) nature. The word 'restite' may not be appropriate for two reasons. Firstly, all light minerals in the gneiss may not have been molten, hence may be also part of the actual restite; secondly, some of the mafic mineral components in the melanosome may be newly formed minerals, crystallized in equilibrium with the melt (this is the case with incongruent melting). The melanosome may not be easy to delineate; it can form mineral aggregates more or less scattered in the leucosome (Fig. 2.2a). This is the reason why Johannes (1983) used the word 'neosome' to refer to both the leucosome and the melanosome.

The use of 'mesosome' instead of 'paleosome' is also recommended by Johannes (1983) to avoid any genetic implication. Indeed, the observed mesosome may be different in composition from the melted gneissic protolith, i.e. the paleosome *sensu stricto*. It is possible that the observed mesosome did not melt, because its composition was not appropriate! Another hypothesis considers the local introduction of a hydrous fluid, that might have favoured local partial melting, leaving the other gneissic layers unchanged (Weber and Barbey, 1986). Finally, the so-called 'injection migmatites' result from the introduction of an elsewhere-formed granitic melt into the gneiss.

Melting reactions and their pressure–temperature (P-T) conditions can be determined in migmatites deriving from *in situ* partial melting. In the Eseka migmatites (Fig. 2.2b), the incongruent melting reaction:

$$\text{biotite} + \text{plagioclase} + \text{quartz} \, (\pm H_2O) = \text{granodioritic melt} + \text{hornblende} \pm \text{garnet} \pm \text{clinopyroxene} \tag{2.1}$$

occurred at $T \geq 750$ °C and $P = 900$ MPa. The protolith is an Archaean orthogneiss of tonalitic to trondhjemitic composition. Such migmatites portray the formation of an I-type granitic melt (where I indeed indicates an igneous protolith).

Migmatites offer the opportunity of a field test for melting experiments. However, their leucosomes are not proper granites. Indeed, granites derive from higher degrees of melting, that favoured melt segregation and transfer to shallower crustal levels. The migmatites often do not represent a thermodynamically closed system: a small percentage of melt may have migrated (hence the origin of the word *mobilisat* in French), leaving a leucosome, whose composition may not truly represent the initial melt (Cuney and Barbey, 1982).

2.2 Experimental data

2.2.1 Early experiments

The first experiments were always performed in water-saturated conditions (P_{H_2O} = total P). They used either quartz–feldspar mixtures or natural gneissic samples.

Tuttle and Bowen (1958) showed that melting of a ternary mixture of quartz, albite and orthoclase, a sort of simple granitic mixture (the so-called 'haplogranitic' system, from the Greek *haplos*: simple), starts at temperatures consistent with conditions prevailing in the lower continental crust. For instance, at a water-saturated pressure (P_{H_2O}) of 200 MPa, melting begins at *ca.* 700 °C (Fig. 2.3). The composition of this first melt always plots on the cotectic line that connects the binary eutectics E1 (for a mixture of quartz and orthoclase) and E2 (quartz–albite mixture). The position of this first melt on the cotectic line depends on the starting proportions of the quartz–albite–orthoclase (Q–Ab–Or) mixture. The lowest point of the cotectic line (M, the cotectic minimum) corresponds to the composition of the most easily molten mixture. At 200 MPa, the composition of the minimum melt is: Q:Ab:Or = 35:40:25, and the melting temperature (T_M) is 680 °C. Notice that this composition only refers to the quartz–feldspar proportions if water is also present.

At higher pressures (P_{H_2O} > 300 MPa), a ternary eutectic E3 replaces the M minimum on the cotectic line. This eutectic corresponds to the composition of the first melt, whatever the proportions of the starting Q–Ab–Or mixture. By contrast, the volume of this first melt depends on the composition of the starting mixture. A mixture with eutectic composition would melt completely at E3 and, in this case, the solidus and the liquidus would be identical. At higher pressures, E3 moves towards more Ab-rich compositions and its temperature is only 635 °C at 1 GPa (Fig. 2.4).

If albite is replaced by a slightly calcic plagioclase in the starting mixture, the cotectic minimum M is replaced by a ternary eutectic, even at low pressure (Fig. 2.5). In this case, melting begins at a slightly higher temperature and the composition of the eutectic melt moves towards the Q–Or line.

Figure 2.6 shows that the compositions of the natural granitic rocks cluster near the experimental eutectic compositions. This observation, already highlighted by Tuttle and Bowen (1958), is a convincing piece of evidence for an origin of granites by partial melting of quartz-feldspathic material in the continental crust.

16 Origin of granitic magmas

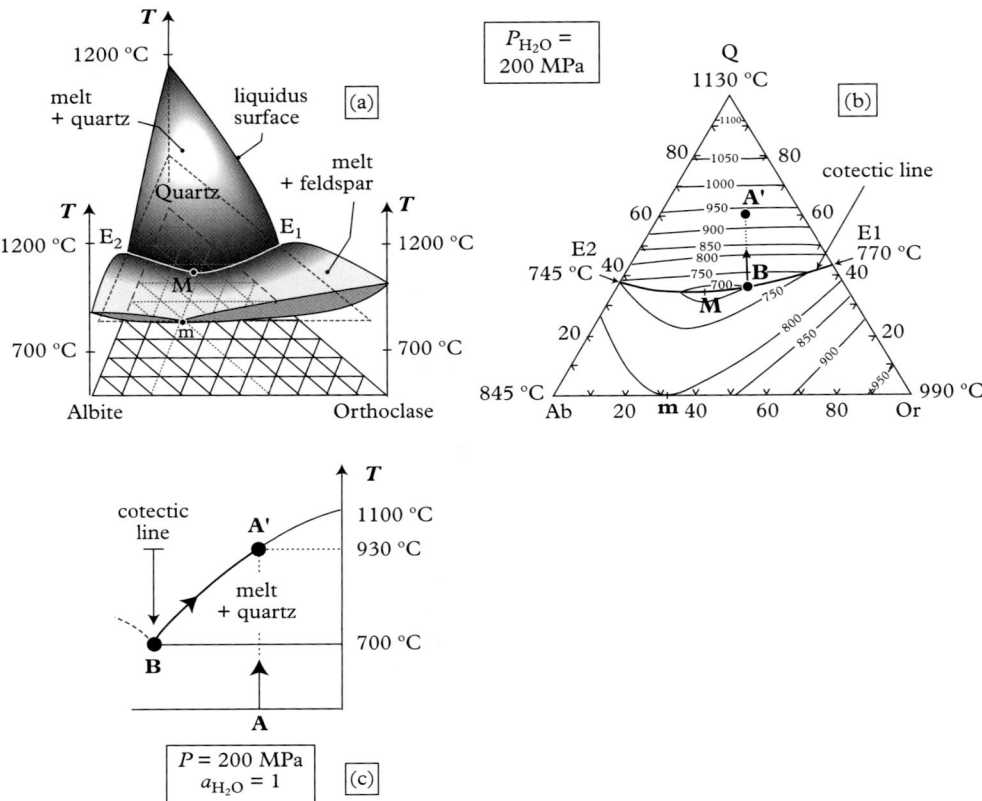

Figure 2.3 *Melting of a water-saturated haplogranitic mixture, after Tuttle and Bowen (1958). (a) 3D phase diagram; E1 and E2: binary eutectic for Q–Or and Q–Ab mixtures; m: minimum melt in the Ab–Or binary system; M: cotectic minimum of the Q–Ab–Or ternary system; the liquidus surface corresponds to complete melting. (b) Projection of the cotectic line and liquidus isotherms on the triangular base of the diagram. (c) Application: melting of mixture A (composition Q:Ab:Or = 62:17:25) in a vertical section of the ternary diagram. Melting begins at 700 °C and yields the melt B (composition Q:Ab:Or = 37:27:36). The volume of melt increases with temperature and the melt composition becomes more quartz-rich following the BA' curve; melting will be complete at 930 °C.*

Wyart and Sabatier (1959) and Winkler and von Platen (1961) performed water-saturated melting experiments with natural samples (pelites and various gneisses) as well. Melting occurs at temperatures very similar to the above experiments (i.e. $T = 650–700$ °C). The resulting melt is granitic to granodioritic in composition, even if the starting material does not contain any potassic feldspar. These authors concluded that micas were likely involved in the melting reaction. The volume of melt is generally abundant, but depends on the starting composition. In some runs, Wyart and Sabatier (1959) observed that new crystals (cordierite or hercynite) formed in equilibrium with

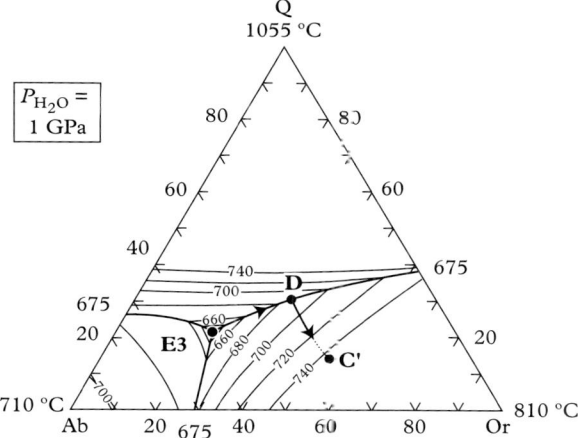

Figure 2.4 *Melting of a haplogranitic mixture at P_{H_2O} = 1 GPa, after Luth et al. (1964); E3: ternary eutectic. Application: melting of mixture C (composition Q–Ab–Or = 15:40:55) yields a first melt E3 (Q–Ab–Or = 23:56:21) at 635 °C. Melt volume increases with temperature and melt composition follows the cotectic line up to D, and then the DC' curve; melting is complete in C' at 760 °C.*

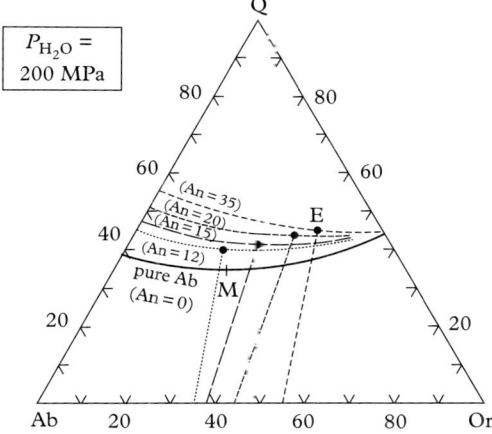

Figure 2.5 *Effect of an additional calcic component at P_{H_2O} = 200 MPa, after von Platen (1965): plots of eutectic points and cotectic lines on basal plane Q–Ab–Or for different plagioclase compositions; M: cotectic minimum; E: eutectic. Notice that the actual mixture is quaternary (Q–Ab–Or–An) and would require a tetrahedral representation.*

18 Origin of granitic magmas

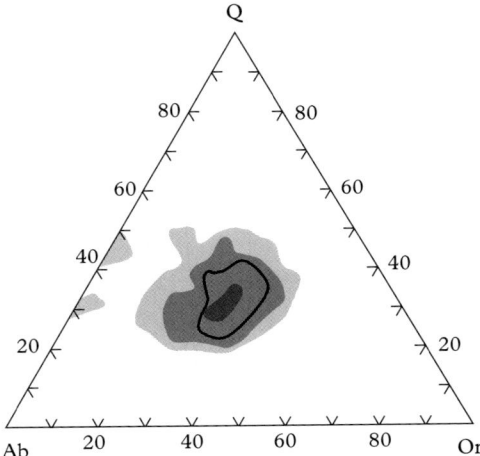

Figure 2.6 *Composition of 1190 granitic rocks after Winkler and von Platen (1961); half of them are enclosed within the solid line.*

the melt. Unfortunately, too many different factors controlled these experiments and no general conclusion could be drawn at that time.

2.2.2 The importance of water

All previous experiments were performed in water-saturated conditions (water activity $a_{H2O} = 1$), meaning that a free hydrous fluid phase was always present. However, this situation is unlikely to exist in the lower crust, the potential site of partial melting, where conditions are either water-undersaturated ($a_{H_2O} < 1$) or dry ($a_{H_2O} = 0$). Implications for the beginning of melting are spectacular (Fig. 2.7). Temperatures as high as 800–1000 °C are required for the beginning of melting of a haplogranitic mixture in these fluid-absent conditions, an unrealistic situation in most natural conditions.

The water content necessary to get water-saturation does not refer to a specific water activity, because water solubility in the melt is strongly dependent on pressure, hence on depth (Fig. 2.8a), and very slightly on temperature (Fig. 2.8b). When solid phases coexist with a silicate melt, the available water dissolves in the melt, until the melt reaches saturation. At that very moment, a free hydrous fluid phase appears in equilibrium with the water-saturated silicate melt. If saturation is not reached, the melt is called 'water-undersaturated' and there is no free hydrous fluid phase (Fig. 2.9a).

The water content influences not only the temperature of the solidus, but also the gap between solidus and liquidus temperatures, hence the volume of melt produced in near solidus conditions. This result is very important for the origin of granite. It has

Experimental data 19

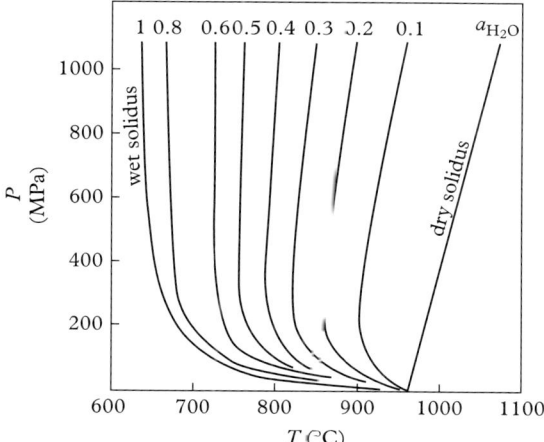

Figure 2.7 *Influence of water activity on the haplogranitic solidus after Ebadi and Johannes (1991): in the experiments, the starting materials have minimal or eutectic compositions and the different water activities are obtained using mixed H_2O–CO_2 fluids (with variable proportions of H_2O and CO_2).*

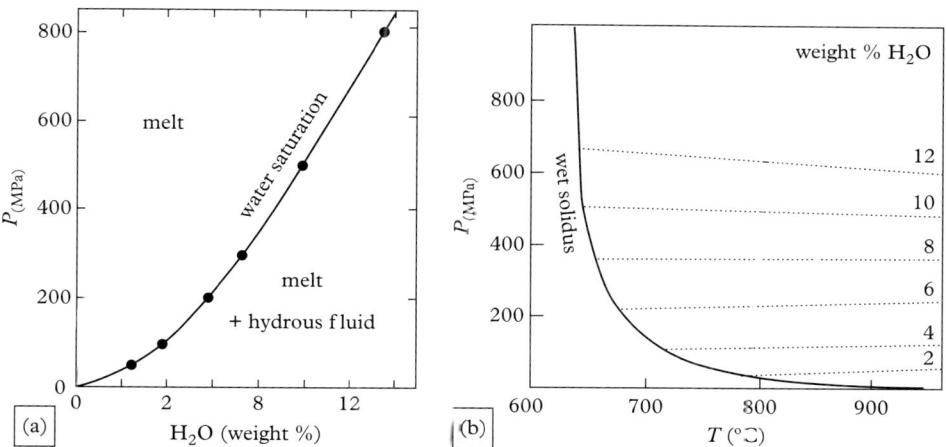

Figure 2.8 *Water solubility in a haplogranitic melt. (a) Effect of pressure. (b) Effect of temperature; dotted curves correspond to the water contents necessary for water saturation of the melt. After Holtz et al. (1995).*

20 Origin of granitic magmas

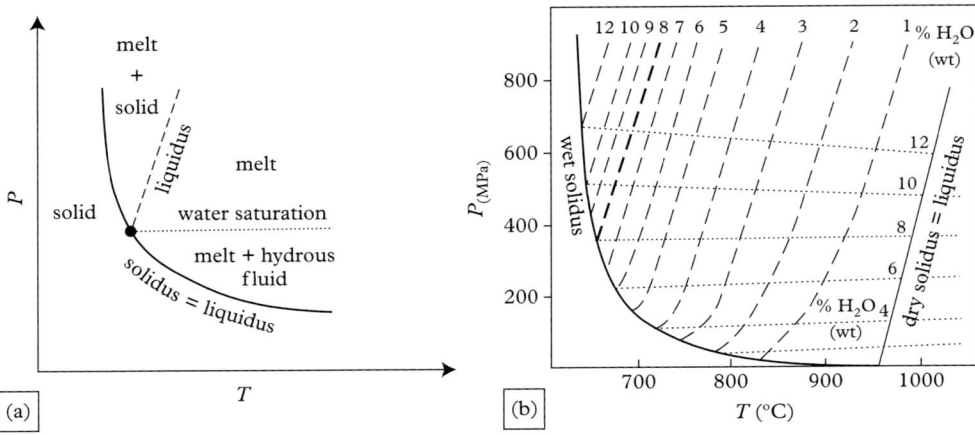

Figure 2.9 *Influence of water content on melting of haplogranitic mixtures of minimum or eutectic composition. (a) General case: solidus (solid curve), liquidus (dashed line) and water-saturation curve (dotted line); solidus and liquidus are identical in case of water saturation, i.e. melting is complete from the beginning. (b) Liquidus for different water contents in water-undersaturated conditions. After Johannes and Holtz (1996).*

implications even for the so-called 'eutectic' compositions, which are regarded as the most easy to melt. In water-saturated conditions, these compositions melt completely (solidus and liquidus are identical). In other cases, melting is incomplete and the temperature difference between solidus and liquidus is always larger when conditions are far from water-saturation. The effect of water content on melting of haplogranitic mixtures of minimal or eutectic composition is summarized in Fig. 2.9b. Obviously, it is difficult to form large volumes of granitic melt in the water-undersaturated lower crust, if only quartz-feldspar compositions are considered.

2.2.3 Melting experiments with hydrous phases

Brown and Fyfe (1970) performed the first experiments that demonstrated the role of hydrous minerals in partial melting. Their experiments were done with no water added, hence in fluid-absent conditions. Water is present only in the crystalline network of the hydrous minerals. The starting mixtures are made of minerals picked from crushed natural samples. In each experimental run including a hydrous mineral, melting begins at a temperature intermediate between the temperatures of the respective water-saturated and dry haplogranitic solidus. Figure 2.10 shows that melting is possible near the base of the continental crust if muscovite is present. If only biotite is present, the solidus is not reached, unless the continental crust is thicker than usual (>30–35 km) or the geothermal gradient is higher than the common value of 20 °C/km.

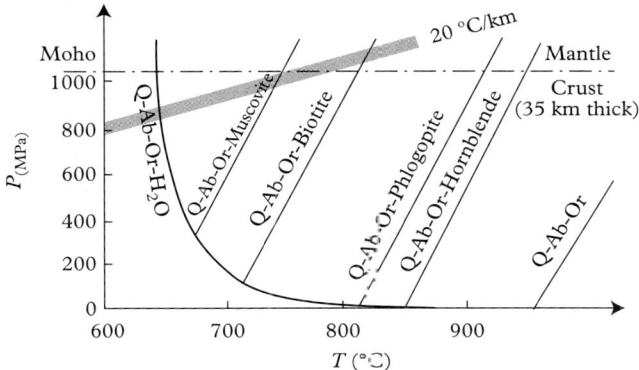

Figure 2.10 *Solidus for quartz–feldspar plus hydrated mineral mixtures, after Brown and Fyfe (1970).*

Such melting reactions with hydrous minerals, but without water added, are called 'dehydration-melting' reactions, an ambiguous expression, because it suggests a two-step process: dehydration first, followed by melting. Actually, there is only one melt-forming reaction and no hydrous fluid phase is released in the system during melting. Breakdown of the hydrous mineral is part of the melting reaction.

The following experiments investigated melting at higher pressures ($P \geq 1.5$ GPa), focussing on the amount of melt produced and the nature of the mineral phases in equilibrium with the melt. Reaching equilibrium is difficult in experimental petrology. Indeed, at temperatures close to the solidus, the reaction rate may not be fast enough to assert equilibrium conditions despite run durations as long as a few weeks. Incidentally, the volume of melt may be small and the new minerals tiny and somewhat difficult to identify. Nevertheless, technical improvements in the past 20 years have made experimental studies much easier (Fig. 2.11).

The volume of melt depends on the amount of water contained in the hydrous minerals of the protolith. It increases with temperature, but not linearly (Fig. 2.12). Experiments used starting materials corresponding to the different crustal protoliths, which are of interest for the origin of granitic magmas. Such protoliths can be derived from either metasediments or former igneous rocks, namely:

- metapelites, i.e. micaschists and gneisses resulting from the prograde metamorphism of clays (pelites);
- metagreywackes, i.e. rocks resulting from the prograde metamorphism of mica- and plagioclase-bearing sandstones (sediments less rich in alumina than pelites);
- biotite (± hornblende)-bearing tonalites or tonalitic orthogneisses; such rocks are common in the Archaean (>2.5 Ga old) terranes.

The following reactions were calibrated using pelitic mixtures:

muscovite + quartz (± plag) = granitic melt + aluminium silicate (sillimanite) (2.2)

22 *Origin of granitic magmas*

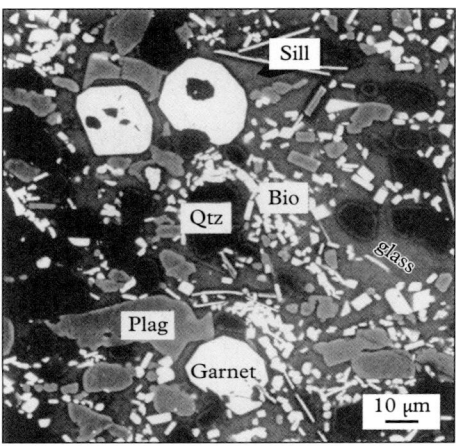

Figure 2.11 *Experimental melting of a quartz–plagioclase–biotite–muscovite mixture at 825 °C, 1 GPa and run duration of 8 days (backscattered electron image). Muscovite disappeared; newly formed minerals are sillimanite (Sill, fine needles) and garnet; garnet crystals display well developed faces, indicating that they grew freely in the melt; quartz, plagioclase and biotite are still recognizable in the melt (now glass). In Gardien et al. (1995).*

$$\text{biotite} + \text{quartz} + \text{sillimanite} (\pm \text{plag}) = \text{granitic melt} + \text{garnet} \, (P \geq 500 \text{ MPa}) \qquad (2.3)$$

$$\text{biotite} + \text{quartz} + \text{sillimanite} (\pm \text{plag}) = \text{granitic melt} + \text{cordierite} \, (P \leq 500 \text{ MPa}) \qquad (2.4)$$

In these three reactions, melting occurs even in the absence of plagioclase, but the participation of plagioclase increases the volume of melt. Such reactions are often observed in migmatites and can explain the origin of peraluminous (S-type) granites. At high temperature, i.e. when all biotite has been consumed ($T \geq 850$ °C), hercynite-rich ($FeAl_2O_4$) green spinel may crystallize. Figure 2.13 displays the stability domains of the main minerals, that may form in equilibrium with the melt in the biotite dehydration-melting reactions.

The following reaction is observed with mixtures richer in silica:

$$\text{biotite} + \text{quartz} (\pm \text{plag}) = \text{granitic melt} + \text{orthopyroxene} \qquad (2.5)$$

This reaction occurs at temperatures slightly higher than reaction (2.4). It can explain the origin of orthopyroxene-bearing granites or 'charnockites' (named after Lord

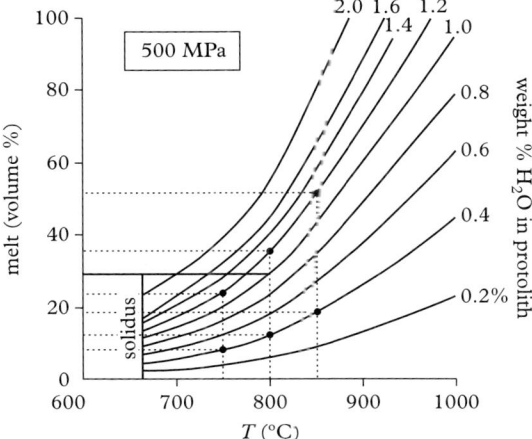

Figure 2.12 *Melt volume versus water content in the silicates of a metapelite (after Clemens and Vielzeuf, 1987). Given that the mica water content is 4 wt %, a protolith with 10 (30)% micas contains 0.4 (1.2)% water. With 0.4% water in the protolith, the melt volume is 7% at 750 °C, 10% at 800 °C and 18% at 850 °C. With 1.2% water in the protolith, the respective melt volumes are 23, 34 and 50% at the same temperatures.*

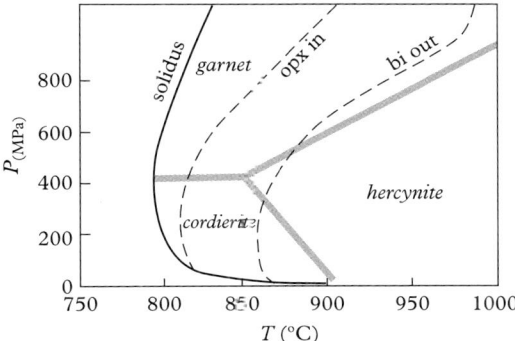

Figure 2.13 *Stability fields of minerals crystallized during incongruent melting of mixtures of metapelite or metagreywacke compositions, after Vielzeuf and Montel (1994); opx in: appearance of orthopyroxene (hypersthene); bi out: disappearance of biotite. The field limits depend on P and T, but also on Fe/Mg, fO_2 (cf. Fig. 2.14) and minor components, such as Zn in spinel, that stabilizes the spinel (hercynite–gahnite solid solution) at $T < 850$ °C.*

Charnock, who founded Calcutta, and whose tombstone in Madras is made of a rock type subsequently called charnockite).

Greywacke-type mixtures give rise to many incongruent melting reactions with the involvement of biotite, such as:

$$\text{biotite} + \text{quartz} + \text{plagioclase} = \text{granitic melt} + \text{orthopyroxene} + \text{garnet}, \qquad (2.6)$$

a reaction that produces more garnet when pressure is higher, because of an increased contribution of plagioclase.

Many factors control the solidus temperature, especially the X_{Mg} ratio (i.e. the atomic Mg/Mg+Fe ratio) in biotite and the oxygen fugacity (fO_2). Strictly speaking, the X_{Mg} values of all ferromagnesian minerals should be indicated in all reactions, and therefore P and T given here are only indicative. The composition of Fe–Mg minerals change during

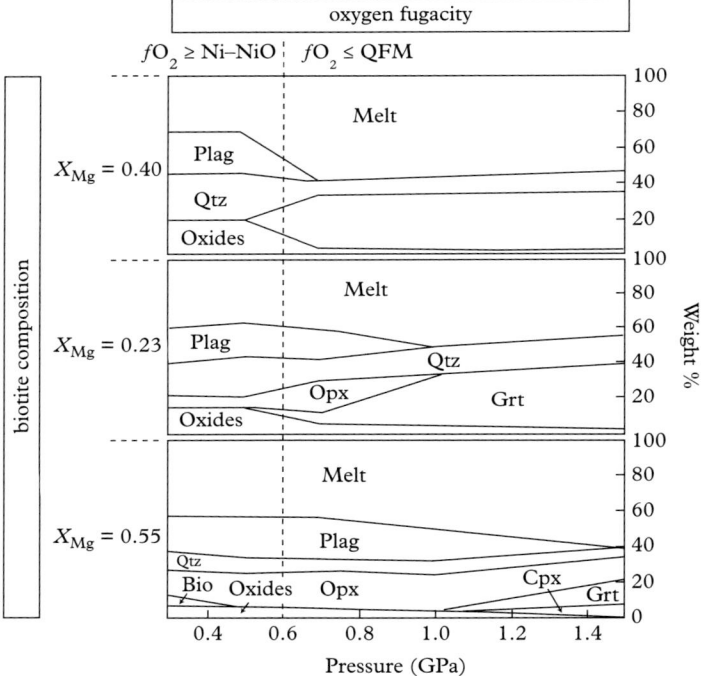

Figure 2.14 *Effect of pressure, biotite composition (X_{Mg}) and oxygen fugacity on the products of incongruent melting of different mixtures at the temperature of biotite destabilization; oxygen fugacity is buffered with Ni–NiO for a high fugacity, or with quartz, fayalite and magnetite, the QFM buffer, for a lower fugacity. Cpx: clinopyroxene; Grt: garnet; Opx: orthopyroxene; Plag: plagioclase. After Patiño Douce and Beard (1996).*

melting reactions, for instance biotite becomes richer in magnesium, because phlogopite is more stable than biotite at higher temperatures.

At high oxygen fugacity, formation of Fe–Ti oxides is favoured, and especially the formation of magnetite, instead of Fe–Mg silicates (Fig. 2.14). The corresponding reaction is:

$$\text{biotite (annite)} + \text{plagioclase} + \text{quartz} = \text{granitic liquid} + \text{magnetite} + \text{ilmenite} \quad (2.7)$$

On the other hand, at lower fO_2 with magnesium-rich biotite, orthopyroxene is formed rather than garnet.

Tonalitic starting materials contain the same minerals than greywacke-type mixtures, but in different proportions (less biotite, more plagioclase). Consequently, they produce less melt at the same temperatures. The newly formed minerals are the same (garnet, orthopyroxene, oxides).

It is worth noting that none of the previous reactions produced hornblende, whereas this mineral is common in some migmatite neosomes (Figs 2.2a and 3.14c). Hornblende crystallizes in water-present (but not necessarily water-saturated) experimental reactions (Gardien et al., 2000), such as:

$$\text{biotite} + \text{plagioclase} + \text{quartz} (\pm H_2O) = \text{granitic melt} + \text{hornblende} + \text{garnet} \quad (2.8)$$

In this reaction, formation of garnet and evolution of the melt towards a granodioritic composition are favoured at high pressure. This is very similar to reaction (2.1), which was recognized in the Eseka migmatites. As a general rule, crystallization of hornblende in natural cases would indicate at least a small water activity.

2.3 Fertility of crustal protoliths and melt compositions

In water-absent conditions, metapelites melt first, but the amount of melt produced remains small and this melt may not easily leave its source (see the segregation problem in Chapter 4). At temperatures ≤750 °C, migmatites will form and not granites. A minimum temperature of 850 °C is required to obtain the volumes of melt necessary to form granitic plutons (Fig. 2.15). At such high temperatures, metagreywackes represent a more fertile protolith, because of their large content of plagioclase, that will contribute to the melting reaction, at least for its albitic fraction.

Whereas melting of metapelites always yields peraluminous (S-type) granitic liquids, melting of greywackes yields peraluminous or metaluminous (S- or I-type) liquids, depending on the starting mineral compositions. Similar results are obtained with tonalitic protoliths. Finally, experiments produce more aluminous and peraluminous melts than metaluminous ones, partly because of the nature of the starting mixtures. Therefore, these experiments are not perfect representatives of the formation of large batholiths composed of calc-alkaline metaluminous granitic rocks. The fact, that these granitoids have a hybrid origin and are not derived exclusively by crustal melting (as will be discussed in Chapter 12), contributes to the discrepancy between experimental and natural melts.

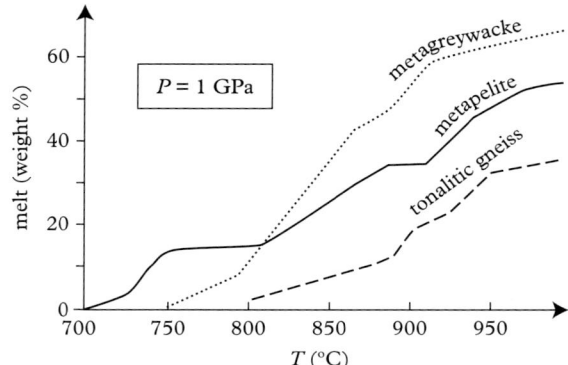

Figure 2.15 *Compared fertilities of a few crustal protoliths after Johannes and Holtz (1996).*

The lower fertility of magmatic protoliths of tonalitic to granodioritic compositions, with respect to metasedimentary protoliths, has already been noted. Indeed, the igneous protoliths display lower contents of hydrous minerals (biotite and hornblende). In addition, water-absent melting reactions involving hornblende require a higher temperature than melting reactions involving biotite. At $T \geq 900$ °C, the disappearance of biotite produces large volumes of melt. As biotite usually contains fluorine, the resulting melt will also be halogen-rich, a typical feature of A-type granites (Skjerlie and Johnston, 1993). Another feature of A-type granites is their high content of high field strength elements (HFSE), which are elements with a large ratio of charge vs. ionic radius, such as Zr, Hf, Ce, Ti or Nb (see Fig. 1.5). These elements are usually contained in refractory accessory minerals (zircon, monazite, . . .), whose dissolution is only efficient at high temperature. These pieces of evidence suggest that A-type granites may form by melting of a magmatic protolith in the continental crust. However, an origin by extreme fractional crystallization of a mantle-derived melt may be argued in other cases on the basis of isotopic data.

2.4 Tracing the sources with isotopes

2.4.1 The Rb–Sr pair and the $^{87}Sr/^{86}Sr$ isotopic tracer

Rubidium (Rb) and strontium (Sr) are lithophilic elements abundant in the outer silicate envelope of the Earth. Rb belongs to the first column of the Periodic Table (see Fig. 14.3): it is an alkaline element, like potassium (K), that it replaces easily in some minerals (micas, alkali feldspars). Its ionic radius (1.68 or 1.81 Å depending on the coordination number) is larger than the ionic radius of K^+ (1.59–1.68 Å). These large ions have a very incompatible behaviour, i.e. they do not fit easily into the crystal lattices, hence they prefer to remain in the melt.

Sr belongs to the second column of the Periodic Table of the elements. It is an alkaline earth like Ca, that it can replace in some minerals (plagioclase, apatite). Sr^{2+} and Ca^{2+} have smaller ionic radii than Rb^+ and K^+; hence, they are less incompatible.

Rubidium has two natural isotopes: ^{85}Rb (the more abundant) and ^{87}Rb, and strontium has four: ^{84}Sr, ^{86}Sr, ^{87}Sr and ^{88}Sr (the most abundant). The radioactive decay of ^{87}Rb produces the so-called 'radiogenic' ^{87}Sr isotope. The period (or half-life) of ^{87}Rb is very long (ca 50 Gyrs), corresponding to a radioactive decay constant λ of 1.42×10^{-11} yr^{-1}.

When a rock becomes partially molten, the resulting melt is always enriched in Rb with respect to Sr, because of the very incompatible nature of Rb with respect to Sr. Partial melting is therefore responsible for elemental fractionation of the Rb–Sr pair. Conversely, when no isotopic fractionation occurs, the isotopic proportions of each element remain unchanged. Note however that it is not the rule for all phase transitions. For instance, ocean evaporation fractionates oxygen isotopes for kinetic reasons, because the lightest ^{16}O isotope is extracted faster from seawater than the heavier isotopes. Water vapour is diluted in the atmosphere, thus the phase transition can be regarded as unidirectional, with no way to reach equilibrium. Returning to the geological processes and to the isotopes of interest, that are much heavier than oxygen, the mass difference between these isotopes is smaller, and solid and melt remain in contact for long enough periods to ensure isotopic equilibrium.

Any magmatic rock crystallized from a rubidium-enriched melt will become enriched in ^{87}Sr due to the decay of the radioactive ^{87}Rb isotope. By contrast, the refractory restite in the parental protolith is depleted in the incompatible Rb element and will contain less ^{87}Sr with respect to the derived magmatic rock. For analytical purposes, the $^{87}Sr/^{86}Sr$ ratios can be measured precisely with the mass spectrometer.

The same argument can be applied to bulk silicate Earth, comprising crust and mantle (the core is not worthwhile considering, because it is mainly made of iron and can be regarded as an isolated reservoir, devoid of Rb and Sr). The continental crust can be considered as mantle-derived through partial melting events in the mantle. Therefore, it is an enriched reservoir, i.e. enriched in incompatible elements such as Rb, when compared to the upper mantle, that was its source and that constitutes a depleted reservoir. The lower mantle is regarded as primitive (neither enriched nor depleted) as a first approximation.

Now, the ultimate question is to know the isotopic composition of the bulk silicate Earth at the beginning, i.e. to know the so-called 'primordial' $^{87}Sr/^{86}Sr$ ratio at 4.5 Ga. By convention, we use the lowest ratio so far determined in specific meteorites, namely the basaltic achondrites. This ratio, called BABI (*basaltic achondrite best initial*) is equal to 0.699 (Papanastassiou and Wasserburg, 1969). From this moment, the silicate Earth, regarded as a uniform reservoir (UR), will become richer and richer in radiogenic ^{87}Sr, ending at a present day $^{87}Sr/^{86}Sr$ ratio of 0.7046 (Fig. 2.16). This value is typical of some basalts from oceanic hot spot volcanoes, likely derived from the melting of primitive mantle material ascended from the depth. By contrast, basalts from mid-oceanic ridges (MORBs) are characterized by a $^{87}Sr/^{86}Sr$ ratio of only 0.702, in agreement with their origin by partial melting of the depleted upper mantle.

28 Origin of granitic magmas

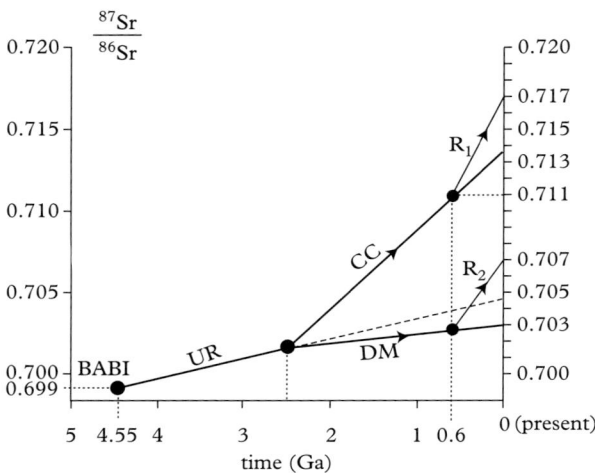

Figure 2.16 *Evolution of the $^{87}Sr/^{86}Sr$ ratio in the main reservoirs of the bulk silicate Earth, after Fourcade (1998). UR (uniform reservoir) is equivalent to the primitive lower mantle. Formation of continental crust (CC) is supposed to occur at 2.5 Ga. DM is the residual upper mantle depleted in incompatible elements. Compare the isotopic compositions of rock R1, derived from the anatexis of a crustal protolith (CC) of late-Archaean age, and rock R2, derived from partial melting of the depleted mantle (DM). Although coeval (600 Ma old), these rocks have not the same initial isotopic ratios.*

The continental crust is characterized by a $^{87}Sr/^{86}Sr$ ratio >0.705, with variable values depending on the age and composition of the constituent rocks. It is not a homogeneous reservoir. The $^{87}Sr/^{86}Sr$ ratios are all the higher as the rocks are older and as their respective Rb/Sr elemental ratios are also high (Figs 2.16 and 2.17). This is the reason why it is necessary to compare the so-called 'initial ratio', Sr(i), i.e. the ratio that would have been measured at the time of crystallization, with the Sr(i) ratio in the upper depleted mantle at this very stage. A high Sr(i) ratio (with respect to the mantle value) points to a dominantly crustal origin. A low Sr(i) ratio is indicative either of a mantle origin (the magmatic rock is then called 'juvenile') or of a crustal origin by melting of young juvenile protolith, i.e. a protolith that was extracted from the mantle less than 0.2 Gyr ago. Finally, hybrid rocks may have intermediate Sr(i) values. Where possible, such ambiguous answers justify the use of several other isotopic tracers.

2.4.2 The Sm–Nd pair and the $^{143}Nd/^{144}Nd$ isotopic tracer

Samarium (Sm) and neodymium (Nd) are two rare earth elements or lanthanides. Their geochemical behaviour is very similar, but Nd is slightly more incompatible

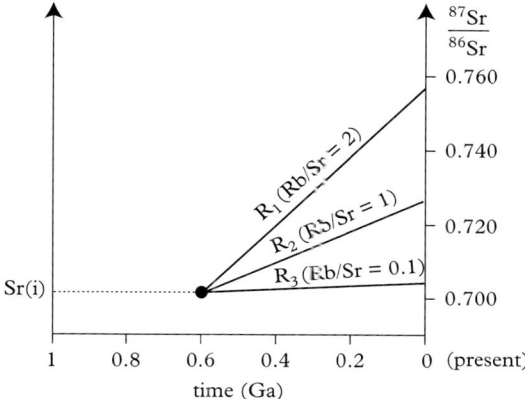

Figure 2.17 *Evolution of the $^{87}Sr/^{86}Sr$ ratio of three coeval rocks derived from the same source, but characterized by different Rb/Sr ratios, after Faure (2001). Rb/Sr ratios of granites are generally >1, whereas intermediate granitoids have ratios <1.*

than Sm. Samarium has seven isotopes, and one of them, ^{147}Sm, is radioactive and yields radiogenic ^{143}Nd. The periodicity of this isotope is extremely long (around 106 Gyrs), corresponding to a decay constant λ of 0.654×10^{-11} yr^{-1}. Neodymium has seven isotopes. The ^{143}Nd/^{144}Nd ratio is used as an isotopic tracer. Compared to the previous isotopic pair, this radioactive isotope is slightly less incompatible than its daughter isotope, explaining the inverse positions of the CC and DM reservoirs in Fig. 2.18, compared to Fig. 2.16.

For the undifferentiated silicate Earth, regarded as equivalent to the lower mantle, the evolution model is supposed to have been identical to the evolution of chondrites (meteorites with silicate nodules). The ^{143}Nd/^{144}Nd ratio corresponding to this reservoir, called CHUR (chondritic uniform reservoir), displays the present value CHUR$_0$ = 0.512638. The range of variations of the ^{143}Nd/^{144}Nd ratio is much smaller than those of the ^{87}Sr/^{86}Sr ratio, because Sm and Nd have rather similar behaviour and because the radioactive decay of ^{147}Sm is extremely slow. Therefore, it is better to use a particuliar indicator, ε, introduced by DePaolo and Wasserburg (1976), resulting in a larger variation range:

$$\varepsilon_{Nd} = 10^4 \left[\left(^{143}Nd / ^{144}Nd\right)_{sample} / (CHUR - 1) \right]$$

The ε_{Nd} can be calculated for the present: $\varepsilon_{Nd}(0)$ or, more useful, at the time t of crystallization of the rock $\varepsilon_{Nd}(t)$.

Rocks formed by partial melting of the continental crust are characterized by negative values of ε_{Nd}, whereas rocks formed by partial melting of the depleted mantle (DM) are characterized by positive ε_{Nd} values.

30 Origin of granitic magmas

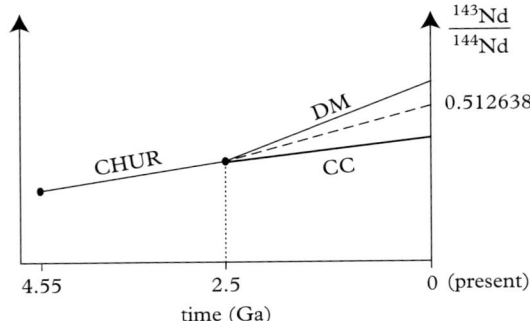

Figure 2.18 *Evolution of the $^{143}Nd/^{144}Nd$ ratio in the main reservoirs of bulk silicate Earth. Here is the case of a continental crust extracted from the mantle at 2.5 Ga. CHUR: chondritic reservoir; DM: depleted mantle; CC: continental crust.*

2.4.3 Coupling Sr and Nd tracers

The contrasting variations of Sr and Nd isotopic ratios found in the complementary reservoirs CC (continental crust) and DM (depleted upper mantle) can be combined in the so-called 'quadrant diagram' (Fig. 2.19). Using the ε_{Nd} indicator and the ε_{Sr} calculated in the same way, the following quadrants can be defined:

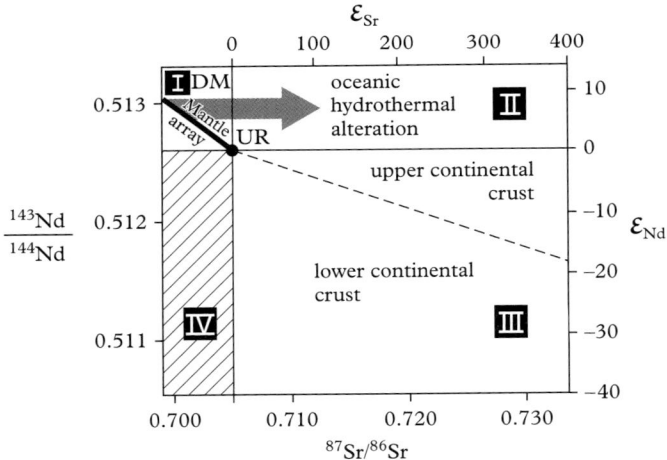

Figure 2.19 *The quadrant diagram. The values of the isotopic ratios are those characterizing the present times. The axes must be translated following the changing ratios of the 'bulk Earth' reservoir (UR or CHUR) according to time. After Fourcade (1998).*

- quadrant I: $\varepsilon_{Nd} > 0$ and $\varepsilon_{Sr} < 0$; magmatic rocks of mantle origin plot roughly along the line connecting the DM point to the origin (UR or CHUR); this line is called the 'mantle array' and represents all possible mixtures of magmas from a source resembling the depleted upper mantle (DM) to a source resembling the lower primitive mantle (origin);
- quadrant II: $\varepsilon_{Nd} > 0$ and $\varepsilon_{Sr} > 0$; magmatic rocks plotting in this quadrant were influenced by oceanic hydrothermal alteration, that modified the Sr isotopic ratio toward the seawater ratio (today 0.709), whereas the Nd isotopic ratio, much less sensitive to alteration, remained unchanged;
- quadrant III: $\varepsilon_{Nd} < 0$ and $\varepsilon_{Sr} > 0$; this is the field of the continental crust divided into two domains respectively corresponding to the lower and upper crust; the former often contains less Rb than the latter, especially if it has a restitic component, hence lower ε_{Sr} values;
- quadrant IV: $\varepsilon_{Nd} < 0$ and $\varepsilon_{Sr} < 0$; no rock plots in this field.

2.4.4 The $^{18}O/^{16}O$ ratio

This tracer uses stable, non-radiogenic isotopes. Theoretically, these light isotopes can fractionate during partial melting, unlike the heavier isotopes discussed above. This fractionation is an inverse function of temperature, and therefore can be neglected at high temperatures of magma formation, implying that the $^{18}O/^{16}O$ ratio of a magmatic rock is similar to the ratio of its source.

Practically, the isotopic ratio of the sample is compared to the isotopic ratio of seawater or standard mean ocean water (SMOW). The deviation is calculated as:

$$\delta^{18}O \text{ sample } (\permil) = 10^3 [[(^{18}O/^{16}O)_{sample} / (^{18}O/^{16}O)_{SMOW}] - 1]$$

The mantle is characterized by low $\delta^{18}O$ values: +5 to +6‰. By contrast, sedimentary rocks, formed in water at low temperatures, registered variable fractionation of the $^{18}O/^{16}O$ ratio, hence the corresponding $\delta^{18}O$ values ranging from +8 to +32‰ (Fig. 2.20). The continental crust has higher $\delta^{18}O$ values than the mantle, because it contains a large amount of recycled sediments. Thus, the use of oxygen tracers helps to discriminate between mantle or crustal sources for granitic rocks (Fig. 2.20). This figure shows that crustal sources have more variable signatures than mantle sources. Moreover, $\delta^{18}O$ values more than +10‰ (S-type granites) always correspond to a sedimentary crustal protolith, whereas values ranging from +8 to +10‰ (I-type granites) may correspond to a sedimentary or an igneous crustal protolith (Fig. 2.21).

Fluid–rock interactions at low to medium temperatures can modify the granite isotope signatures in a significant way. Indeed, hydrothermal fluids (either magmatic or metamorphic) are characterized by $\delta^{18}O$ values ranging from +5 to +8‰, whereas meteoritic waters are characterized by negative values. Therefore, using oxygen isotopes as tracers of the granite sources requires collecting samples devoid of any hydrothermal

32 Origin of granitic magmas

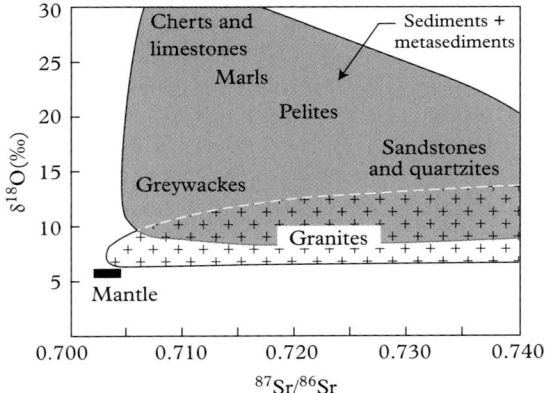

Figure 2.20 *Fields of (Sr, O) isotopic compositions for the mantle, granites and sediments. After Taylor and Sheppard (1986).*

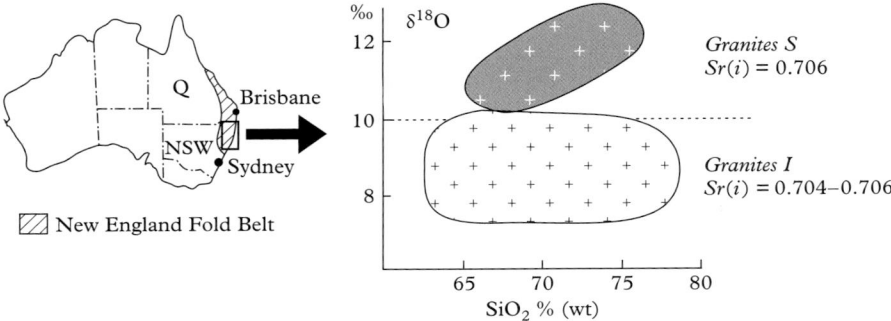

Figure 2.21 *(Sr, O) isotopic compositions of granites from New England. The New England batholith is located in eastern Australia and straddles the boundary between Queensland (Q) and New South Wales (NSW). It is made of 80% of I- and S-type monzogranites. Both types can be discriminated by their oxygen isotopic composition. Higher $\delta^{18}O$ in the S-types than in the I-types witness to a greater amount of sedimentary component in their source. After Shaw and Flood (1981).*

alteration or weathering. In addition, it is also possible to trace the pathway of magmatic fluids in the country rocks using oxygen isotopes (see Fig. 7.11).

2.4.5 Hf isotopes in zircon

Hafnium (Hf) is an element that chemically resembles zirconium (Zr). Owing to technological analytical progress, a new isotopic tracer has been introduced recently: the $^{176}Hf/^{177}Hf$ ratio measured *in situ* in single zircon grains using LA-ICP-MS.

> **Box 2.2** Oxygen isotope palaeothermometry
>
> Application of oxygen isotopes as a thermometer is a common tool in palaeoclimatology, for instance using $\delta^{18}O$ variations measured in ice cores or in tests (shells) of marine foraminifera (unicellulars). These applications are detailed in the book *Isotopic Geology* (Allègre, 2005). Thermometry is also possible in the metamorphic and igneous domains. In this case, it is necessary to consider mineral pairs and not the whole rock (as in tracing the sources of magmas). The difference between $\delta^{18}O$ of each mineral is denoted as $\Delta^{18}O$ and is an inverse function of the squared equilibrium temperature (as a first approximation). As an example, here is the relation yielding T, using the quartz–muscovite pair:
>
> $$\Delta^{18}O(\text{quartz} - \text{muscovite}) = \left(2.2 \times 10^6 / T^2\right) - 0.6 \text{ with } T \text{ in K (kelvin)}$$
>
> If the rock is in isotopic equilibrium, the temperatures calculated from $\Delta^{18}O$ of different mineral pairs (quartz–magnetite, quartz–biotite, quartz–alkali feldspar, etc.) must be identical. Indeed, there should be no isotope fractionation between minerals during cooling of a magmatic rock, because solid-state diffusion is extremely slow, thus preserving high-temperature isotopic fractionation. Actually, high-temperature fractionation is very small and, because of the slow cooling of granitic rocks, the inferred temperature may not be the crystallization temperature, but a lower one. Conversely, the calculated $\delta^{18}O$ may indicate an isotopic disequilibrium due to interaction between the rock and a late fluid, whose origin and temperature remain to be determined by the method.

Radiogenic ^{176}Hf results from the decay of ^{176}Lu, a radioactive isotope of the rare earth element lutetium, whose half-life is around 35 Gyrs. During magma generation, fractionation of Lu from Hf occurs, as Hf is the more incompatible element. The chondritic initial Lu/Hf ratio of the Earth has been progressively modified by episodes of partial melting in the upper mantle. The Hf isotopic compositions of the depleted mantle and of the enriched crust diverge from unfractionated (chondritic) material in the same way as their Nd isotope composition (Fig. 2.22, to be compared with Fig. 2.18). As in the Nd isotope system, deviations of Hf isotopic composition from chondritic values at time t are expressed in epsilon units (parts per ten thousand) as given by the formula:

$$\varepsilon_{Hf} = \left[\left(^{176}Hf/^{177}Hf\right)_t - \left(^{176}Hf/^{177}Hf\right)_{\text{chondrites}} - 1\right] \times 10^4$$

Zircon hosts Hf at the % level, but has Lu/Hf very low, typically 0.002, hence $^{176}Lu/^{177}Hf$ is usually less than 0.0005, meaning that changes due to *in-situ* decay of ^{176}Lu proceed at a negligible rate. Hence, zircons preserve the initial $^{176}Hf/^{177}Hf$ ratios acquired at the time of their magmatic crystallization (Fig. 2.22).

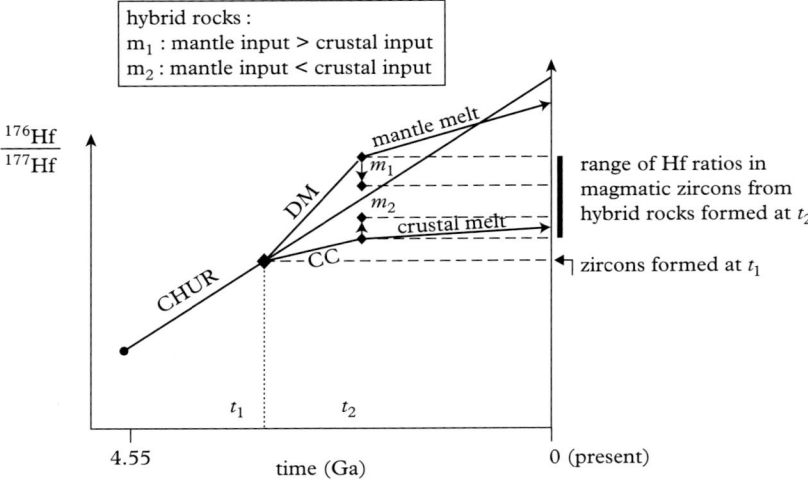

Figure 2.22 *Evolution of Hf isotope composition in the main reservoirs: CC (continental crust), CHUR (bulk silicate Earth as a chondritic uniform reservoir) and DM (depleted mantle). Two successive melting events are considered: partial melting of the primitive mantle at t_1, coeval partial melting of the depleted mantle and the continental crust at t_2. Zircons preserve their initial Hf ratios, because they contain negligible Lu unlike their whole-rock hosts. Magmatic mixtures would be characterized by a large range of Hf ratios in their zircons, depending on the proportions of their components. After Kinny and Maas (2003).*

In situ Hf studies are a most efficient tool for the following reasons:

1. The *in situ* technique is able to detect isotopic differences between grains, not just producing average values for macroscopic samples.
2. Zircon is more robust to post-magmatic alteration than its whole-rock host.
3. The Lu–Hf system has a shorter half-life than the Sm–Nd system and will produce a larger range of variations.
4. As zircons are often zoned minerals, they may also yield different Hf ratios corresponding to newly added rims, whereas their cores preserve an unmodified Hf isotope composition.

2.5 Crustal vs. mantle contributions

Most of the S-type, or peraluminous, granites are derived from partial melting of metasediments. From the isotopic point of view, their Sr(i) and $\delta^{18}O$ are expected to be high, and their ε_{Nd} to be negative. Let us consider the case of the Himalayan leucogranites (also studied in Chapters 6 and 12) and, more precisely, the example of the Manaslu

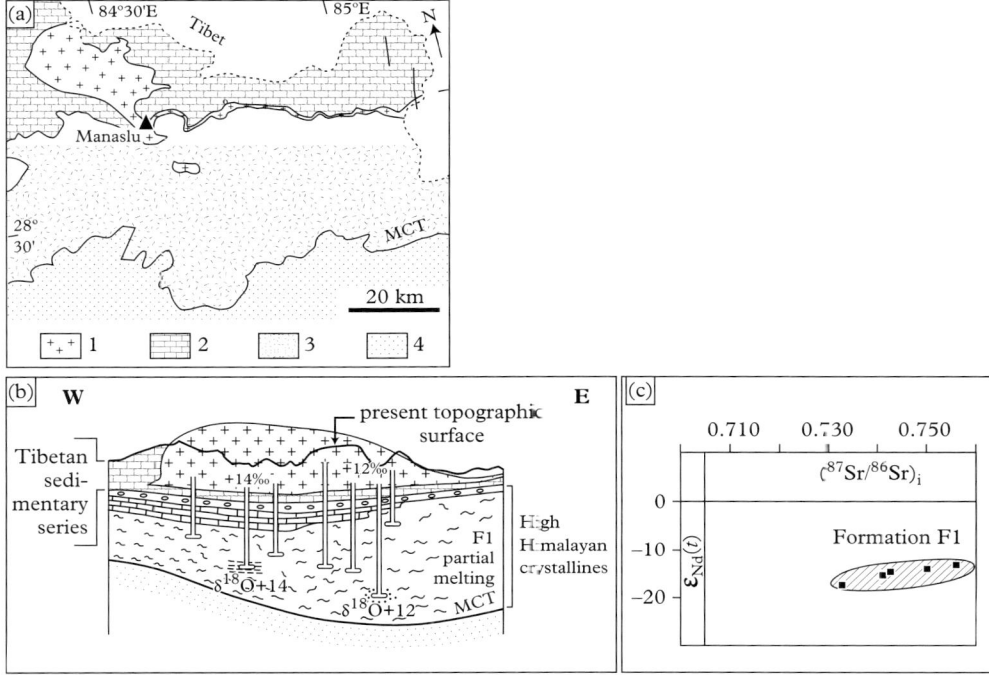

Figure 2.23 Nd and O isotopic signatures of the Manaslu granite. (a) Geological map; 1: granite (Manaslu: 8.125 m); 2: Mesozoic and Palaeozoic Tethyan sedimentary series; 3: High Himalayan crystallines; 4: Lesser Himalayas metasediments. (b) E–W vertical section at the latitude of the Manaslu granite (after France-Lanord et al., 1988). (c) Quadrant diagram (at 20 Ma) for the Manaslu granite and the F1 formation at the base of the High Himalayan crystallines, after Vidal et al. (1984).

granite (Fig. 2.23). This Miocene granite is characterized by very high and heterogeneous Sr(i) ratios (0.730–0.760). The $\varepsilon_{Nd}(t)$ calculated at the time of granite formation are negative (−13 to −17). These (Sr, Nd) signatures are consistent with the isotopic composition of the High Himalayas crystalline rocks (also called the Tibetan slab, although the outcrops are in Nepal), and especially with the composition of the F1 migmatitic gneisses situated immediately above the Main Central Thrust (MCT). The granites also display a perfect fit of $\delta^{18}O$ values (between +12 and +14) with the gneisses of the F1 formation located immediately below. The initial heterogeneities of the gneissic protolith are preserved in the granite. The Sr, Nd and O isotopic tracers confirm the crustal origin of the Himalayan leucogranites, as first suggested by field observations, petrologic evidence and comparison with melting experiments.

The origin of I-type granitic rocks *s.l.* (actually granodiorites are the most abundant lithology among I-types) is less easy to determine. The experiments previously reported show that a crustal protolith of either sedimentary or igneous nature is likely. In addition, mingling of felsic and mafic (basaltic *s.l.*) magmas are often observed in the field exposures of these granitoids, as will be seen in Chapter 4. Only the isotopic data from the Sierra Nevada batholith (California) will be presented here. The same batholith will also be used as an example in Chapters 10 (magnetic fabrics) and 12 (plate tectonics). The Sr, Nd and O isotopic data of DePaolo (1981) points to a mixed origin for the Sierra Nevada granitoids (Fig. 2.24). Two major components, or mixing end-members,

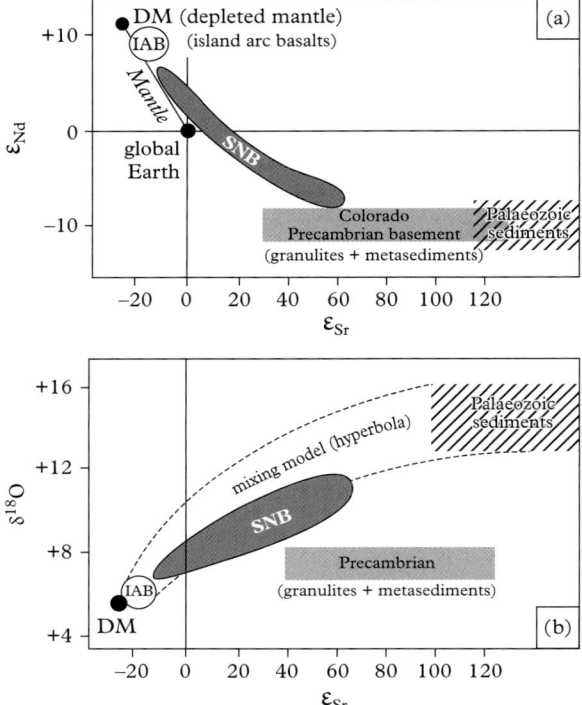

Figure 2.24 *Sr, Nd and O isotopic signatures of the Sierra Nevada batholith (SNB). (a) Sr and Nd data point to a mixed origin for the granitoids of this batholith. One of the component corresponds to island arc basalts (notice the location of these basalts slightly shifted to the right with respect to the depleted mantle, likely due to an early enrichment of Sr in their protoliths as a consequence of the oceanic hydrothermalism). The other component is crustal, but the diagram does not permit one to choose between the Colorado-type Precambrian rocks (granulites and metasediments) and the Palaeozoic sediments. (b) Sr and O data demonstrate that Palaeozoic sediments play the dominant role, but a small contribution of the Precambrian basement cannot be excluded, because the data plot slightly under the two-component mixing hyperbola. After DePaolo (1981).*

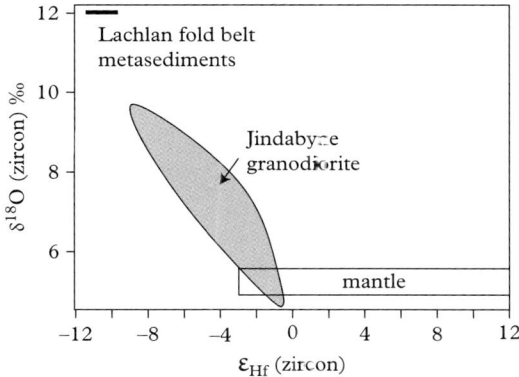

Figure 2.25 *Plot of $\delta^{18}O$ vs. ε_{Hf} for zircons of the I-type Jindabyne suite of granodioritic composition (SiO_2 = 60–67 wt %), after Kemp et al. (2007).*

are an island arc basalt and a magma derived from partial melting of nearby Palaeozoic sediments. The unsolved question is the 'mixing' process. Answers will be suggested in the next chapter using physical, petrological and experimental data. Kemp et al. (2007) studied the Hf and O isotope compositions of I-type granitoids from the Lachlan fold belt type area in eastern Australia and demonstrated their mixed and partly supracrustal source (Fig. 2.25). Their interpretative model proposes the generation of silicic melts at depth through interaction between residual liquids from basalt crystallization and melts derived from the overlying supracrustal assemblage. These hybrid magmas ascended in the shallow crust where they crystallized zircons whose isotopic signature records the progress of supracrustals incorporation at depth. Due to increased crustal anatexis promoting higher rates of metasedimentary rock assimilation by basalts with time, any incrementally built pluton would precipitate zircons with progressively higher $\delta^{18}O$ and lower εHf.

Finally, A-type granites (defined in Chapter 1) are characterized by their high contents in silica and in alkalis, as well as in some trace elements (Zr, rare earths, ...). Their origin is a matter of debate: either mantle-derived, through fractional crystallization of a basaltic magma variably contaminated by the crust (van Breemen et al., 1975), or crustal (Fig. 2.26).

In the hypothesis of a crustal origin, two types of sources are suggested: a granodioritic to tonalitic protolith, possibly enriched in fluorine (Creaser et al., 1991; Skjerlie and Johnston, 1993), or a restitic granulite protolith (Collins et al., 1982) The latter hypothesis is the more unlikely, as a crustal protolith already depleted by a previous anatectic event would not be able to yield a magma with high contents of silica and incompatible elements, that are the characteristics of A-type granites. The two other hypotheses are not mutually exclusive, as can be seen from Nd and O data (Fig. 2.27). The Sr tracer is not considered, because it is less adapted for these Sr-poor rocks. Two groups of A-type

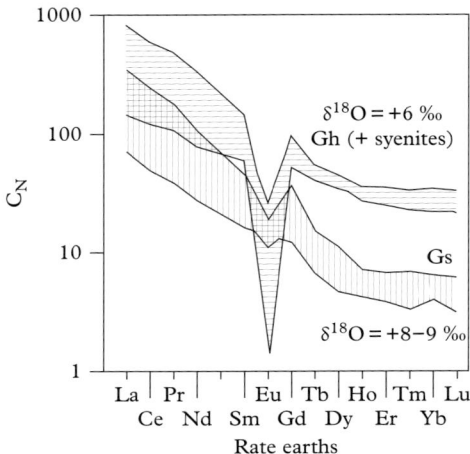

Figure 2.26 *The two different types of Neoproterozoic A-type stratoid granites from Madagascar: hypersolvus granites (and cogenetic syenites) display mantle-type $\delta^{18}O$ values, whereas subsolvus granites display crustal values. The Eu negative anomaly is much pronounced in the REE distribution patterns of the first group (Gh + syenites) and results from the fractional crystallization of a huge amount of feldspar in a mafic magma. The REE patterns of the Gs granites point to a different origin: they are very similar to the REE patterns of the migmatitic country rocks, a possible analogue of the granite protoliths. C_N: chondrite-normalized concentration. After Nédélec et al. (1995).*

granites are thus identified owing to their $\varepsilon_{Nd}(t)$ and $\delta^{18}O$, in agreement with their respective mineralogy and with their geochemical differences, regarding their major and trace element contents. The first group is metaluminous to aluminous and includes subsolvus (two-feldspar) granites; its origin could be purely crustal. The second group is metaluminous to peralkaline and includes hypersolvus (one perthitic alkali feldspar) granites, that display more pronounced A-type characters; their origin is mantle-derived or mixed. Hf isotope data confirm the dichotomy of A-type granites. Kemp et al. (2005) obtained the first Hf data from zircons of the Narabura peralkaline granites from SE Australia, demonstrating their depleted-mantle derivation. By contrast, zircons of the associated metaluminous A-type granites display a large range of ε_{Hf} values reflecting a crustal source that variably interacted with juvenile magmas.

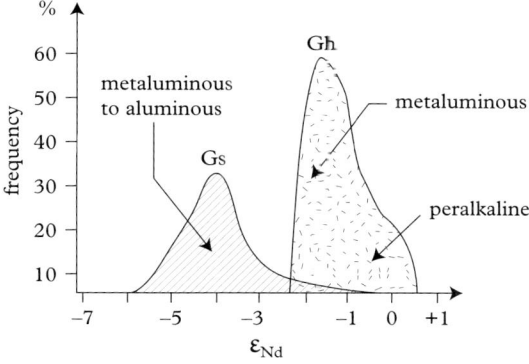

Figure 2.27 *Distribution of $\varepsilon_{Nd}(t)$ values in Permian A-type granites from Corsica; subsolvus granites (Gs) and hypersolvus granites (Gh) are clearly distinct. After Poitrasson et al. (1995).*

2.6 Conclusions

Formation of granites by partial melting of the continental crust has been a matter of debate for a long time. Experimental studies prove that it is a valuable process, even in water-absent conditions, because of partial melting reactions involving hydrous minerals (muscovite, biotite, amphibole). These reactions have their natural analogues in migmatites. However, it is necessary to reach temperatures of 800–900 °C to form large volumes of granites, implying specific geodynamic settings.

The nature of the crustal protoliths, either sedimentary or igneous, and an eventual mantle contribution, can be identified from Sr, Nd and O isotopic tracers. It must be remembered that S-type granites have a purely crustal origin by partial melting of metasediments, whereas I- and A-type granites have either a crustal origin (by partial melting of metasedimentary or metaigneous protoliths in the first case, and by partial melting of only metaigneous protoliths in the second case), or a mixed origin (both crustal and mantle derived). Chapter 13 will present the case of Precambrian granitoids that were formed by partial melting of an igneous protolith, whose composition was similar to that of the oceanic crust.

3
Segregation of granitic melts

This chapter will examine what happens in migmatites during deformation of continental crust undergoing partial melting at depths ranging from 10 to 40 km depending on the geothermal gradient. The composition of granitic melts in migmatites was examined in Chapter 2. It is important to note that the source rock (protolith) from which the granitic magma is derived never melts entirely. Complete melting in a migmatitic protolith would require special conditions such as an exact eutectic composition and water saturation (see Chapter 2). Not only would such a composition melt completely, but it would melt almost instantaneously at an appropriate temperature, between 650 °C and 700 °C. Partial melting is therefore commonplace in the geological record.

The granitic liquid resulting from partial melting begins to collect over short distances of transport (from cm to dm) by a process called segregation. This process is worth distinguishing from a situation where certain amounts of crystals derived from the source, called restites, become entrained in the melt over greater distances. Magma transport is discussed in Chapter 5. Likewise, composition and behaviour of a granitic liquid remaining in a magma at the end of its crystallization is not exactly the same as the melt that forms at the beginning of crystallization: this question will be examined further in Chapter 9 (microstructures and fabrics) and Chapter 11 (zoned plutons).

The main factors that govern the movement of a granitic melt, namely 'viscosity' and 'wetting', are considered first. Based on experiments and observations of migmatites, the distribution of melt can be scrutinized under the microscope as well as in hand specimen. Rates of segregation, volume change from solid to melt as well as various physical mechanisms responsible for melt concentration in layers, tension gashes or shear planes will be also discussed. Then, it will be time to observe migmatites as they appear in the field.

3.1 Viscosity of granitic melts

Viscosity measurements of silicate liquids have spectacularly progressed due to experimental studies using specific techniques adapted to magnitudes of viscosity (see Dingwell, 1999, for a review). Viscosity of a granitic melt varies with composition and,

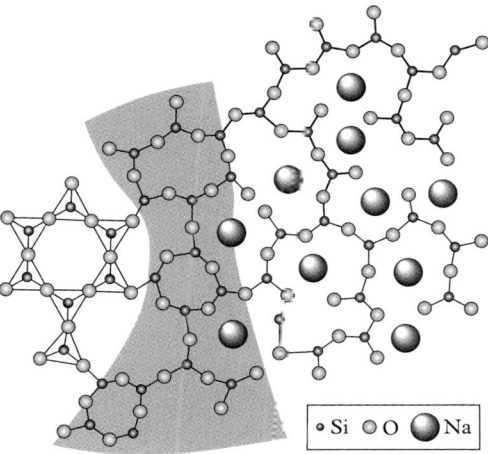

Figure 3.1 *Solid–liquid transition. Change from an ordered tetrahedral structure (quartz) to a disordered one (liquid) is accomplished through the 'vitreous transition' (grey zone) whose kinetics depends on the nature of the atoms. Here, sodium acts as a 'melter', i.e. breaks the Si–O bonds and occupies the large cavities left by the broken bonds. A similar role, with different kinetics however, is played by K, Fe, ..., or by anions such as OH, F and B, common in crustal fluids.*

above all, with temperature and water content. Unexpectedly, pressure has a tendency to decrease the viscosity, not particularly for a mechanical reason, but because pressure plays a dominant role on the amount of water accepted into solution by the melt (see Chapter 2).

Let us examine the effect of composition. Silica is well known to be a strong polymerizer or Si–O chain-former. Conversely, Na, K and Fe are the main Si–O chain-breakers (Fig. 3.1). However, if the content of chain-formers or chain-breakers is decisive for viscosities (see Box 3.1) in a broad range of compositions of magmatic liquids (dry felsic melts are known to be more viscous than mafic ones), then it has a minor effect on the narrow compositional range that concerns granitic to tonalitic melts.

3.2 Melt behaviour during partial melting

At the onset of partial melting, melt films are observed to develop at contacts between crystals, and to migrate along crystal edges then to pool at triple junctions (Fig. 3.3). At this stage, the higher the wetting capability of the melt, the faster its mobility. 'Wettability',

Box 3.1 Viscosity as a function of composition, temperature and water content

From a practical point of view, Bottinga and Weill (1972) calculated the viscosity (η) of anhydrous melts as a linear function of their composition, $\eta = x_i D_i$ using molar fractions (x_i) of oxides present in the liquid (SiO_2, TiO_2, ... , obtained from standard chemical analyses) and empirical constants, D_i, valid for a given domain of SiO_2% (65%–75% for granites s.l.) and a given temperature (1200 °C and above, by 50 °C steps). D_i are adjusted with respect to experimental viscosity measurements.

Decrease in viscosity with increasing temperature reflects the increase of the atomic or molecular mobility with temperature. To the first order, viscosity follows an Arrhenius equation: $\log \eta = a + b/T$, with η in Pa s and T in kelvin (Fig. 3.2a). Water content is a crucial factor for the viscosity of a melt: a viscosity decrease of 3–6 orders of magnitude is observed between 2 and 6% water content (Fig. 3.2b). Based on their data concerning haplogranitic compositions, Hess and Dingwell (1996) propose the following relation for $\eta(T,w)$, where w (0–8 wt %) is the water content of the melt:

$$\log \eta = \{[-3.545 + 833 \log w] + [9601 - 2368 \log w]\} / [T - 195.7 + 32.25 \log w]$$

Such a strongly nonlinear behaviour versus water content reflects the complexity of the transition that exists in silicates between the crystalline state, which has an ordered three-dimensional structure, and the liquid state, which has a disordered Si–O structure. The intermediate status between solid and liquid behaviour, called the vitreous transition (Fig. 3.1), highly depends on the rate of formation of Si–O bonds in the presence of cations (see above), or anions such as OH^-, B^- and F^-. OH^- depends on the water content, which greatly influences the viscosity of the melt. F^- and B^- may be common in the most differentiated granites and their associated fluids (see Chapter 14).

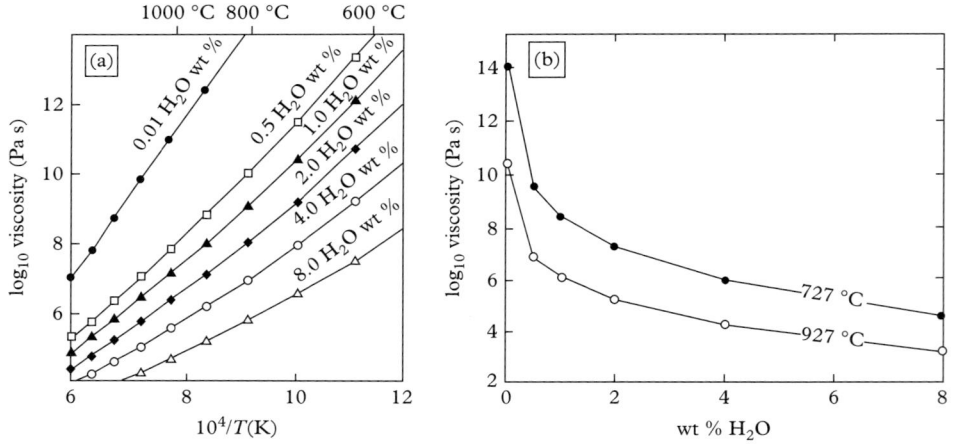

Figure 3.2 *Viscosity of a haplogranitic melt as a function of temperature and water content, after experiments by Hess and Dingwell (1996). (a) Viscosity vs. temperature curves (in H_2O wt %); this representation in $1/T$ emphasizes the changing slope of the curves with increasing water content, hence their departure from Arrhenius linear behaviour. (b) Viscosity vs. water content (curves labelled after temperatures in °C).*

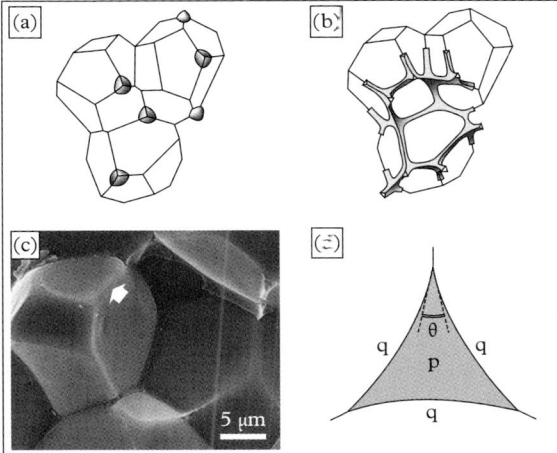

Figure 3.3 *Spatial distribution of the melt at the onset of melting. (a) First formed melt (in grey) gathers at triple grain junctions. (b) Network of minute interconnected channels along edges. (c) Electron microscope view illustrating channels of melt along grain edges (around the missing grain in the centre) of a quartz aggregate containing 0.2% of hydrated granitic melt (vitrified after the experiment); white arrow shows the aspect of a grain edge after accidental removal of its covering channel of glass. (d) Section of a channel filled with glass (l), perpendicular to an edge separating three grains (q), and showing the dihedral angle* θ. *After Laporte et al. (1997).*

or aptitude of a melt to spread over a solid surface, is quantified through the dihedral angle θ, or equilibrium angle between liquid-solid interfaces of a solid-solid-liquid triple junction. θ depends on the ratio between the solid–solid (γ_{ss}) and solid–liquid (γ_{sl}) surface energies, according to $\cos(\theta/2) = \gamma_{ss}/2\gamma_{sl}$. The dihedral angle is rather small for granitic melts, never larger than 60°, and usually between 10° and 30°. A granitic melt is therefore a wetting liquid that easily becomes interconnected within the solid crystal framework.

A few per cent of melt is sufficient to wet most of the crystals. The rock begins to weaken, and eventually the melt starts segregating. However, approximately 20–25% of melt is necessary to reach the critical melt fraction, a threshold above which an abrupt viscosity decrease is observed (Fig. 3.2). Segregation becomes effective when the melt exceeds this critical fraction.

3.3 Progressive partial melting experiments

This section examines the way in which the melt tends to collect before eventually escaping from its solid matrix. Experimental melt appears as glass because of the abrupt

44 *Segregation of granitic melts*

cooling (quenching) of the specimens at the end of each run. In static experiments, the only variable is temperature, making geometry and behaviour of the draining sites easier to define with confidence. In dynamic experiments, addition of a differential stress helps to evaluate the role of deformation in the process of segregation.

3.3.1 Melt location in a static aggregate

In experiments involving several species of crystals and different volumes of melt, it is observed that at low melt fractions, on the order of one to a few per cent, the melt sticks to the grains and nestles in the pores located along grain edges and at grain junctions (Fig. 3.3). Note in addition that γ_{sl} (hence θ) depends upon the orientation of the considered crystalline surface. Such an anisotropic behaviour is observed for feldspars, amphiboles and biotites: the melt better wets some surfaces than others.

In the case of partial melting of an amphibolite (Fig. 3.4), a rock mostly made of amphibole and plagioclase, it is observed that the melt: (i) has a low viscosity (10^3–10^4 Pa s), and (ii) better wets surfaces that are parallel to the prism than the basal ones. Considering that (iii) amphibole preferentially defines a lineation, and that (iv) amphibole and plagioclase define more or less continuous chains of crystals, all these observations explain why melt connectivity across the rock is reached for melt percentages as low as 5% or less. In conclusion, the melt may escape from its matrix at very low melt fractions.

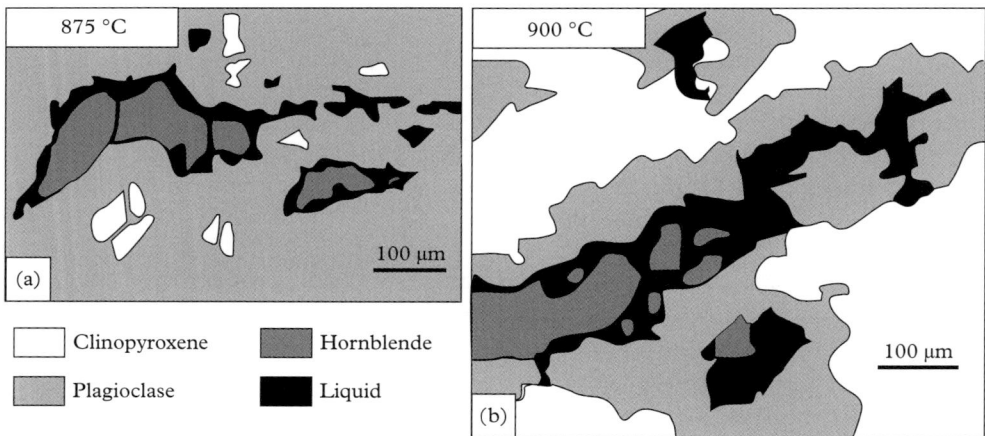

Figure 3.4 *Melt-pocket morphologies resulting from dehydration–melting at 1 GPa of a natural amphibolite. Hornblende–plagioclase contacts are impregnated with melt with preferential wetting along the prismatic faces of hornblende. (a) 875 °C, 14 days, liquid ~2%. (b) 900 °C, 14 days, liquid ~5%. After Wolf and Wyllie (1995).*

3.3.2 Dynamic melting of a granite

An experiment of granite 'dry' melting (no water added) under uniaxial compression at 900 °C is now considered (Rutter and Neumann, 1995). In fact, this experiment is unrealistic since granites rarely form granitic magmas. However, such experiments provide useful observations with respect to the rheology of a partially molten material. The main role of stress is to build up a network of both intra- and intergranular microfractures, a realistic process since any rock containing less than 10% of melt remains essentially brittle. It is also observed that the melt preferentially migrates towards the fractures under tension (the material is dilatant) and towards relay shear fractures (Fig. 3.5a). Since the specimen is constrained by the experimental device, the melt cannot escape from the experimental assemblage. As a consequence, volume increase due to the melting

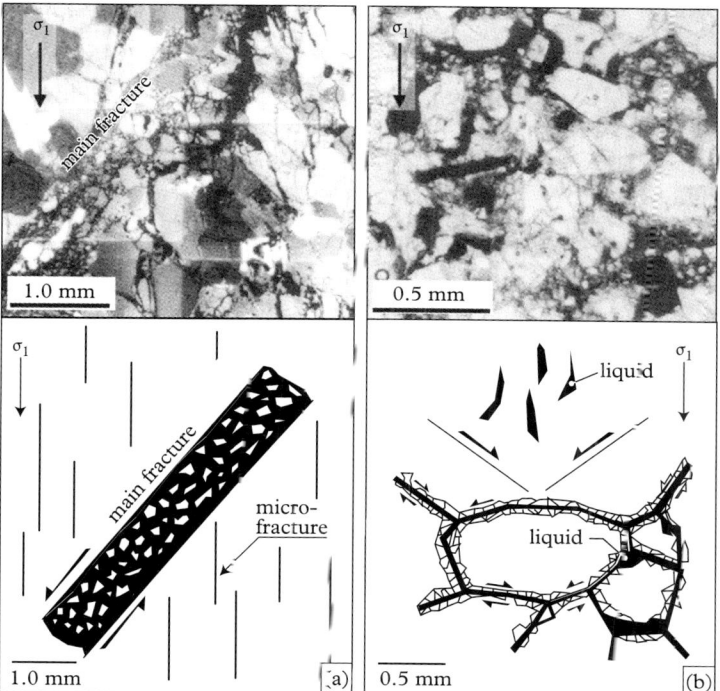

Figure 3.5 *Experimental partial melting of the Westerly granite under uniaxial stress (σ_1 vertical). Scanning electron microscope images (SEM, backscatter mode) and interpretative sketches. Melting was obtained by heating during a few hours (a), and a few days (b), at 8×10^{-5} s^{-1} strain rate under a confining pressure ($\sigma_1 + \sigma_2 + \sigma_3/3$) of 250 MPa in Paterson's apparatus (ANU-tech, Australia). Strain magnitude: 10–15% in (a) and 15–20% in (b). After Rutter and Neumann (1995).*

reaction (see Fig. 5.3) reinforces the fracturing process: this is the main drawback of such melting experiments.

When the melt fraction reaches high enough values (10–45%), microfracturing becomes widespread and, along with the interconnected conjugate fractures, melt migration is made possible through compression of the melt pathways and collapse of the matrix. Such a mechanism of deformation-assisted compaction, illustrated in Fig. 3.5b, is the cause of layering in migmatites, sometimes referred to as the 'filter-press' mechanism (see Box 3.2).

Finally, when the melt exceeds 45% in volume, it is no more restricted to pathways. Rather, it soaks the whole aggregate which then deforms viscously. The melt cannot be extracted any more from the solid matrix. In nature, this stage represents the magmatic stage in which the melt, together with a solid fraction (made of restites and eventual newly formed crystals), is susceptible to migration (see Chapter 5).

3.3.3 Dynamic partial melting of an amphibolite

Rushmer's (1995) experiments, also performed without added water, between 650 °C and 1000 °C at 1.5 GPa of confining (isostatic) pressure, conditions which are typical of the lower thickened crust, and under uniaxial compression (using a Griggs apparatus) at a (fast) strain rate of 10^{-5} s^{-1}, show that different micro-mechanisms operate according to the temperature of the experiment. (i) Below the solidus (650–800 °C) solid-state plastic deformation is apparently observed, giving a calculated bulk viscosity (η_{eff}) around 1.5×10^9 Pa s. While amphibole remains brittle, intragranular slip operates in plagioclase, as confirmed by undulose extinctions and slip along cleavages. (ii) At a temperature close to the solidus (850–900 °C), the specimen begin to weaken ($\eta_{eff} = 10^9$ to 5×10^8 Pa s) and, in addition to plastic deformation of plagioclase, several melt-infilled tension microfractures (parallel to the applied load) appear, affecting mostly the amphibole, but plagioclase is also affected (Fig. 3.6). (iii) Above 900 °C, the melt fraction increases up to 15% at 935 °C, along with development of microfractures and cataclastic micro-shear zones, as illustrated in Fig. 3.5b. (iv) Finally, at approximately 1000 °C, the specimen continues to weaken ($\eta_{eff} = 2.6 \times 10^8$ Pa s) with a melt percentage reaching 20%; the deformation becomes more homogeneously distributed giving a behaviour close to viscous, as in the experiments recorded in Section 3.2 but for a much lower melt fraction.

These experiments, performed at high strain rates in order to obtain high enough strain values at each experimental run, favour the development of microfractures. Nevertheless, their main virtue is to enlighten the importance, at low melt fractions, of the mechanical anisotropies that favour connectivity of the melt-forming sites.

Under differential stress, the most common situation in nature, the first percentage of melt is collected into micro tensional sites. Gathering of the melt is realised through both tension- and shear-microfractures. The high permeability of microfractures is due to their cataclastic microstructure, i.e. to the tiny pieces of broken crystals that fill the fractures, hence allowing a highly connected porosity. The role of volume change associated with the melt reaction remains to be discussed (see Section 4.2).

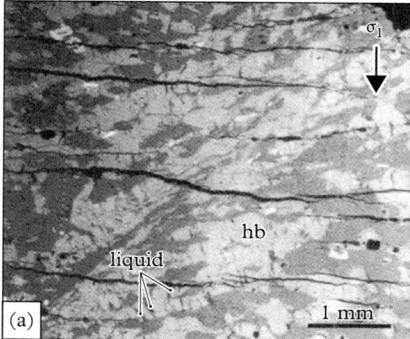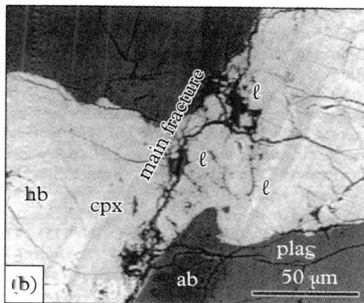

Figure 3.6 *Experimental incipient melting under uniaxial stress (σ_1 vertical) of an amphibolite at 850 °C. Scanning electron microscope backscatter images. (a) Low magnification: note the melt-infilled tension microfractures (~N–S features) mostly inside amphibole crystals; no melt is present in the microshear zone (NE–SW feature). The horizontal fractures (perpendicular to σ_1) result from decompression at the end of the experiment. (b) High magnification: the fracture line (at the centre) and the microfractures are filled with melt (l); note that, close to the main fracture, hornblende (hb) and plagioclase (plag) are transformed (melting reaction) respectively into clinopyroxene (cpx) and albite (ab), as indicated by the different grey tones. In Brown et al. (1995).*

3.4 Melt distribution in migmatites

If you observe a migmatite in the field, a part of its melt fraction forming the leucosome has already escaped from the rock. Except in the exceptional case of partial melting in a metamorphic aureole where the thermal peak is transient, the onset of segregation is never observed, hence justifying the importance of these experiments. Therefore, in the usual case of a long cooling time, the observed distribution of the melt characterizes the end of the anatectic event.

At the grain scale, melt is observed to spread along grain boundaries but hardly along fractures, contrary to the experiments. The reactant minerals have curved boundaries and often show corrosion embayments at triple grain junctions (Fig. 3.7a). The melt films gather into pockets (Fig. 3.7b) that are elongate parallel to the local grain shape anisotropy (foliation) or perpendicular to the rock lineation where tensional domains form.

At the outcrop scale, geometry of the draining network, although mostly driven by strain, is also conditioned by local mineralogy which determines the most adapted sites in the rock. Two main types of drains can be distinguished: (i) the local drain network, markedly branched, that distributes the melt in its site of formation (Fig. 3.8a and b), and (ii) the transfer network, significantly dictated by the local grain anisotropy (foliation or layering), which allows the melt to escape from its source (Fig. 3.8c). Highly oblique

48 *Segregation of granitic melts*

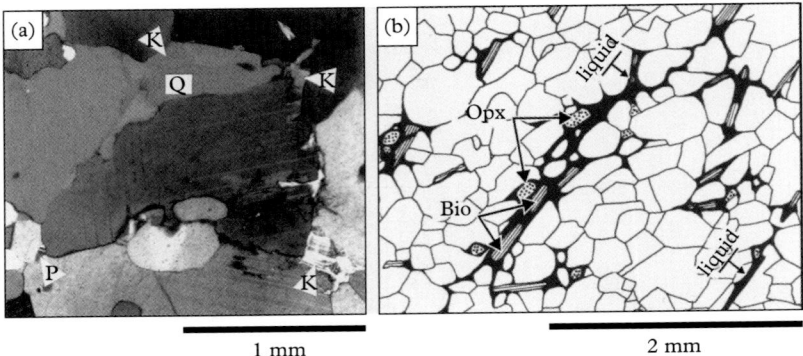

Figure 3.7 *Experimental incipient melting at the grain scale. (a) Microstructures witnessing the presence of melt; large (residual) plagioclase crystals with typical lobate and corroded boundaries, host small pockets of melt from which interstitial neoblasts of K-feldspar (K) and quartz (Q) were formed. (b) Melt pockets (in black) developing along the rock foliation, itself marked by alignments of biotite crystals (striped); melting begins at quartz–plagioclase–biotite contacts, and (in this example) neoblasts of orthopyroxene (dotted) develop during partial melting. In Sawyer (2001).*

channels are observed to develop slightly later (Fig. 3.8d), mostly favouring extraction of the foliation-parallel accumulation of melt (belonging to the transfer network) without increasing substantially the matrix drainage.

When melting stops, the small branch-like networks mostly disappear, leaving a few per cent of melt trapped in the matrix: this medium is called the melanosome (coloured grey in Fig. 3.8d). A number of transfer channels are entirely drained, but the late melt remains rooted in most of them along with newly formed crystals: this medium, so characteristic of migmatites, is called the leucosome.

3.4.1 Melt segregation rate

Residence time of the melt at contact with its source, hence the level of chemical equilibrium between melt and source, is controlled by the rate of melt segregation. We know that trace element contents in the melt are determined by the rate of dissolution of the accessory minerals (which concentrate these elements) that are in contact with the melt (Fig. 3.9a). For crustal melts, the content in Zr depends on the dissolution rate of the zircons that are contained in the protolith.

It has been shown experimentally that dissolution of zircon in a granitic melt, hence the amount of Zr in the melt, depends on the temperature and the water content of the melt. Dissolution in the melt proceeds up to the saturation in Zr. The empirical law of Watson and Harrison (1983) gives the content in Zr corresponding to saturation, calculated at the temperature of melt formation for any granitic melt of known composition:

$$\ln\left(D_{Zr}^{zircon/melt}\right) = \left[-3.8 - (0.85(M-1))\right] + 12{,}900/T$$

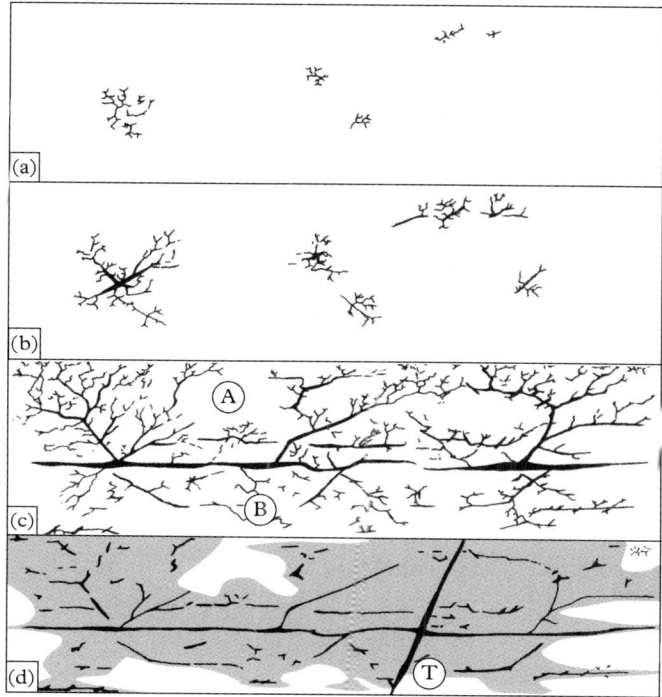

Figure 3.8 Sketches representing the evolution of drainage (a–c) and transfer (d) networks in a 50 cm-thick migmatitic layer. (a) Onset of partial melting. (b) Development of small branching networks. (c) Maximum development of the drainage network. Note that some areas are not affected by melting (A) while others have undergone complete melting. Melt was drained and its draining network has almost disappeared (B). (d) Transfer network stage: a dilational fracture (T), supplied with melt from the horizontal layers, favours the final collapse of the drainage network. The residual rock (melanosome) is represented in grey. After Sawyer (2001).

where D is the ratio of Zr concentrations (in ppm) in zircon (497,600) and melt, M is the cationic ratio $(Na + K + 2Ca)/(Al + Si)$ of the granite and T the temperature in kelvin. The calculated content in Zr is then compared to the real value analysed in the granite (Fig. 3.9b). If the analysed content is lower than the calculated one, it is concluded that the melt was not long enough in contact with its source to reach saturation in Zr. This melt is called a 'disequilibrium melt' with respect to its source and for the studied element. In this case, extraction of the melt took place before Zr saturation was reached. Applying the known dissolution rate of zircon in the melt, the maximum residence time of the melt in contact with its source can be derived. By contrast, if the

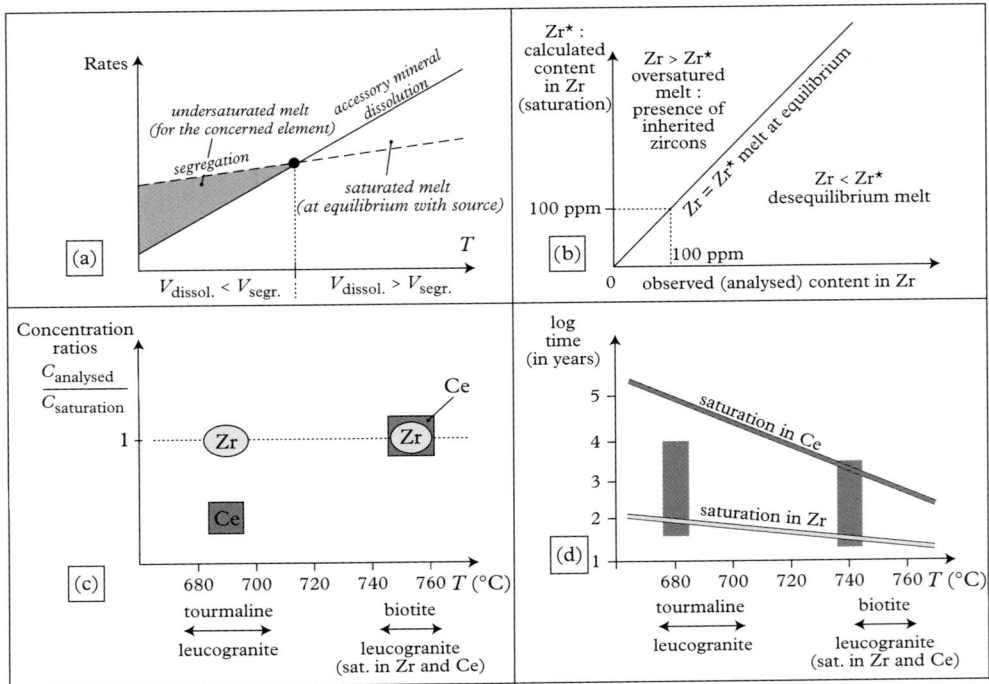

Figure 3.9 *Melt segregation rate. (a) Saturation and undersaturation domains of a melt as a function of temperature and rates of segregation and dissolution of the accessory mineral. (b) Comparison between analysed and calculated contents in Zr. (c) Analysed and calculated concentration ratios in trace elements (Zr and Ce) in leucogranites from Zanskar (Himalayas). (d) Residence times of leucogranitic magmas from Zanskar at contact with their source. After Ayres et al. (1997).*

melt is saturated in Zr, the known dissolution rate of zircon gives a minimum residence time before melt removal from the source.

The same reasoning is valid for the content in Ce through dissolution of monazite, a rare earth element-bearing phosphate mineral, that is an accessory mineral common in S-type and some A-type granites. Since the dissolution rate of monazite is two to three orders of magnitude lower than dissolution rate of zircon, the two different chronometers may therefore help to bracket the residence time of a magma in contact with its source. These chronometers were used by Ayres et al. (1997) for leucogranites from Zanskar (NW Himalaya) and allowed the authors to calculate a residence time not exceeding a few thousand years (Fig. 3.9c and d). Such residence times might be overestimates since accessory minerals may be entrained in the melt during segregation and migration, hence increasing the trace element content of the melt. In conclusion, magma segregation may, indeed, be a quick process, strengthening the idea of its association with some rapid tectonic events.

3.4.2 Role of the melting reaction

At a low melting rate, volume change, ΔV, associated with the melting reaction constrains the early appearance of melt mobility in the rock, hence its segregation. Melting reactions without added water, called 'dry', have positive volume changes ($\Delta V > 0$) hence positive slopes in (P,T) graphs. It is the case of melting reactions called 'dehydration melting', a common situation where melting is accompanied by destabilization of a hydrous phase, such as biotite and/or amphibole, and in which ΔV of the reaction products may reach 20%. By contrast, melting reaction in the presence of a water-rich fluid phase, an uncommon case in nature, has $\Delta V < 0$ (see Chapter 2).

In the previous examples (Figs 3.7 and 3.8), the increase in melt pressure due to the $\Delta V > 0$ melting reactions ($P_{melt} > P_{lithostatic}$) plays a key role for the localization of early melting sites, and for melt migration across the grain boundary network. Equalization of local pressures is the main driving process.

3.4.3 Role of gravity-driven compaction

In static conditions, i.e. in absence of deformation, gravity resulting from density contrast ($\Delta \rho$) between melt and crystals is the main driving force for melt migration, crystals being heavier than melt. The rate of ascent v_l of the melt by gravity-driven compaction, as deduced from porous media equations (McKenzie, 1985), is:

$$v_l = k(1-\phi)\Delta\rho g / \eta\phi$$

where k is the matrix permeability, ϕ its porosity (or melt fraction), η is the viscosity of the melt and g the gravity constant. Taking a partly molten 'dry' granite at $T = 900$ °C, $k = 2.5 \times 10^{-12}$ m^{-2}, $\phi = 10\%$ and $\Delta\rho = 500$ kg m^{-3}, the ascent rate v_l becomes 2.5 μm/yr for a viscosity of 10^5 Pa s. Accepting this figure, 10 million years would be required to obtain a 60 m-thick layer of melt resulting from the compaction of one km-thick layer of matrix + melt (Rutter and Neumann, 1995). In other words, this mechanism cannot be invoked for the formation of a whole pluton. By contrast, a few mm-thick migmatitic layer may form in a few thousand years by gravity-driven compaction (Fig. 3.10; see also Box 3.2), considering a melt fraction larger than 10%, preferably at $\Delta V > 0$, and supposing a large enough matrix permeability and melt connectivity (Brown et al., 1995). As already noticed, the addition of differential stress greatly facilitates melt mobility.

In conclusion, gravity-driven compaction is very efficient in the mantle due to the low viscosity of mafic melts, but is only a short-range process for granitic melts, for instance explaining the formation of layering in migmatites.

3.4.4 Fracturing assisted by melt pressure

In presence of a differential stress, a very common situation if all matrix grains are in contact with the melt, fracturing assisted by fluid pressure can be invoked as an efficient

Box 3.2 Filter-press mechanism: a dynamic compaction mode

Let us examine the filter-press mechanism, frequently evoked by geologists. The term filter press comes from industry (mining industry, metallurgy, ceramics, water industry ...), which uses compaction of muds of different kinds into low water content 'cakes'. In migmatites, a similar mechanism is acting, in a more efficient way than gravity, to enrich the melt in light-coloured layers at the expense of the dark-coloured ones that form the 'cake'. In order to take into account the alternating light- and dark-coloured layers, Robin (1979) proposes the following explanation (see Fig. 3.10). Consider a vertical force (overburden) applied to an aggregate of refractory grains forming a (dark) layer. Since this layer is more rigid and less deformable than a nearby aggregate containing a larger melt fraction, the differential stress that applies to the dark layer will increase. The pressure of the melt trapped in the dark layer will also increase. Therefore, the melt will tend to migrate toward lower pressure sites, i.e. toward light-coloured layers which become richer in melt fraction.

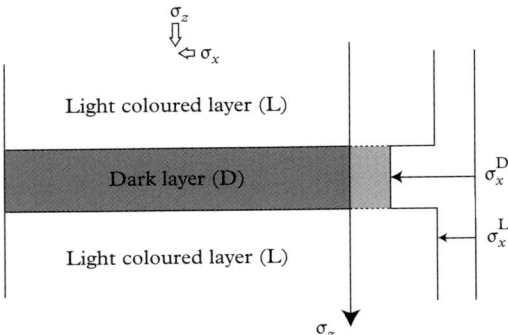

Figure 3.10 *Theoretical sketch of the filter-press mechanism. The dark layer (D, richer in Fe–Mg minerals) and the light-coloured layer (L, quartz and feldspar) have the same vertical stress σ_z (it is supposed that $\sigma_z \sim \sigma_1 > \sigma_x \sim \sigma_3$). The dark layer, the easiest to deform, experiences a horizontal stress σ_x^D larger than the light-coloured layer ($\sigma_x^D > \sigma_x^L$). The small volumes of melt that formed in the dark layers suffer a mean normal stress ($\sigma_n^D = (\sigma_z - \sigma_x^D)/2$) larger than that the one suffered by the melt contained in the light-coloured layers ($\sigma_n^L = (\sigma_z - \sigma_x^L)/2$). The difference is at the origin of the melt pressure gradient. After Robin, 1979.*

process. Mohr space (τ, σ_n) analysis of this process stipulates that the circle representing the state of stress in absence of melt will be translated toward the left (i.e. toward lower pressures) in presence of melt. The amount of translation is equal to the melt pressure (Fig. 3.11). At this stage, the Mohr circle centre represents the isostatic pressure, called 'effective pressure', suffered by the system. During its movement toward lower pressures, the Mohr circle may cross-cut the intrinsic curve characterizing the matrix, causing fracturing to develop. Note that rupture usually takes place in the form of shear fractures, particularly if differential stress is large ($\alpha \gg 0°$). Due to the draining action of the fractures, as soon as fracturing takes place, a drop in melt pressure is expected. The early drainage network is hardly preserved in the rock (Fig. 3.8) since fracture closure takes place as soon as drainage ends. Fracturation assisted by melt pressure therefore acts as a control valve mechanism. Melt segregation, whether continuous or not, has been numerically modelled by Vigneresse and Burg (2000) using a cellular automaton, i.e. a mathematical model to investigate self-organization in statistical mechanics (Wolfram, 1986).

3.5 Migmatites in the field

In the field, migmatites showing various stages of melt segregation can be observed. In fact, structures related to segregation can be seen only when some melt

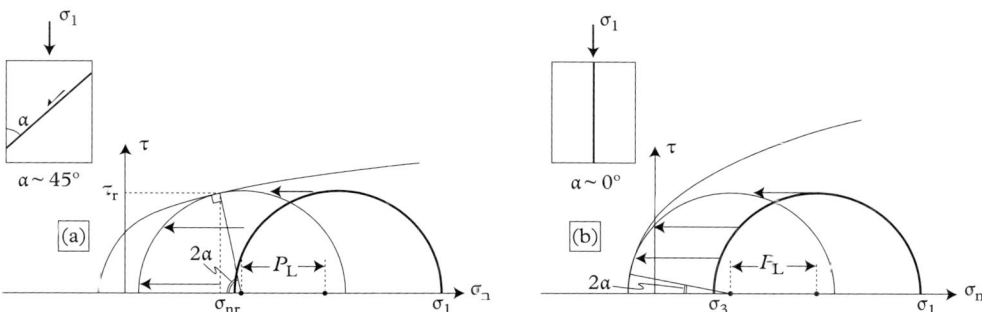

Figure 3.11 *Effect of the partial pressure of fluid, P_L, on fracturation of the host rock. For a given initial state of stress (circle in bold print) and for the same fluid pressure, a shear fracture will appear when the intrinsic curve is flat ((a): high 2α value; ductile material), otherwise a tensional fracture will appear ((b): small 2α; brittle material). In this Mohr diagram, τ represents the tangential, or shear stress, and σ_n the normal stress applied to the material; the circle printed in bold represents the state of stress ($\sigma_1 - \sigma_3$; $\sigma_1 > \sigma_3$ belong to the σ_n axis; $\tau = 0$); centre of circle, at $\sigma_n = (\sigma_1 + \sigma_3)/2$, represents the lithostatic pressure; when translated to the left by P_L, $(\sigma_1 - P_L + \sigma_3 - P_L)/2 = (\sigma_1 + \sigma_3)/2 - P_L$ gives the effective pressure in the presence of a fluid pressure P_L. The intrinsic (or rupture) curve of a material is the envelop of all the Mohr circles when rupture occurs. For a given state of stress at rupture (Mohr circle is tangent to the intrinsic curve): $\tau_r(rupt) = [(\sigma_1 - \sigma_3)/2] \sin 2\alpha$, and $\sigma_{nr}(rupt) = [(\sigma_1 + \sigma_3)/2] - [(\sigma_1 - \sigma_3)/2] \cos 2\alpha - P_L$. After Nicolas (1989).*

54 *Segregation of granitic melts*

is still present in the matrix. In summary, a migmatite exists when it has not been transformed into a granite or into a restite, or a rock composed only of refractory minerals.

3.5.1 Layering in migmatites

Layering, a feature characterizing stromatic migmatites, often derives from the compositional layering of the protolith. The latter, however, cannot be easily distinguished from layer-parallel melt segregation. The light-coloured layers are mm- to cm-thick and their spacing may reach a few centimetres (Fig. 3.12a).

Melt concentration within layers, forming leucosomes, or neosomes if new Fe–Mg bearing crystals are present, are favoured by the strong planar anisotropy of the foliated gneiss from which migmatites are often derived. Accordingly, darker layers (melanosome) are located immediately above or below the light-coloured ones, due to their depletion in quartz and feldspar.

The frequent almost horizontal attitude of migmatitic layers strongly suggests that gravity plays a role through matrix compaction and melt migration from one layer to the other. However, some features (in the field), or microstructures (under the microscope), confirm that such a mechanism was operating in exceptional circumstances. Such features are often erased by deformation following segregation. Nonetheless, in the Velay migmatites (French Massif Central), where subsequent deformation is only incipient, Burg and Vanderhaeghe (1993) showed that gravity-driven compaction played a role

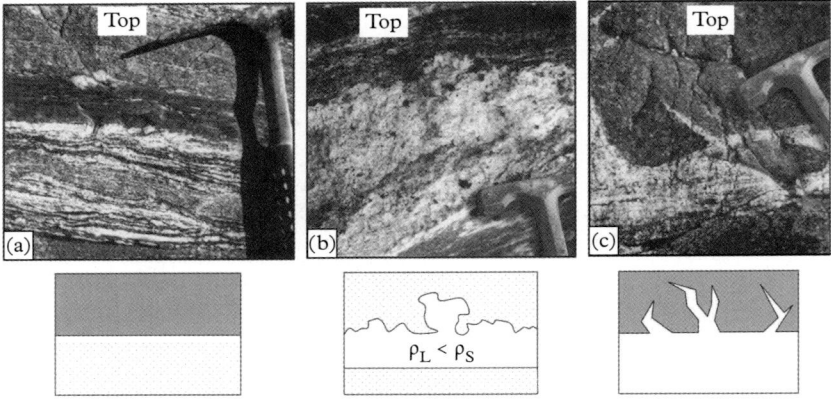

Figure 3.12 *Polarity criteria suggesting gravity-driven compaction in a migmatite from the Velay massif (French Massif Central). (a) The light-coloured layer (enriched in melt) at the image centre, becomes clearer upwards, at the contact with the dark layer, which is more rigid and impermeable. Due to the ascending pressure of the light-coloured layer, its upper contact reveals a cauliflower-shape (b), or fractures of the dark layer (c). After Burg and Vanderhaeghe (1993).*

for melt segregation, and established several polarity criteria based on the outline of the leucosome–host interface (Fig. 3.12).

3.5.2 Other migmatitic structures

Once a given percentage of melt is reached, the role of deformation during partial melting is illustrated by melt migration along shear zones (Fig. 3.13). Dilatant sites, or sites into which melt is supplied by a local pressure gradient, are commonly observed in the field. Such sites correspond to local geometries such as anticlinal fold hinges, shadow tails around enclaves, or inter-boudin volumes of a less deformable amphibolite layer within a gneissic matrix (Fig. 3.14b). Here again, the theory of fracturing assisted by melt pressure may be applied. The slope of the intrinsic curve being steeper in a brittle material, the Mohr circle at contact with the intrinsic curve yields low α values close to rupture in tension (Fig. 3.11b). In practice, for almost identical values of differential stress and melt pressure, tension fractures or shear fractures will appear in the gneissic or amphibolitic material, according to the nature and distribution of these lithologies.

Nebulites (Fig. 3.15), or diatexites, represent cases where the melt fraction, which has reached rather high values (>>20%), could not be extracted from its matrix due to insufficient differential stress, insufficient anisotropy or $\Delta V < 0$ melting reaction. In such cases, the bulk composition of the migmatite is close to that of its protolith. Hence nebulites have a viscous behaviour, the bulk viscosity being very low compared to that of the protolith. In the extreme case, the material under consideration is no more a migmatite but rather a granitic magma that contains a certain fraction of restitic material which may begin to separate during transport (see Chapter 5).

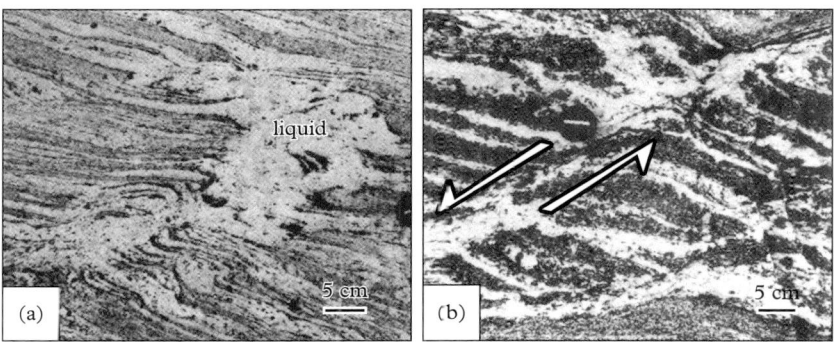

Figure 3.13 *Drainage of melt in migmatites along shear zones. (a) A folded layer in a migmatitic gneiss evolves into a shear zone. (b) Sinistral, tensional shear zone in a layered amphibolite. Lower Proterozoic of Karelia (Russia). From Brown et al. (1995).*

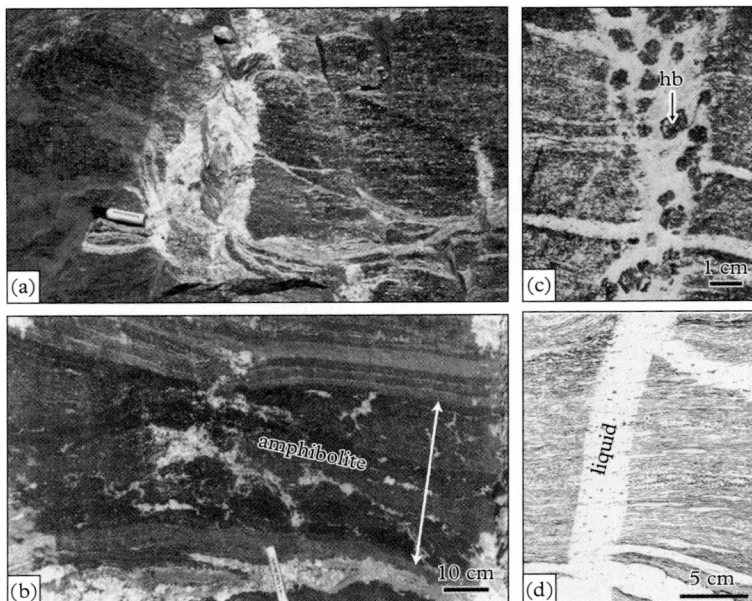

Figure 3.14 *Melt collection in tensional domains. (a) In a tension site of sillimanite-bearing micaschists from the Main Central Thrust pile, Himalayas. (b) In a volume under tension of amphibolites undergoing boudinage in the Grenville Front (in Sawyer, 1991). (c) The neosome of this tension gash contains, in addition to quartz and feldspar, large neoblasts of hornblende (Todtnauberg migmatite, Black Forest; in Mehnert, 1968). (d) Preservation of the fine structure of the host-gneiss in this tension gash, or dyke, suggests its progressive feeding from the dyke walls (migmatite from Urenkopf, Black Forest; in Mehnert, 1968).*

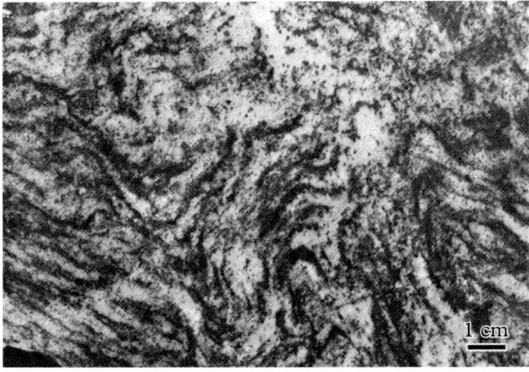

Figure 3.15 *Nebulitic migmatite or migmatitic granite. The material is greatly weakened by the amount of mobile fraction. Origin: Black Forest (Germany). From Mehnert (1968).*

3.6 Conclusions

Segregation is the process helping extraction of granitic melt from its protolith. Segregation is facilitated by a low melt viscosity, hence by a high temperature and even more by a high water content. On the one hand, the melt water content is limited by the rather small volume of water brought by the hydrated phases (principally biotite and amphibole) participating to the dehydration–melting reactions. On the other hand, since the resulting water-undersaturated melt occupies a volume larger than its solid counterpart, the increase in melt pressure will facilitate fracturing of the source rock. Finally, mobility of the first formed melt is greatly facilitated by the application of differential stress, which increases the number of melt collection sites, and hence prepares the process of magma transport (Chapter 5).

I and S granitic melts never contain 100% liquid. They carry restitic crystals inherited from the protolith, or newly formed crystals resulting from the melting reaction.

Geochemical tools that use, for example, trace element contents derived from dissolution of accessory minerals, allow segregation rates of crustal magmas, hence residence times at contact with protoliths, to be obtained. Residence times are thus revealed to be rather short. Magma segregation therefore appears to be a transient process (at least at the geological scale) in accordance with the common thought that segregation is facilitated by deformation, itself usually related to tectonic processes.

4
Genesis of hybrid granitoids: mingling and mixing

Coexistence of granitic and basaltic magmas is commonly observed in the field. Geochemical, and especially isotopic, data suggest that these magmas are able to mix (see Chapter 2 and Section 4.5). At this stage, two different cases must be distinguished: either mingling, i.e. incomplete mixing, in which both components are still recognizable by their colour or texture, or proper mixing, resulting in a hybrid rock.

Granitic rocks in the broadest sense may be hybrid, i.e. produced by mingling and/or mixing of magmas derived from mantle and crustal sources. For US geologists, who are familiar with magma production in an active margin setting, such as the North American cordillera, hybridization represents an important process of production of granitic rocks. By contrast, European geologists often favour a dominantly crustal origin for these rocks. Influence of the geological setting on the nature and origin of granites will be examined in Chapter 12.

4.1 Field observations

Conspicuous magma mingling features are commonly observed in granitic massifs. Usually, it is the occurrence of dark enclaves, varying from a few centimetres to a few metres in size. Their shapes are rounded, elliptical or lobate, suggesting mingling between coeval felsic and mafic magmas, the felsic one being the host and the mafic one constituting the enclaves. Contrast in colour between both rocks is amplified by the difference in grain size: the enclaves, that always display a finer grain-size, appear darker than their host. Enclave–host contact is not always clear-cut; it may be progressive, indicating that a limited contamination between both magmas took place (Fig. 4.1a and b). Locally, a rock with an intermediate colour may be observed, as a result of some hybridization between both magmas. However, more careful observation often helps to recognize heterogeneous domains within the hybrid magmas, for instance clots, or aggregates, of ferromagnesian minerals. Contamination may also be shown by crystals originating from the host, such as orthoclase megacrysts, that are observed in the enclave. These

Granites. First Edition. Anne Nédélec and Jean-Luc Bouchez. © Oxford University Press 2015.
English edition published in 2015 by Oxford University Press.

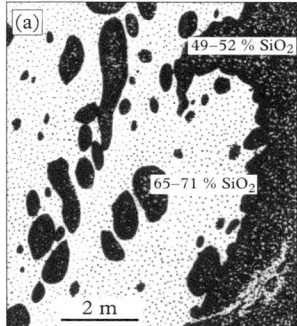 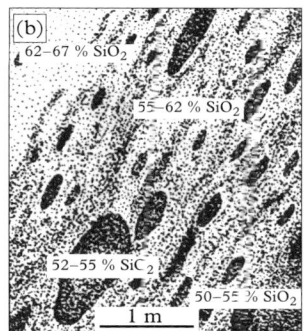 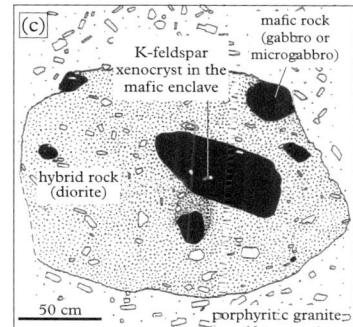

Figure 4.1 *Diversity of enclave–host relationships. (a, b) Lamarck granodiorite (Sierra Nevada batholith, California); note the compositional heterogeneities pointing to incipient hybridization during mingling. After Frost and Mahood (1987). (c) Double enclave (Shap Fell granite, Lake District National Park, Cumbria, UK); orthoclase phenocrysts, similar to those of the host granite, are also present in the (hybrid) diorite and in the microgabbro. After Pitcher (1993).*

megacrysts appear to have been mechanically introduced while the enclave was soft and incompletely crystallized (Fig. 4.1c).

Elsewhere, enclaves with chilled margins may be observed with a finer grain size, indicating rapid cooling in the host magma. In other cases, enclaves display fractures infilled by their host granite. Such a variety of situations can be observed in the same granite massif, and even locally in the same enclave! They record different crystallization histories. According to its more mafic composition, the enclave magma crystallizes earlier than the host magma. In the case of mingling between both magmas, mixing appears limited. Indeed, although two silicate liquids of different compositions can theoretically mix, their different crystalline fractions, hence their different effective viscosities, actually prevent or hinder mixing.

4.2 Influence of viscosity on magma mingling or mixing

First of all, viscosity depends on the composition of the silicate liquid, on its temperature and on the content of water, fluorine, boron ...; all of these factors are possibly responsible for viscosity changes up to several orders of magnitude. But a magma is not only a silicate liquid, it also contains a variable amount of crystals: it is a suspension. These crystals may have been extracted from the protolith, therefore representing either restitic material or products of the melting reaction (Section 2.2.3). The viscosity of a magma is therefore dependent on its chemical composition, but even more on its crystalline load, that increases during cooling (see Box 4.1).

The curve representing the evolution of viscosity vs. solid load in Fig. 4.2 corresponds to a granitic magma. For a basaltic magma, the curve has the same sigmoidal

Box 4.1 Viscosity of a suspension: Einstein's model

The viscosity of a suspension consisting of liquid + crystals (L + S) is called 'effective' when it is normalized to the viscosity of the liquid, emphasizing the role of the crystal load. One of the first attempts to model the viscosity of a suspension was reported by Einstein (1906) and further developed by Roscoe (1952). It is known as the Einstein–Roscoe (E-R) equation:

$$\eta = \eta_0 (1 - RS)^{-2.5}$$

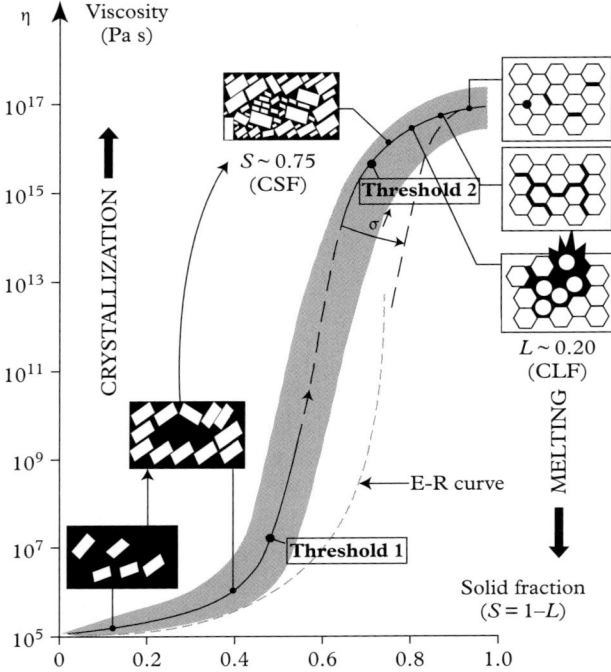

Figure 4.2 *Evolution of magma viscosity versus solid fraction (S). The dashed Einstein–Roscoe (E-R) curve to the right of the diagram, asymptotic to an S value of ~0.75, results from the E-R equation at R = 1.35 and is valid for low concentrations. The large grey zone corresponds to more realistic curves. Small sketches to the left represent liquid–solid transition due to crystallization; CSF: critical solid fraction corresponding to the appearance of rheological behaviour close to that of a solid. To the right, the small sketches represent solid–liquid transition due to partial melting; CLF: critical liquid fraction corresponding to liquid segregation. The percolation thresholds can be slightly modified, for instance CLF decreases when the applied stress increases. After Nicolas (1992) and Vigneresse et al. (1996).*

continued

> **Box 4.1** *continued*
>
> where η_0 stands for the viscosity of the suspending liquid alone and R is a constant based on the volume ratio of the solid fraction. For a stacking of identical spheres of maximum compactness, $R = 1.35$. In this case, the ratio $S/(S + L)$ for a stacking of rigid spheres (S) is 0.75 (causing a 25% porosity, hence the large retaining capacity of sand reservoirs), leading to a value of 1 for the product RS, and consequently to an infinite viscosity. This is the reason why this equation is realistic only at low concentrations. This approach was refined on many occasions (taking account of the particle sizes and distribution, of the temperature dependence of η_0, etc.). The resulting variation of viscosity with the crystal load is highly nonlinear, as can be seen by the sigmoidal curve representing the log of the viscosity vs. S (Fig. 4.2).
>
> The viscosity is increased by about ten orders of magnitude when the crystal load changes from 50% to 75%, with conspicuous consequences on the magma rheology. Indeed, for $S > 50–55\%$ (the rigid percolation threshold of Vigneresse and Tickoff, 1999), the crystals come in contact and build a framework able to support some stress, although the interstitial liquid is still able to filter through this crystalline framework. For $S > 70–75\%$ (the locking framework threshold or critical solid fraction), the crystalline framework becomes rigid. At that time, and despite the fact that some residual liquid is still present, the magma behaves like a solid, i.e. it can support shearing stress and may fracture. The location of the thresholds may vary, but the sigmoidal shape of the curve remains unchanged.

shape, but begins at viscosities *ca* five orders of magnitude lower. The basalt viscosity follows the same evolution, but not at the same temperature: increasing viscosity due to crystallization occurs at much higher temperature for a basaltic magma than for a granitic magma (Fig. 4.3a and b). Below the so-called inversion temperature T_i, the viscosity of the basaltic magma becomes higher than the viscosity of the granitic magma. Mixing is only possible if $T \geq T_i$. Below this temperature, mixing is difficult or impossible. This is the common situation of basaltic enclaves preserved in their granitic host, because the latter, also formed at lower temperatures, represents the larger magmatic volume. Indeed, when two different magmas come in contact, they do not have the same temperature. Thermal equilibrium is reached fast enough, at least in the contact zone, and corresponds to an intermediate temperature T_e, determined by the respective volumes of the coexisting magmas. Then, both magmas cool at the same time with the same temperature decrease, but each magma follows its own viscosity evolution, hence the variety of contacts observed (Fig. 4.1), corresponding to the different stages of this evolution. It ensures that complete hybridization can only occur between magmas with similar compositions, and that carry only small solid charges when they come in contact.

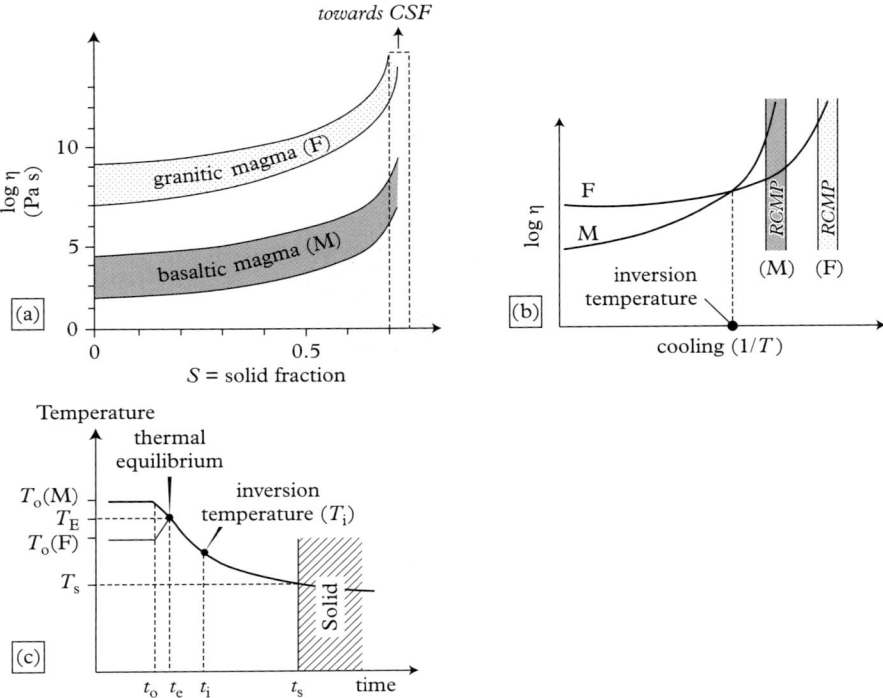

Figure 4.3 *Comparison of the evolution of two magmas: felsic (F) and mafic (M). (a) Viscosities vs. solid fraction; CSF: critical solid fraction. (b) Viscosities vs temperature; RCMP = rheological critical melt percentage, equivalent to CLF in Fig. 4.2. (c) Evolution of temperature (T) vs. time (t). After Fernandez and Gasquet (1994).*

4.3 Characteristics of hybrid rocks

Examples cited in Fig. 4.1 correspond to cases where the granitic magma is more important in volume and where mainly mingling occurs. Consider a different situation that will end in mixing. Rocks resulting from such a mixing process display specific characters typical of hybrid rocks.

4.3.1 Mineralogy

As discussed previously, only magmas containing a small percentage of crystals are able to become hybridized. At the very instant of mixing, these early crystals are submitted to physical and chemical conditions different from those prevailing during their growth in the basic or in the granitic magma. They sometimes record this perturbation. This is the case with plagioclase, a mineral that does not re-equilibrate quickly with its environment; hence it has a zoned appearance. Without any perturbation, zoning is normal with

Characteristics of hybrid rocks 63

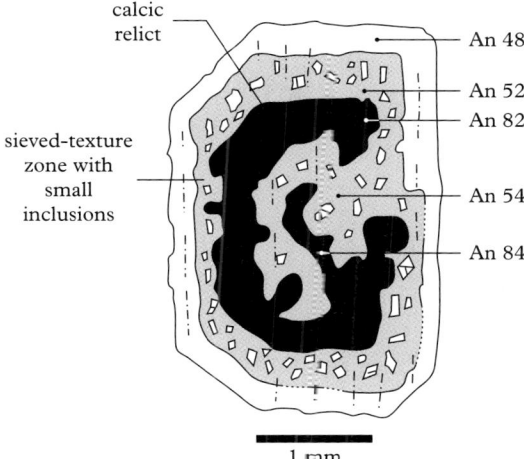

Figure 4.4 *Plagioclase with corroded core and sieve texture from an enclave in the Bono granite (Sardinia, Italy). The calcic core (An84) crystallized in a mafic magma that was suddenly cooled at the contact with a granitic magma. The core was then partly resorbed and small crystals of pyroxene and Fe–Ti oxide nucleated in the middle zone of composition An54–52. The outermost zone (An48) corresponds to the final crystallization of the hybrid magma. After Cocirta (1986).*

a calcic core surrounded by successive growth zones of a more and more sodic composition. Meanwhile, if these crystals are transported in a different magma resulting in a disequilibrium situation, their zoning will be modified. Discontinuous or complex zoning may be observed, with a corroded core, resorption zones or sieved zones characterized by small mineral inclusions nucleated during the sudden temperature decrease of the host silicate liquid (Fig. 4.4). The acicular shape of apatite crystals, a common accessory mineral, is also typical of a fast cooling stage or undercooling. Such textures are frequently observed in granitic rocks of intermediate composition and are pieces of evidence suggesting magma mixing. Finally, alkali feldspars typically surrounded by a plagioclase rim in the rapakivi granites are also regarded as the result of magma mixing (see Chapter 13).

4.3.2 Major and trace element chemistry

As expected, hybrid rocks are often tonalitic to granodioritic in composition, because they display compositions intermediate between their parental (granitic and basaltic) magmas. In binary diagrams of the type % oxide = f(% SiO_2), or Harker diagrams, hybrid rocks plot along a straight line, whose two end compositions correspond to the parental magmas, either postulated or confirmed in the field. This straight line is called a mixing line (Fig. 4.5b).

64 *Genesis of hybrid granitoids: mingling and mixing*

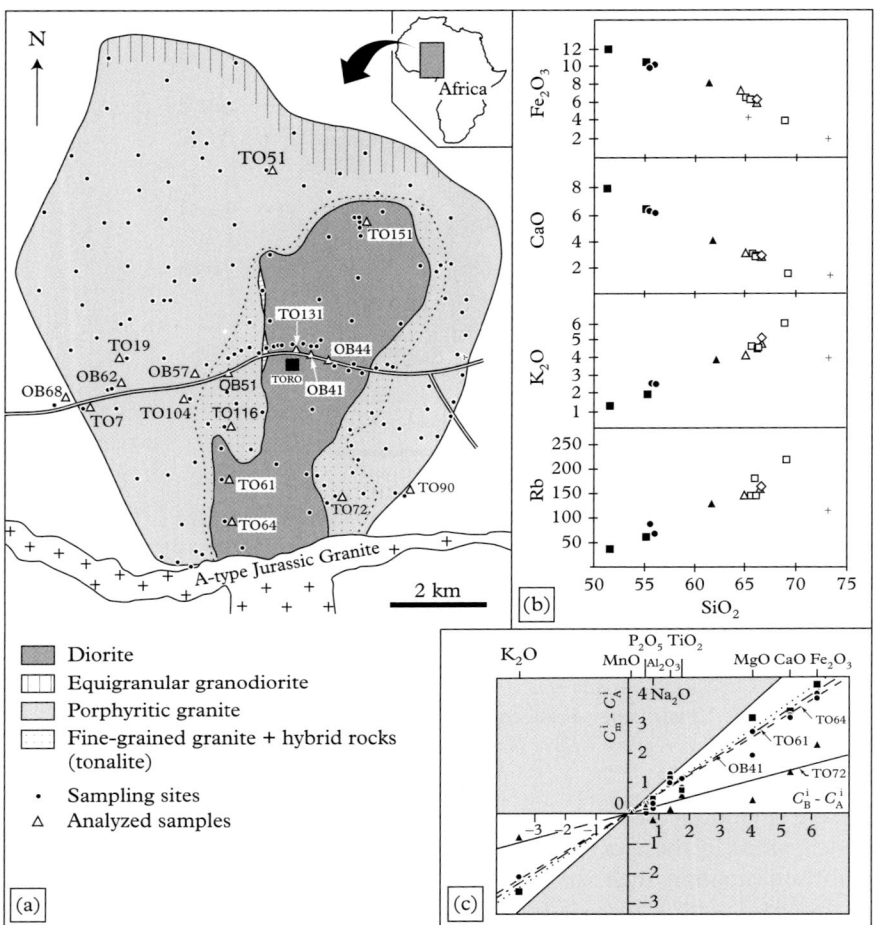

Figure 4.5 *Magmatic interactions in the bimodal plutonic complex of Toro (Nigeria). (a) Petrographic map with location of analysed samples. (b) Mixing lines in Harker diagrams. (c) Test of Fourcade and Allègre (1981) considering OB44 diorite sample as the mafic (B, basic) parental magma and a granite average composition as the felsic (A, acid) parental magma; TO72 tonalite sample displays mineral features suggesting a hybrid rock and corresponds to magma mixing case with a mafic fraction of 0.32; some diorites (TO64 and TO61) also correspond to hybrid rocks with a mafic fraction of 0.76 to 0.70. After Déléris et al. (1996).*

The mixing test of Fourcade and Allègre (1981) enables one to quantify the mixing proportions. Let A be the felsic (acid) parent, B the mafic (basic) one and M the tested hybrid mixture. Concentration C of a given element in the hybrid is related to the concentration of the same element in each parent magma by the equation:

$$C_M - C_A = p(C_B - C_A),$$

where p, with a value between 0 and 1, is the fraction of mafic magma (f_B) in the tested hybrid (M). For perfect mixing, all points corresponding to specific elements would be aligned and p would be the slope of this line. Practically, the best line passing through the origin is calculated (linear regression). The test can be used with major elements (as oxide weight percentages) as well as with trace elements (in ppm). An example of test application is provided in Fig. 4.5c.

4.3.3 Isotopic signatures

When we consider the isotopic ratios of an element, mixing of two components A and B results in a hyperbola and not a straight line, as is shown below for Sr. Element concentrations in mixtures M are given by the equation:

$$(Sr)_M = [(Sr)_A f_A] + [(Sr)_B (1 - f_A)] \text{ with } f_A = 1 - f_B.$$

If A and B do not display the same isotopic signature, their respective contributions must be multiplied by their concentration ratios:

$$(^{87}Sr/^{86}Sr)_M = (^{87}Sr/^{86}Sr)_A f_A [(Sr)_A/(Sr)_M] + (^{87}Sr/^{86}Sr)_B (1-f_A)[(Sr)_B/(Sr)_M]$$

This equation is of the type: $(^{87}Sr/^{86}Sr)_M = [a/(Sr)_M] + b$, which corresponds to a hyperbola (Fig. 4.6). The hyperbola can be replaced by a straight line considering the concentration inverse ratio (1/Sr). The hyperbola concavity depends on the concentration ratios $(Sr)_A/(Sr)_B$. Thus, in the case of a granitic magma (A) that is less Sr-rich than a basaltic magma (B), the concavity is upward (Fig. 4.6a). Notice that the isotopic ratio of the mixture rapidly decreases with respect to the isotopic ratio of magma A, when a small quantity of B is added to A. In

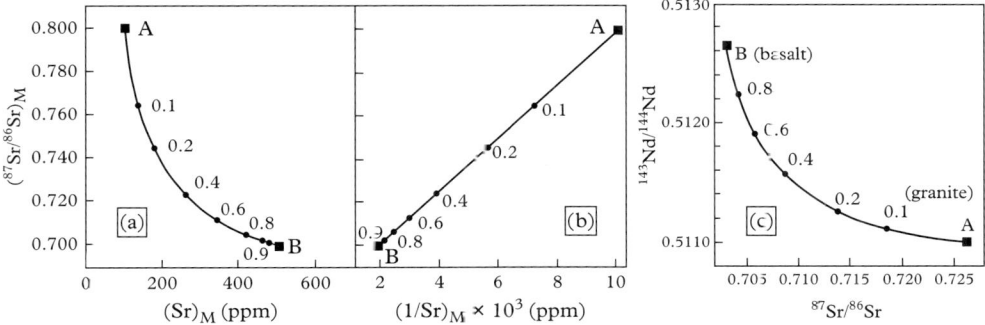

Figure 4.6 *Isotopic ratios resulting from mixing of granitic (A) and basaltic (B) magmas. Numbers along the curves correspond to the basaltic fraction (f_B). (a) Example of mixing hyperbola. (b) Conversion of the hyperbola to a straight line in the Sr ratio vs. 1/Sr diagram. (c) Combination of two isotopic pairs in another example. After Faure (2001).*

66 Genesis of hybrid granitoids: mingling and mixing

the same way, it is possible to calculate the evolution of the Nd or O isotopic signature in magma mixing. Combining both isotopic ratios, still results in a hyperbola (see Figs 2.23 and 4.6c).

Note that the isotopic evidence alone is not sufficient to prove that a magmatic rock was derived by magma mixing. Sometimes, the so-called 'hybrid' signature of the magma is inherited from a mixed source (or mixed protolith). Finally, a basaltic magma can assimilate (i.e. dissolve) a small volume of crustal rocks, without these rocks forming a distinct crustal magma. In the case of very old crust (with a high $^{87}Sr/^{86}Sr$ and a very negative ε_{Nd}), the isotopic signature of the contaminated magma will be noticeably modified. Petrological evidence with additional field observations is required to prove the occurrence of magma mixing.

4.4 Processes of magma mingling and mixing

To have mixing when two contrasting magmas mingle not only requires transfer of crystals from one magma to the other, but also some process involving the liquid phases. Let us consider first the interdiffusional process of chemical elements.

4.4.1 Element interdiffusion in silicate liquids

Diffusion of an element i from one liquid to another in a sense opposite to the concentration gradient can be defined by Fick's first law:

$$\mathcal{J}_i = -D\nabla C_i,$$

where $\mathcal{J}$ is the flux of the element i during a time unit through an unit area perpendicular to the transport direction, D is the diffusion coefficient (or diffusivity) and ∇C_i is the concentration gradient in all space directions. Fick's second law enables one to calculate the evolution of concentration C with time (Fig. 4.7).

The diffusion coefficient depends on the temperature T following the Arrhenius law:

$$D = D_0 \exp(-E/RT),$$

where E represents the activation energy and R is the universal gas constant. In addition, D is inversely proportional to the liquid viscosity η and to the radius r of the element of interest, in agreement with the Stokes–Einstein relation and enabling the diffusivities of cations with similar charges to be compared:

$$D = kT/6\pi\eta r,$$

where k is the Boltzmann constant. Diffusion coefficients can be determined experimentally. They are always low in granitic liquids, with values between 10^{-16} and 10^{-12} m^2 s^{-1} for most major and trace elements, except alkalis (Na^+ and K^+) that diffuse faster

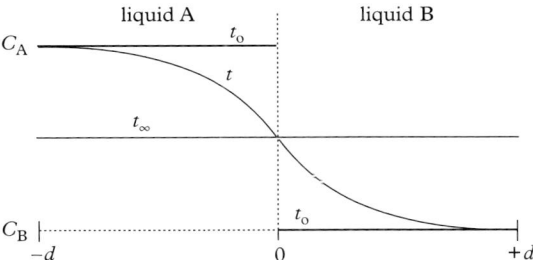

Figure 4.7 *Evolution of the concentrations (C) of an element versus time (t) as the result of an interdiffusion between two liquids A and B; t_o: initial situation; t_∞: uniform concentrations on both sides.*

($D \sim 10^{-12}$ to 10^{-10} m^2 s^{-1}). D is increased by one or two orders of magnitude when the silicate liquid contains water. Water itself, or rather the hydroxyl ion OH$^-$, diffuses nearly as quickly as the alkalis.

Actually, the average diffusion length of an element in a granitic liquid is only a few metres during one million years! According to the fact that a granitic pluton can crystallize in a much shorter time, a hybrid rock cannot result from the sole interdiffusion mechanism. By contrast, this process can play a significant role in basaltic magmas, which are both hotter and less viscous than granitic ones, at least regarding the diffusion of water and potassium. This is a likely explanation for the common occurrence of biotite in mafic enclaves hosted in granitic rocks.

4.4.2 Transfer of granitic liquid into mafic magma enclaves

Diffusion of water and potassium is not sufficient to explain the hybrid compositions of some mafic enclaves. By making a series of thin sections, it is possible to reconstruct the three-dimensional network of granitic liquid in a mafic enclave (Fig. 4.8a). The network is connected to the host granite magma and represents the open porosity that persisted during crystallization of the mafic enclave. In thin sections, the infiltrated granitic liquid is recognized as interstitial quartz and K-feldspar. Therefore, granitic contamination of a mafic enclave initially results from fast infiltration of granitic liquid along distances up to a few metres, eventually followed by a much slower chemical process due to element diffusion at the scale of individual grains (Fig. 4.8b).

4.4.3 Dispersion of mafic magma into a granitic magma host

Formation of hybrid granitoids with intermediate compositions requires the kinetic dispersion of mafic components into the original granitic host. The corresponding conditions are ruled by fluid mechanics. Dynamic conditions are considered, namely the

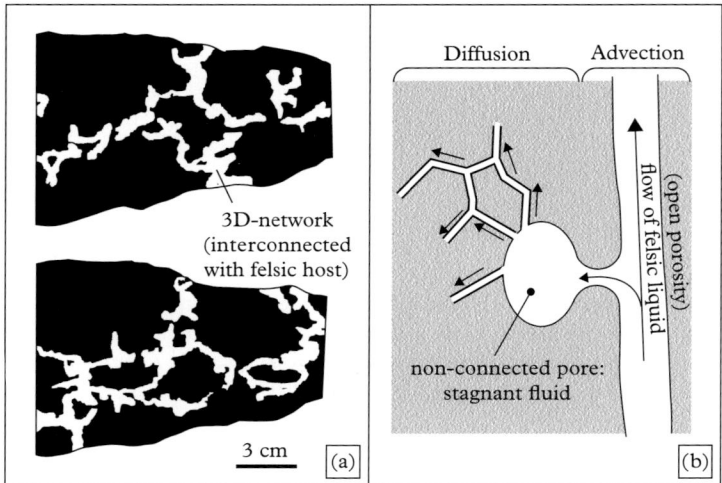

Figure 4.8 *The contamination process of a basaltic enclave. (a) Successive thin sections of a series cut from the same enclave; the enclave host is in black; the white network corresponds to infiltrated granitic liquid. (b) At a non-connected pore, liquid advection is followed by liquid–solid diffusion at grain boundaries. After Petford et al. (1996).*

injection of mafic magma B (density ρ_B, viscosity η_B) coming from a conduit of width (w) at speed (u), into felsic magma A (ρ_A, η_A). The density of B being greater than the density of A, it is expected that magma B will remain stagnant on the floor of the magma chamber without mixing with A. However, if the initial speed of B is sufficient, B can ascend through the magma chamber.

The nature of the flow of mafic magma B into felsic magma A is determined by its Reynolds number:

$$\mathrm{Re} = \rho_B Q / \eta = Q / \nu_B,$$

where ν_B is the kinematic viscosity ($\nu_B = \eta/\rho_B$) of magma B and Q is the input rate, calculated as follows:

$$Q = [(g\Delta\rho)/(f\rho_B)]^{1/2} w^{3/2},$$

in which $\Delta\rho$ is the density difference between B and A, f is a friction coefficient and w is the width of the input conduit. Increasing Re values successively first result in laminar flow of the mafic magma without any mixing or mingling with the felsic magma, then instabilities occur and form enclaves of B in A. Finally a turbulent plume of B in A appears if Re > 400 (Fig. 4.9). To reach this condition of turbulent flow or fountaining of magma, the mafic magma input must occur through a conduit, whose minimum width is one metre.

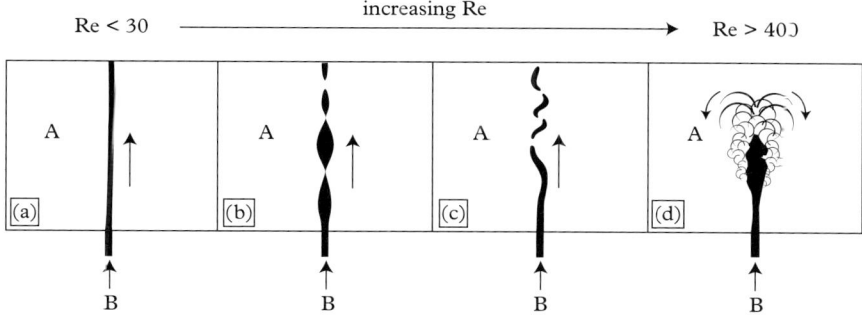

Figure 4.9 *Analog models of the injection of a magma B into a magma A. In order to obtain appropriate Reynolds numbers (Re), both magmas are simulated by variable proportions of water and glycerine mixtures in experiments (a–c), and by the injection of coloured salty water in a tank of fresh water in experiment (d). (a) Laminar flow. (b) Varicose vein-like instabilities. (c) Meander-like instabilities. (d) Fountaining (turbulent) plume. After Huppert et al. (1986).*

Formation of a turbulent plume or fountain is a necessary condition to obtain a well-mixed hybrid magma, but it is not sufficient. Campbell and Turner (1986) demonstrated that the viscosity of magma A is a critical parameter. Indeed, if the ratio η_B/η_A is greater than 70, complete mixing is theoretically possible. Conversely, for a ratio less than 7, there is no mixing at all. Intermediate values correspond to mingling. as calculations show that mixing is impossible, if viscosity η_A is greater than viscosity η_B by more than two orders of magnitude, even if magma B formed a turbulent plume fed by a 500 m wide conduit! Consequently, only magmas with rather similar viscosities can form hybrid compositions.

4.5 Where do hybrid magmas form?

Following the review of mingling and mixing processes between mafic and granitic magmas, places should be sought where hybrid magmas may form. There are three possibilities: at the granite–magma source, in a crustal magma chamber or in feeding dykes.

Mixing at the granite–magma source is often suggested in geodynamic situations where large volumes of basaltic magmas accumulate at the base of the continental crust (magmatic underplating) or in the continental lower crust (magmatic intraplating), namely in active continental margins and at continental hot spots or continental rifts. These hot magmas modify the continental geotherm, resulting in partial melting of the continental crust. However, segregation mechanisms are fast enough (see Chapter 3) to prevent basaltic magma and granitic crustal magma from remaining in contact for a sufficiently long time to lead to their hybridization. Moreover, there is no geological evidence for such a situation to have existed.

By contrast, the coexistence of different magmas in the same magma chamber is often observed. This is the case in a pluton emplacement site, which can precisely be regarded as a high-level magma chamber. It is also possible to imagine the same situation in an intermediate magma chamber at a deeper level in the continental crust. Field observations point to the predominance of mingling conditions (Section 4.1). Whereas the contamination of mafic enclaves of metric sizes can occur *in situ*, production of large volumes of hybrid magmas appears much more difficult.

Stratified chambers with mafic magma ponding below granitic magma without any magma mixing (because their respective viscosities are too different) are commonly observed. Fast thermal equilibration will result in cooling and crystallization of the mafic magma, whereas the granitic magma is heated, hence decreasing its solid load and its viscosity. Mingling (and locally mixing) of the granitic magma with the evolved and residual (not yet crystallized) mafic liquid is still possible. Examples of outcrops showing vertical or oblique sections of mafic magmas contaminating granitic plutons enable this model to be tested in the field (Fig. 4.10).

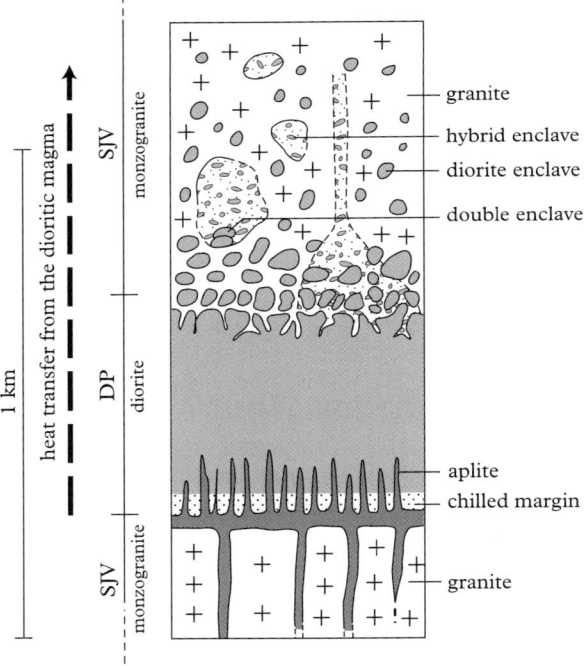

Figure 4.10 *Contacts between the Piolard diorite (DP) and the Saint-Julien-la-Vêtre granite (SJV), Massif Central, France, after Barbarin (1988). Notice the asymmetry on both sides of the dioritic sill: a chilled margin formed on the floor of the mafic intrusion with hybrid enclaves in the roof.*

Where do hybrid magmas form?   71

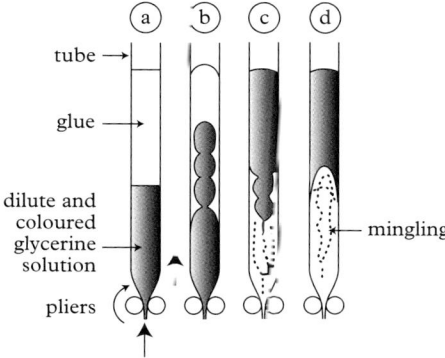

Figure 4.11 *Analog models of magma mingling (and mixing) in a conduit. (a) Starting conditions with glue representing the felsic magma (in white), on top of a diluted coloured solution of glycerine representing the mafic magma (in black); the tube base is compressed at a speed of 6–7 cm s^{-1}. (b) The 'mafic magma' is injected toward the top through the 'felsic magma'. (c) A portion of the 'mafic magma' is mixed with the felsic one. (d) At the end of the experiment, some mafic 'magma' rest on top of a mixture. After Koyaguchi (1987).*

Heat transfer from the mafic magma upwards, i.e. towards the roof of the magma chamber, is responsible for an asymmetric distribution of the structures at the contact between both magmas. Some of these structures can be used as polarity criteria pointing to the top of the magma chamber (see Fig. 11.10c). Even in the case where significant volumes of hybrid magma can be produced, numerous enclaves are still observed. This is the evidence of restricted duration and location of potential mixing processes in the magma chamber. It is also possible that the crystallizing and differentiating magma chamber received several magma inputs corresponding to different mixing or mingling stages (see Fig. 11.6).

The highest level of mingling and potential mixing is in dyke conduits used by magmas coming from different sources. Such a context corresponds to dynamic conditions, which might favour magma mingling and mixing despite the eventual contrasted natures and physical properties of the ascending magmas. Analog models illustrate that a mafic magma can ascend through a felsic magma, a common observation in bimodal dykes. Part of the mafic magma may become mingled or even mixed with the felsic one in the experiment (Fig. 4.11). The final result will depend on the pressure of the magma, hence on its ascending rate, but also on the width of the conduit. These experiments suggest that mingling (and possibly mixing) may be an efficient process in narrow dykes, even for relatively slow ascending rates.

4.6 Conclusions

Field observations often show variable mingling and mixing stages between coexisting magmas. The variety of cases is the consequence of a large range of viscosity contrasts between coexisting magmas, mostly depending of their respective crystal loads. Limited mingling results in the formation of dark enclaves of mafic or intermediate composition, commonly observed in I-type and A-type granites. Mingling followed by mixing results in the formation of hybrid rocks characterized by intermediate chemical compositions, that can be represented by mixing lines (for major and trace element contents) or mixing hyperbolas (for isotopic ratios). Some minerals (for instance plagioclase) may have registered the mingling or mixing event. Formation of a hybrid magma generally requires restricted viscosity contrasts between the potentially mixing magmas. It is favoured by dynamic conditions, for instance during ascent of coeval mafic and granitic magmas in dykes or in the case of magma fountaining in a host magma chamber derived from a different source.

5
Transport of granitic magma

Once formed and separated from their source regions, granitic magmas travel through the crust, up to *ca.* 30 km vertically, before assembling as a high-level magma chamber. It is worth saying that mass transport of magma as a diapir (Fig. 5.1), travelling like a hot-air balloon that gets flatter against the country rocks when densities equilibrate, is now an old-fashioned idea based on Ramberg's laboratory experiments. An exception is when the continental crust was hotter (i.e. thermal gradients were much steeper) in ancient geological times, such as during the Palaeoproterozoic/Archaean eons, hence when the lithosphere was softer than later in Earth's history (see Chapter 13). Nevertheless, granitic magmas are always hotter and lighter than their country rocks, and are normally subjected to rising gravity forces.

Ramberg-type diapirism, i.e. balloon-like magma ascent, is impeded by the physical strength of the crust. The brittle crust may be easily fractured, thus promoting upwards transportation granitic magmas to high-level magma chambers though dyke-like networks and fissures (Petford et al., 1994). Magma transport, via ductile or brittle wrench faults and shear zones, or through tensional fractures creating dykelets and, at higher levels, forming dyke swarms, is a modern alternative to diapirism, as illustrated in Fig. 5.2.

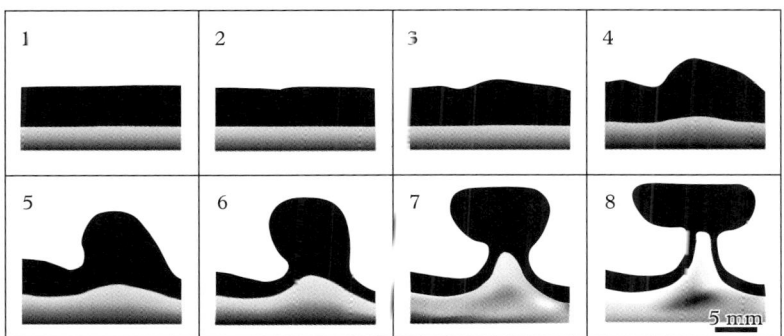

Figure 5.1 *Ramberg's experiments, showing successive evolution stages of an oil layer capped by a syrup layer, which led to the idea of granite diapirism. After Ramberg (1981).*

74 Transport of granitic magma

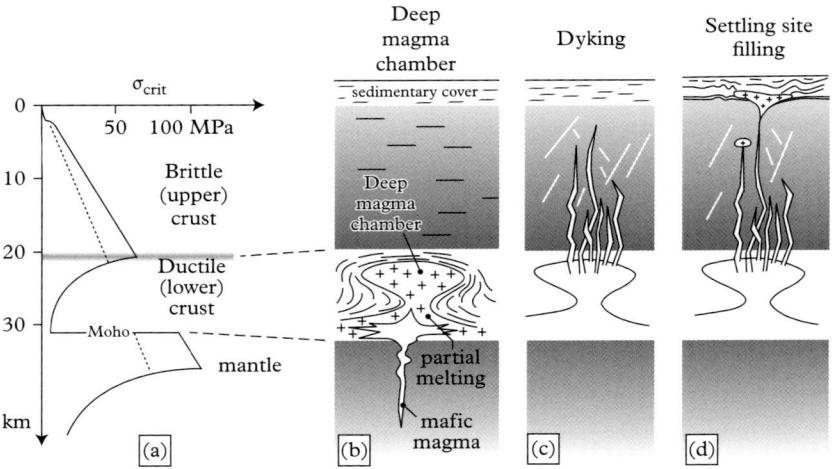

Figure 5.2 *Granitic magma transport at the light of a rheological profile of the crust. (a) This 'Christmas tree' represents the maximum differential stress, $\sigma_{crit} = (\sigma_1 - \sigma_3)_{crit}$, that can be sustained by rocks without failing. For $\sigma > \sigma_{crit}$ fracturing takes place for the linear parts of the profile (Byerlee's law, 1978; in compression—full line—or in tension—broken line), and replaced by plastic deformation along the curved portions (Kuznir and Park, 1986). (b) Short-range vertical magma transport, within the ductile crust. (c, d) Brittle crust fracturing, dyke-assisted transportation and magma settling aided by discontinuities. After Brun et al. (1990).*

5.1 Diapirism

The rate of ascent v (m s^{-1}) of a spherical diapir, supposedly at a constant temperature and travelling through a constant viscosity medium, is given by the Stokes' law:

$$v = 2\Delta\rho g r^2 / 9\eta,$$

where r is the radius of the sphere (in m), g the gravity constant (9.8 m s^{-1}), $\Delta\rho$ the density contrast between the magma and its host rocks (e.g. 300 kg m^{-3}) and η the viscosity of the host (Pa s). This rate derives from the ratio between the upward directed Archimedes force, which is proportional to volume, hence to r^3, and the drag force proportional to r, thus explaining the r^2 term in the numerator.

Viscosity is indeed the sensitive factor of the ascent rate. Since crustal materials have a high viscosity, at least when in the solid state, say 10^{22}–10^{24} Pa s, the ascent rate of a diapir will always be very slow, on the order of 10^{-8}–10^{-10} m s^{-1}, and will decrease rapidly with time due to heat loss. Under such conditions, a diapir would hardly reach maturity, i.e. rise then flatten at gravity equilibrium, just like oil within syrup (Fig. 5.1). This cartoon, showing the spreading of a light and viscous liquid, and its flattening or ballooning

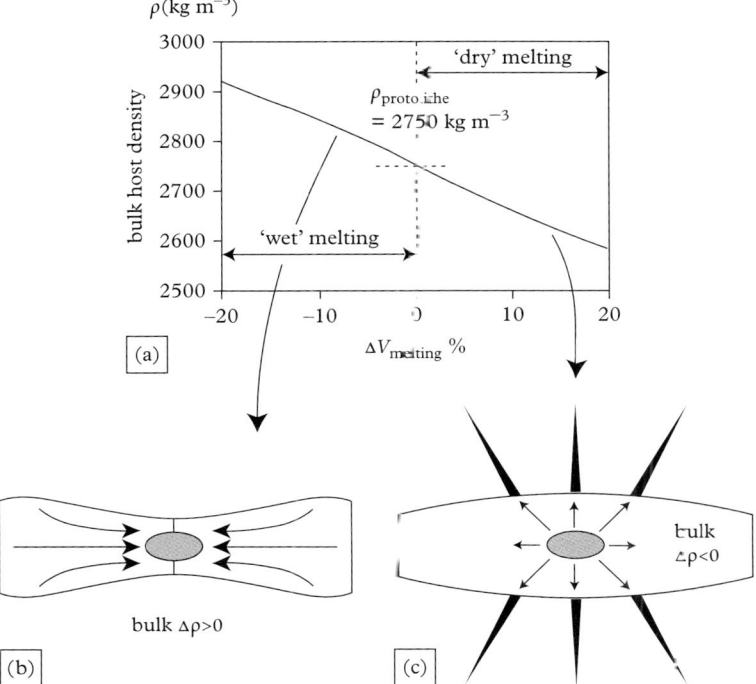

Figure 5.3 Density and volume changes of a rock mass due to melting reaction. (a) Bulk density response to volume changes. (b) Fluid-present ('wet') melting leads to negative volume change, hence a density increase of the whole (host + melt) system. (c) Fluid-absent ('dry') melting leads to a positive volume change, hence a density decrease of the system, and favours brittle failure and melt expulsion into fractures. Shaded area: melting site. After Clemens and Droop (1998).

at the end of its ascent into a denser and viscous medium, is a realistic image of a granitic reservoir once emplaced in the crust. It explains the original success of granite diapirism.

Partial melting of crustal material is itself unfavourable for a diapiric ascent of the melt. If partial melting takes place in presence of free water (wet melting), for example in a warmed upper crust, the melt will tend to decrease in volume hence the whole medium will tend to contract (Fig. 5.3). If, on the contrary, the melting process is anhydrous (dry melting), the increase in volume of the melt, hence its density decrease due to the melting reaction, will increase its pressure. The melt will tend to fracture its host and to feed dyke systems (Fig. 5.3).

Consider the hypothesis of an ascending diapir. Even if slightly elongate like a hot-air balloon rather than spherical, with a narrow head helping its upward penetration, a diapir is conceivable only in a hot and soft crust, close to its melting point such as in migmatites. In the brittle crust, a diapir would rise only if it were able to transfer enough heat

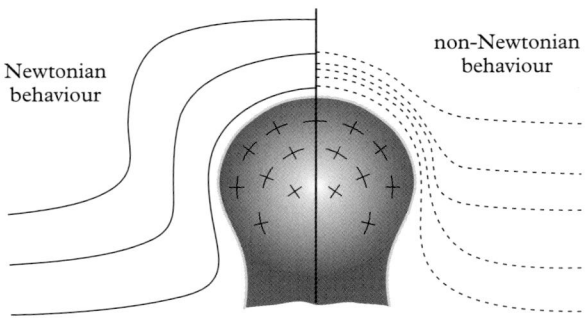

Figure 5.4 *Flow trajectories in the country rocks surrounding the head of a diapir. To the left: Newtonian behaviour (no strain localization). To the right: power flow behaviour; strain is highly localized along the diapir contact, facilitating its ascent. After Weinberg and Podlachikov (1994).*

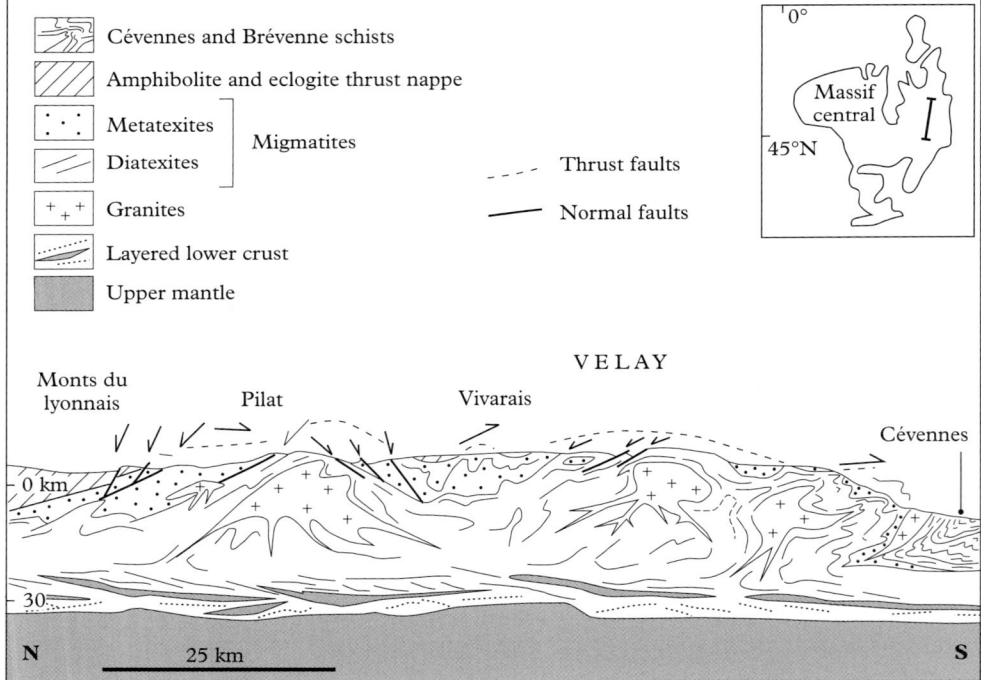

Figure 5.5 *Schematic cross-section of the migmatitic Velay dome (French Massif Central). Thickness of the layered lower crust and depth of the Moho are determined from seismic data. Upwelling of the partially molten layer in the middle crust is coeval with the Late Variscan regional extension, itself responsible for the development of Stephanian sedimentary basins in the upper crust of this region. After Burg and Vanderhaeghe (1993).*

into its country rocks to weaken a path at least as large as the diapir diameter. In such a case, the diapir would rapidly loose its heat. Its ascending force would die out. The most favourable calculations, that take into account a strongly non-Newtonian viscosity of the host rock at contact with the diapir (Fig. 5.4), do not manage to exceed 5×10^{-8} m s^{-1} in rate of ascent, a figure giving about 6 m per year. Such a rate would not last long due to the unavoidable cooling of the magma.

Therefore, the huge volumes of granite that abound in the upper crust, which is brittle down to 15–20 km, cannot represent frozen ascending diapirs. In fact, as clearly shown by geophysical data (Chapter 6), the width of granite plutons as mapped in the field is usually much larger than their thickness. In addition, field observations do no militate in favour of diapirism. For example, analysis of granite–host contacts hardly reveal any deformation that could justify the transit of a diapir. On the contrary, structures on the margins of a pluton help to reconstruct the final part of magma emplacement history which rarely represents a diapiric ascent into the crust.

Nevertheless, diapirism is correctly invoked when important masses of migmatites appear as domes rooted in amphibolite facies country rocks. Such domes are frequently coeval with extensional tectonic events, marked by normal faulting and detachments which were active during unroofing of the dome. This is precisely the situation of the Late Variscan Velay dome in the French Massif Central (Fig. 5.5).

5.2 Transport by dykes

Magma transport through dykes is also due to gravity, the Archimedes' force or buoyancy that results from the density difference between the magma (usually lighter) and the country rocks. Magma pressure may eventually be added to this force. At sites of mantle upwelling (ridges, plumes), buoyancy and magma pressure help the basaltic melt to escape from the asthenospheric mantle and to travel through the lithosphere. Magma transportation is achieved via dyke complexes that operate at ridge axes, as classically described in the ophiolitic sections. However, basaltic magmas have a low viscosity (10^2 Pa s), making their transit easy through narrow conduits. The situation for the transit of granitic magmas via dyke swarms is examined below.

5.2.1 Analysis and orders of magnitude

Following Poiseuille analysis, the upward velocity (v) of a fluid flowing in a laminar mode in a conduit (fracture) having smooth walls separated by a distance l, is not controlled by the viscosity of the host, but by its own viscosity (η_m) and (of course) by its density contrast with the host: $v = \Delta \rho g l^2 / 12 \eta_m$.

As illustrated in Fig. 5.6, the hypothesis of laminar flow is fully justified since the Reynolds number remains far below the critical value for turbulent flow whatever the width of the conduit. Because natural fractures are not smooth, the pressure loss appearing during fluid transportation increases the apparent viscosity of the fluid, lowering the

78 *Transport of granitic magma*

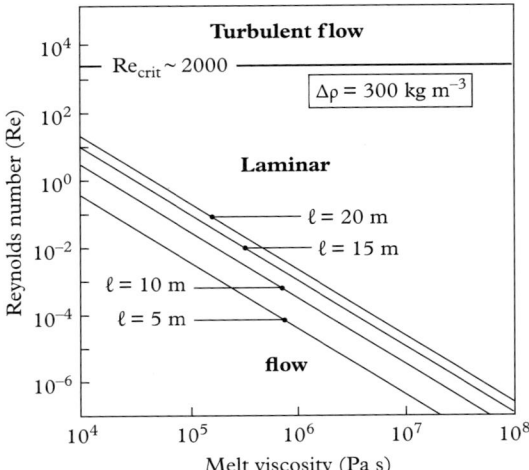

Figure 5.6 *Reynolds number (Re) versus viscosity of a melt flowing in a dyke (width l), for a constant Archimede traction ($\Delta \rho = 300$ kg m^{-3}). Re is always far below its critical value Re$_{crit}$ separating laminar flow from turbulent flow. After Petford (1996).*

Reynolds number even more. Note in addition that magma pressure at the source P_s, which is called 'overpressure', whatever its origin (melting process or tectonic compression of the reservoir), further adds to the gravity force and increases the melt velocity in the fracture.

Supposing no overpressure, taking $\Delta \rho = 200$–400 kg m^{-3}, magma viscosity between 10^5 and 10^7 Pa s, a water content from 2% to 4% at 900 °C and a solid fraction of about 30%, the ascent velocity of such a magma would vary from 2 mm to several metres per second for a fracture width of 3 to 13 m. Therefore, a presumed magma velocity of 0.1 m s^{-1} in a conduit (Clemens, 1998) would be 10^7–10^9 times faster than the ascent velocity of a diapir.

The magma infilling a fracture must not freeze during its ascent as a result of conductive heat transfer through the fracture walls. The minimum width of a fracture beyond which the magma would crystallize is proportional to $l^{1/4}$, $1/\Delta \rho$ and depends on the temperature difference between magma and host rocks, on the latent heat for crystallization and on the specific heat and thermal diffusivity of the magma. Minimum fracture widths resulting from these figures remain on the order of a few metres (Petford et al., 1994).

Finally, the time span to fill up a pluton should be much shorter than the time needed to complete its crystallization. For the slowest ascent velocities considered above (a few mm s^{-1}), a magma flowing in a single dyke, 3 m in width and 1 km in length, continuously fed from its source, should be able to travel though the whole crust in a few weeks and to feed a 1000 km³ pluton in a few thousand years. Such a filling time is much shorter than the time necessary for the pluton to crystallize. Thermal models indicate that 30,000 years are needed to obtain a temperature decrease from 850 °C to 700 °C (~ crystallization) of a 3 km-thick granite layer setting at a depth of 3 km (100 MPa) in a crust at 100 °C (Clemens, 1998).

5.2.2 Transfer from the source

The capacity of a source to generate a volume of magma adapted to the potential transfer rate out of this protolith remains poorly known. The output flow from the source depends on the efficiency of partial melting and segregation, and eventually on the presence of a reservoir located above the source and devoted to temporary magma storage before final emplacement (Fig. 5.2b). Continuity of magma supply into fractures depends on the nature of the rocks between source and dykes. Presence of an intermediate magma reservoir that could discharge progressively or episodically may also account for continuous or discontinuous magma differentiation, a feature often observed in granite plutons as petrographic zoning (see Chapter 11).

5.2.3 Role of magma pressure

Magma pressure at the source is naturally considered to be as large as the lithostatic pressure at the source: $P_m(h_{source}) = \rho_{crust} g h_{source}$. An overpressure at the source (P_s) may eventually be present and added to P_m. Provided that the connection inside the magma column is perfect from source to surface, magma pressure above the source ($h < h_{source}$) is equal to magma pressure at the source *minus* the pressure due to rising magma segment ($h_{source} - h$). Therefore, in absence of overpressure at the source (Fig. 5.7), we can write:

$$P_m(h)_{ideal} = \rho_{crust} g h_{source} - \rho_{magma} g (h_{source} - h) = \Delta \rho g h_{source} + \rho_{magma} g h$$

According to the same principle as the artesian well, a basic concept for hydrogeologists here applied to volcanism, magma pressure above the source is larger than the lithostatic pressure. This is due to $\rho_{magma} < \rho_{crust}$ and to magma pressure at the source being equal to the lithostatic pressure (Fig. 5.7). Overpressure P_s, either at the source or during magma ascent, will eventually be added to magma pressure. Tectonic compression, which may represent a large fraction of total magma pressure, may also be invoked. In all these cases, a rapid pressure loss is expected as soon as connection to the surface is realized in the plumbing system.

There is nothing to indicate, however, that the rising column is full of magma and that magma transport is frictionless. On the contrary, the most common situation is probably a segmented column whose length is tortuous and rugose, hence responsible for pressure loss during magma transport (a phenomenon well known to hydraulic engineers). As a consequence, the upwards magma flow will be reduced, due to magma

80 *Transport of granitic magma*

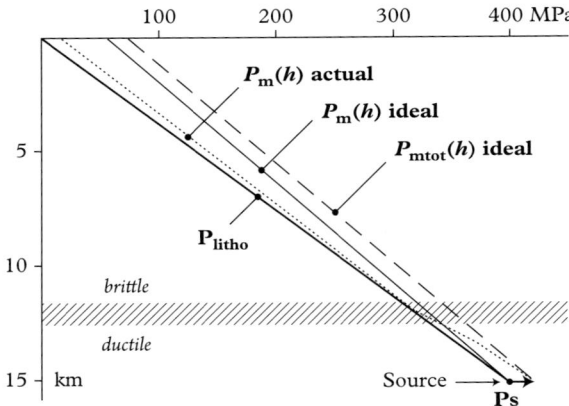

Figure 5.7 *Magma pressure versus depth. Straight line 'P$_{litho}$' represents the lithostatic pressure for crustal density of 2.7. Hatched: brittle–ductile transition. Straight line 'P$_m$ ideal' represents magma pressure for magma density of 2.3, magma being ideally connected from source (depth: 15 km) to surface. P$_s$ represents the eventual overpressure at the source, from which the line 'P$_{mtot}$ ideal' is derived. 'P$_m$ actual': general case with pressure loss during transport, for a brittle–ductile transition around 12 km (see also Fig. 6.1).*

pressure lower than ideal, the actual $P_m(h)$ curve ranging between the ideal $P_{mtot}(h)$ and $P_{litho}(h)$ (Fig. 5.7).

5.2.4 Fracturing the crust

Let us imagine the way by which fracturing operates, either in tension or in shear mode. Close to the source where the protolith is impregnated with melt, fracturing assisted by local stress concentration at solid contacts and eventually by melt pressure may be invoked (see Fig. 9.6). Far from the source, in the ductile domain as well as in the brittle crust, propagation of fractures soaked with melt will depend on magma pressure, intensity of the differential stress ($\sigma_1 - \sigma_3$) and orientation of the principal stress components. This point will be discussed in Chapter 6.

In the brittle domain, fracture propagation will also depend on the pre-existing fracture system, on the elastic constants of the host rocks and on the shape of the fracture tips (Fig. 5.8). This last factor, which relates to the lithological nature of the host rock, deserves some attention. A brittle host, made of quartzites for instance, will show acute fracture tips, a feature favouring fracture propagation (Fig. 5.8a). By contrast, a soft rock, for instance rich in clay minerals, will plastify the fracture tips (Fig. 5.8b), thus impeding fracture propagation. The magma will then be able to gather, and eventually to form a sill (Fig. 5.8c).

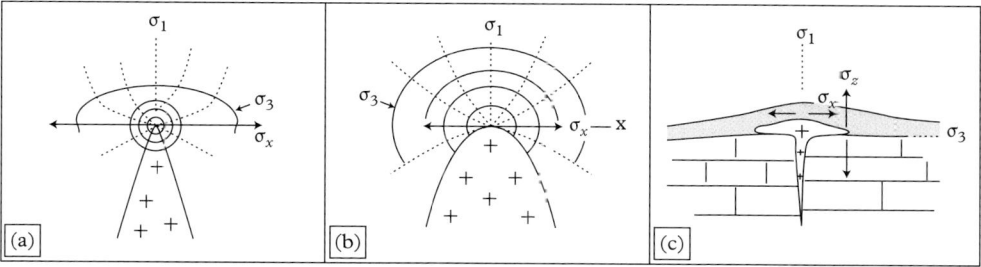

Figure 5.8 *State of stress at a fracture tip. Traction acting perpendicularly to fracture (σ_x) increases with the sharpness of the tip. Stress concentration tends to open the fracture and is all the more important when the radius of curvature is small (Baïlon and Dorlot, 2000). (a) Since σ_x largely surpasses the tectonic (far-field) stress in magnitude, a sharp fracture tip will easily propagate upwards. (b) If a fracture tip abuts a plastic medium, it will become more rounded and, by the way of a horizontal fracturing at the interface, a sill may develop (c). Far-field stress: the maximum principal stress σ_1 is supposed to be vertical and the minimum principal stress σ_3, horizontal. Near-field stress (around fracture tip): dotted and full lines represent, respectively, σ_1 and σ_3 trajectories at fracture tips.*

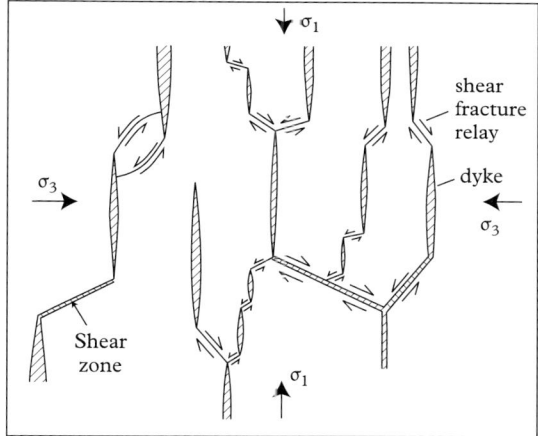

Figure 5.9 *Magma transportation in the crust through a succession of dykes, tension fractures, sills or layers, relayed by shear-faulting. Viewed in cross-section, the maximum principal stress (σ_1) is vertical, in the case of an extensional tectonic setting; tension fractures being relayed by normal faulting. In compression (σ_3 vertical: rotate the figure by 90°), horizontal layers, relayed by thrust-faults will dominate. In strike-slip setting (σ_2 vertical), the figure is to be seen in map view. Inspired from Shaw (1980).*

82 Transport of granitic magma

The fracture network allowing magma transit is more complex than the host and is mechanically heterogeneous. It can be seen as a succession of tension and shear fractures (Fig. 5.9), but the actual network will depend on the regional state of stress. If the principal stress component (σ_1) is horizontal, shear fractures will prevail, magma transfer taking place along shear zones becoming steeper in dips with σ_2 approaching vertical (strike-slip tectonic regime). On the contrary, in case of tectonic extension (σ_1 vertical), normal faults relayed by vertical tension fractures will prevail. Such a complex geometry explains why it is not realistic to expect that magma transport be vertical everywhere and at every time within the fracture network.

Finally, tectonic regimes are usually accompanied by specific thermal regimes. Depth of the brittle–ductile transition will vary accordingly, thus reducing or increasing the part of the crust active for magma transfer through fractures. Taking 300 °C as the temperature for the brittle–ductile transition (based on quartz behaviour), Fig. 5.10 shows that the depth of this transition, located at 15–18 km for a 'normal' crust (geothermal gradient 17.5 °C/km; thermal flux $Q = 60$ mW m^{-2}), will move upward or downward by about 10 km according to the context, HT/LP (crustal thinning;

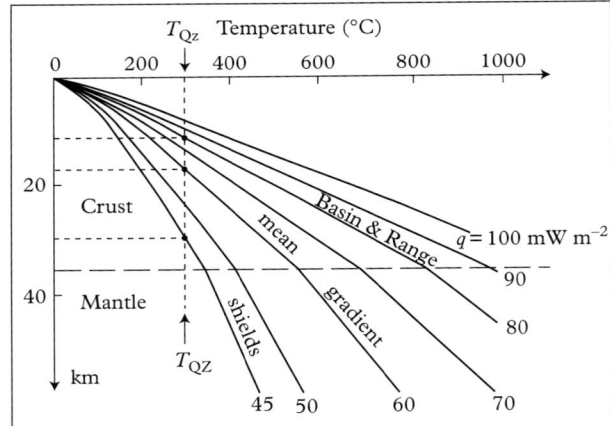

Figure 5.10 *Geotherms and brittle–ductile transition in the crust. q: heat flux (surface measurements). T_{Qz}: temperature above which quartz plasticity becomes efficient for 'usual' stress values, i.e. a few 10^1 MPa. The brittle–ductile transition (horizontal dashed segments) is ≤20 km in depth in average, but may be as high as 10 km in the Basin and Ranges Province ('hot' gradient), and reaches 30 km in shields ('cold' gradient). In the mantle, olivine determines the rheology of the brittle–ductile transition, with T_{ol} at 800–900 °C, giving variable depths for the base of the lithosphere according to geological contexts. After Kuznir and Park (1986).*

$Q = 75$–80 mW m^{-2}) or LT/HP (cratons; $Q = 45$–50 mW m^{-2}). If the rheological behaviour of the crust is dominated by feldspar (granulites), the temperature of the brittle–ductile transition will be higher (450 °C) and dykes may originate from a deeper level.

5.2.5 Magma ascent rate

The viscosity of a magma depends on the cooling rate undergone during ascent. Figure 5.11 shows that the solid fraction of a granitic magma greatly depends on the followed P,T path. For example, the adiabatic ascent (path ④), the fastest, will cause a partial resorption of the newly formed crystals, conformably to the observation of 'corroded' quartz phenocrysts in rhyolites. Decrease of the solid fraction, hence decrease in magma viscosity, will make the ascent easier. On the contrary, increasing the solid fraction (paths ①–③) will lower magma ascent rates. Therefore, for a rough and segmented fracture network, the ascent rate will decrease, due to both pressure loss and increase in magma viscosity. We easily understand that the ascent rate influences the level

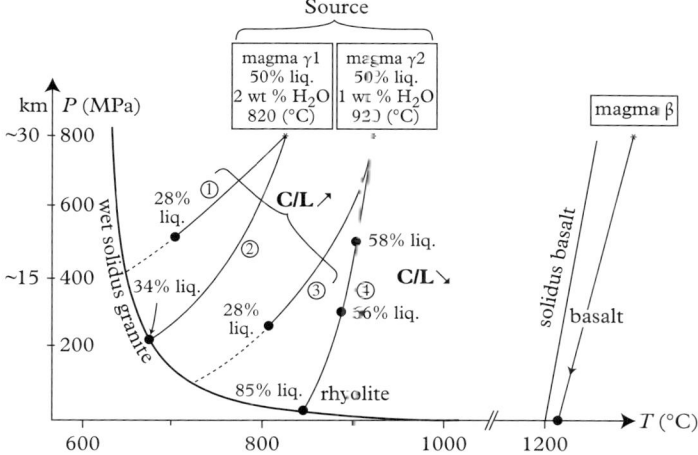

Figure 5.11 *Solid fraction of a granitic magma in relation to its ascending path. The critical parameter is the cooling rate of the magma which depends on the velocity of its ascent. For two haplogranitic magmas formed at different temperatures but at the same depth: for ①, rapid cooling (60 J/g kb), the increasing solid fraction stops magma ascent before reaching its solidus; ② is a more realistic cooling path (40 J/g kb); ③ has the same cooling rate as in ①, but due to its higher initial temperature, the magma reaches a higher level in the crust; ④ shows quasi-adiabatic cooling (rapid ascent) of subvolcanic granites and rhyolites. C/L: crystal/melt ratio. Note that the slope of the basalt solidus favours its emplacement up to the surface. After Holtz and Johannes (1994).*

84 Transport of granitic magma

of magma settling in the crust (Chapter 6) as well as the properties of the crystalline fabrics at the emplacement site (Chapters 9 and 10).

5.3 Field data

Since a pluton cannot be totally observed in three dimensions, it is difficult to confirm that a network of dykes at its floor was responsible for magma infilling. The opportunity to observe a few metres-wide feeder dykes to the Himalayan leucogranites (Fig. 5.12) does not help to infer their extension in map-view nor their density.

The Dharwar craton (peninsular India) offers an oblique section of the Archaean crust through more than 300 km from north to south, as shown by the mineral parageneses in the country rocks recording increasing pressures from 300 MPa (to the north) to 600–700 MPa (in the south). A spectacular example of a feeder dyke system (Fig. 5.13) is provided by the Closepet batholith. The southern end of the batholith, monzonitic to granodioritic in composition, is fed by magma from a 'root' zone made of a set of deformed dykes and shear zones, heterogeneously mixing mafic magmas of mantle origin and magmas

Figure 5.12 *Magma transport through dykes. (a) In vertical section, below the floor of the Himal Chuli (Himalaya) leucogranite pluton, the coarse-grained light-coloured granite cross-cuts the sillimanite-bearing mica schists of the Tibetan Slab (photo: J.L. Bouchez; note the scale). (b) Horizontal section, peraluminous granite, with or without garnet, cross-cutting Breidvagnipa migmatites, East Antarctica (Shimura et al., 1998).*

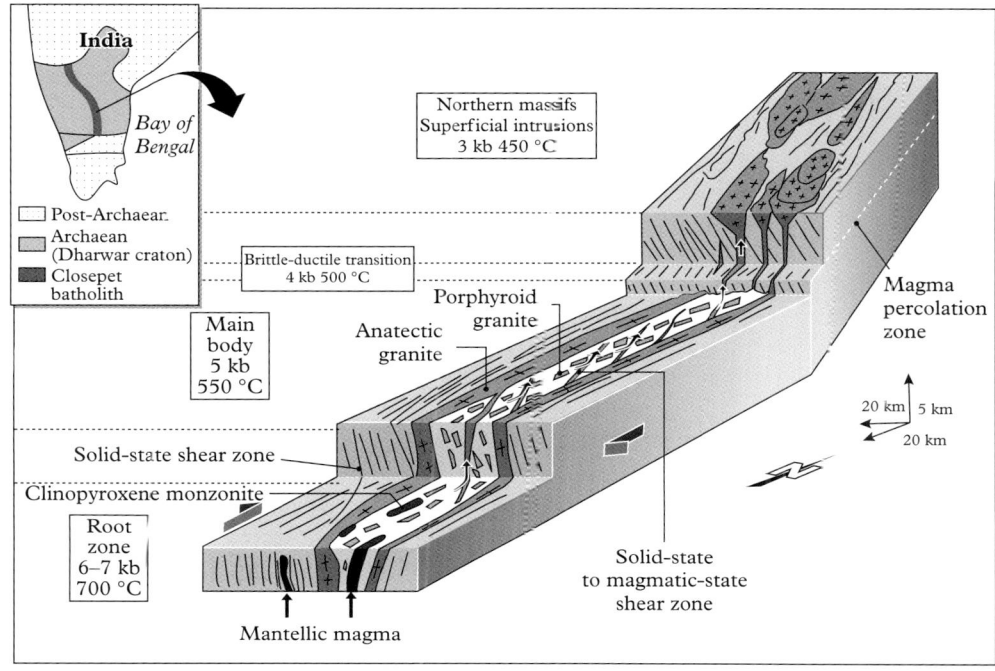

Figure 5.13 *The Closepet batholith, Archaean Dharwar craton. The oblique section of the crust (2° of tilt over 250 km) helps to reveal: the root zone to the south of the pluton (heterogeneous migmatites and melange of mantle and crustal magmas); the elongate main body; the 'gap zone' (where magmas are present only inside narrow dykes); and finally the surficial intrusions to the north. A large part of the magma remains in the main body; only the most evolved melts rise above the rheological brittle–ductile transition through dykes feeding surficial intrusions. After Moyen et al. (2003).*

derived by the partial melting of the lower crust. To the north of the batholith, the northern Massifs, monzogranitic in composition, are fed by a specific dyke system. In map view, this dyke system appears as a few narrow (<<1 km) and very elongate domains (15 km), themselves corresponding to a swarm of granitic dykes more-or-less elongated parallel to the batholith. Due to the different outcropping levels of the batholith with respect to the original horizontal level, the transfer zone takes place in the ductile crust to the south, while to the north it begins at the brittle-ductile transition and then ends in the brittle crust.

Detailed gravity field measurements confirm that plutons generally reveal narrow and slightly elongate roots, attributed to feeding zones, themselves guided by a few sets of fractures (Fig. 5.14). The question of the three-dimensional (3D) shapes of plutons is the subject of the next chapter.

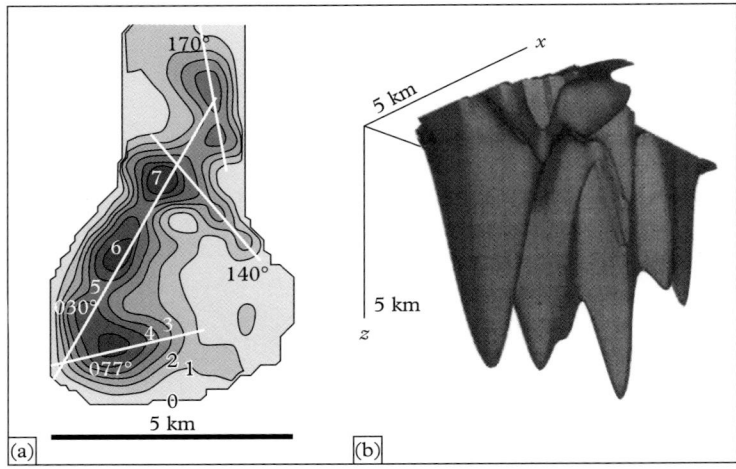

Figure 5.14 *Gravity map and corresponding 3D model of the Ulu pluton, a small, 2.6 Ga old leucogranitic body belonging to the Canadian Archaean craton, north of the Slave Province (115°E, 65°N). (a) The iso-depth values, in kilometres, are aligned according to four main trends, and are also characterized by satellite imagery. (b) A 3D view of the aligned roots, viewed from below. After Dehls et al. (1998).*

5.4 Conclusions

The difficulty in developing concepts on magma transport comes from the lack of direct observations of this phenomenon. Instead, physical principles must be examined which, however, remain insufficient. Diapirism, in the sense of bulk transport of a large quantity of magma through the crust between the source region and the emplacement site, is not realistic. However, diapir-like short distance movement of magmas from the source region remains possible if the host material is hot and easily deformable (migmatitic domes). The recent idea that magma is transported from depth through a network of dykes up to the site of accumulation to form a pluton proves to be a reasonable alternative. In summary, the analysis of magma transport needs the evaluation of pertinent parameters concerning the magma (viscosity, solid fraction, magma pressure, etc.) and their integration into the local and regional, lithological and tectonic setting.

6
Emplacement and shape of granite plutons

This subject is still a hot topic of discussion among geologists, each having their own point of view based on their favourite plutons and inherited from field experience. Emplacement modes are indeed multiple and observations obtained from a pluton, already emplaced into the crust, may be subject to various interpretations. The sensitive parameters that govern magma collecting at the pluton's emplacement site are examined here in detail. Magma transport being now better understood, the crucial question of magma accommodation with respect to its crustal environment, known as the 'room problem' (Read, 1948), seems to be less critical than six decades ago. Magma emplacement into the brittle part of the crust, the most common ultimate destination of granitic magmas, will be considered first. The resulting plutons are easy to identify, having well-defined outer limits on geological maps. Then we shall examine granitic magma emplacement at deeper levels in the ductile crust.

6.1 Cessation of magma ascent

Pluton settling begins at the end of upward transportation of magma. It is largely due to the drastic increase in magma viscosity as a consequence of decreasing temperature and increase in the solid fraction. It is often said that abundant granites but few rhyolites, and plenty of basalts but scarce gabbros are observed at the surface of the Earth's. Indeed, contrary to basaltic liquids that easily reach the surface, granitic melts do not behave in the same way. For a given composition this is explained by the crystal load of the melt as a function of the cooling path, and by the shape of the solidus. Basalt, with the steep positive slope of its solidus, rapidly travels through the crust according to a quasi-adiabatic path, leading to volcanism, provided that the crust is already fractured or can be easily fractured. This is the case in extensional settings where basalts often erupt onto the surface. By contrast, a silicic rhyolite will reach the surface only if the granitic magma is very hot, a rather uncommon condition that would allow a rapid ascent (see Fig. 5.11). Under usual conditions, granitic magma settles at depth and its load in crystals, hence its viscosity, quickly becomes so high that upward continuation to the surface is no longer possible.

88 Emplacement and shape of granite plutons

The upward traction of the magma results from the Archimedes' force, proportional to magma volume and to its density difference with respect to the host rocks. This upward body force is also called buoyancy. Since the density of a granitic magma increases as its temperature decreases, from about 2.2 for a rhyolitic melt to about 2.6 for a solid granite, it must not be concluded that the magma will gather and settle at gravity equilibrium, i.e. when magma density equals that of its country rocks. This conception is reasonable if only body forces are acting, as recognized in migmatitic (see Figs 5.2b and 5.5) or anorthositic domes (see Fig. 13.19), made of light and hot magma within rather low viscosity host rocks. Note, however, that magma density is not a linear function of depth since it depends not only on the competition between its isothermal compressibility and thermal expansion coefficients, but also on its crystal load and dissolved water content, factors depending themselves on the initial water content and P-T path of the magma during its ascent (see Fig. 5.11). In fact most granite plutons are far from being at gravity equilibrium with respect to their host, because granitic magmas are less dense than their host. The resulting gravity anomalies justify gravity measurements in order to define the shape of the plutons (see Section 6.5).

As the magma leaves the ductile crust toward the surface, the increasing brittleness of the enclosing crust has to be taken into account. From the brittle–ductile transition, at

---

Box 6.1 Emplacement in an extensional context: a simple exercise of mechanics

The vertical stress due to the weight of the overlying strata as well as magma pressure (P_m) increases with depth. This is represented by the straight line S_V crossing the origin and whose equation is $S_V = \rho_{crust} g h$ (with ρ_{crust} = 2700 kg m^{-3}), and by the straight line P_m starting from the source and whose slope is $\rho_{magma} g$, with ρ_{magma} on the order of 2300 kg m^{-3}. The equation $P_m = S_V$ at the source is justified by the fact that a static magma cannot be distinguished from its neighbouring rocks in terms of pressure. Under the unrealistic supposition that magma undergoes no pressure loss on its way up to the surface (plumbing system perfectly smooth between source and surface), $P_m = \rho_{magma} g h$ is the maximum pressure of a static magma due to density, but degassing may largely increase the volume of the whole magmatic system, hence P_m, at least close to the surface, reinforces the eruption of a volcano. In the case of a partially clogged plumbing system, P_m may get close to S_V at various depths. Consider now the strength profile of the crust (σ_{crit}) in an extensional environment; the depth of the brittle–ductile transition (~300–350 °C) is around 10 km (high thermal gradient). σ_{crit}, which represents the differential stress needed to cause rupture, is close to $S_V - S_H$ if the principal stress is close to vertical as in an extensional tectonic setting. This helps to trace the profile of $S_H = \sigma_{crit} - S_V$ in Fig. 6.1.

The pressure necessary to keep a vertical fracture opened must be at least equal to S_H. It can be written as $P_m = S_H + P_{sm}$, P_{sm} being the driving pressure for fracturing. This formulation helps to trace the driving pressure profile $P_{sm} = P_m - S_H$ (dotted in Fig. 6.1). We observe that, above 6.5 km depth, dykes may be maintained open ($P_{sm} > S_H$), and, above 3.5 km, large dilatant volumes are made possible, P_{sm} being larger than both

continued

Box 6.1 *continued*

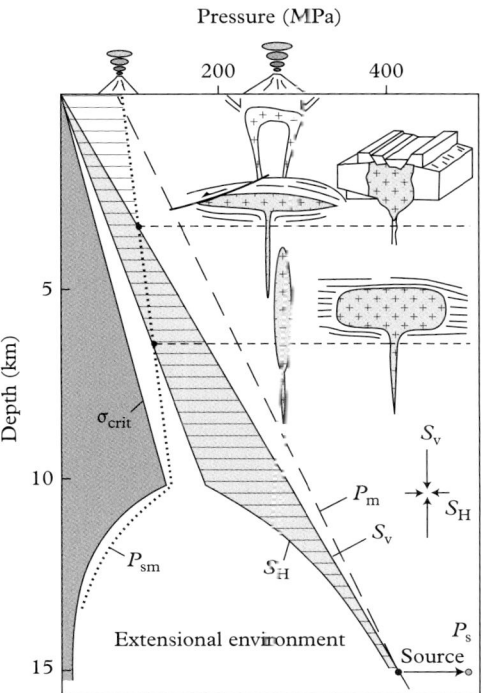

Figure 6.1 *Pluton emplacement in extensional regimes, after Hogan et al. (1998).*

S_H and S_V. These critical depths depend on the location of the source (15 km in the present case) and of the brittle–ductile transition (here 10 km), on the rheological profile, on the intensity of magma pressure, but also on the tectonic overpressure (P_s) which may not be negligible in compressive situations (which translates P_m to the right side of the diagram by a value P_s), and finally on the presence of dilatant sites created by crustal deformation, particularly under extensional and strike-slip tectonic environments. For compressive tectonic environments ($S_H > S_V$), S_V and S_H must be exchanged, and σ_{crit} versus depth must be increased by at least 50 MPa at the brittle–ductile transition.

depths between 15 and 20 km for 'normal' geothermal gradients (see Fig. 5.10), up to the surface, fracturing takes place as soon as the critical stress for rupture σ_{crit} is reached. Since lithostatic pressure tends to prevent the opening of fractures, c_{crit} decreases linearly with decreasing depth (Fig. 5.2; Byerlee law). Letting S_V and S_H be, respectively, the vertical and horizontal stress (supposed to be close to the principal stress components), two main tectonic settings can be distinguished, extensional (for which $S_V > S_H$) and compressional ($S_H > S_V$). The slope of $\sigma_{crit}(h)$, steeper in extension than in compression (see Fig. 5.2a), reflects the fact that rock fracturing is easier in tensional settings than in a compressional environment. As a consequence, granite magma will have the ability of ascending higher into the upper crust in extensional environments than in compressive regimes.

Magma pressure (P_m) has been evoked to be a necessary condition for magma transport (see Fig. 5.7). The part of magma pressure that helps to keep a fracture opened, or driving pressure P_{sm}, is the difference between magma pressure and the least stress component, i.e. $P_{sm} = P_m - S_H$ (Box 6.1) in extension (where $S_V > S_H$), or $P_{sm} = P_m - S_V$ in compression ($S_H > S_V$). Hence, the upward transport of magma ends as soon as P_m becomes lower than S_V or S_H depending on the tectonic stress conditions. This is followed by the collection of magma at its emplacement site.

6.2 Meeting dilatant sites

The brittle crust is dilatant, i.e. it increases in volume while deforming, like a set of dominoes initially standing on their edge that are subjected to a tilting deformation. If the material has no ability to seal the resulting porosity (by plastic deformation, crystal deposition, etc.), open spaces or volumes, also called 'dilatant sites', will appear. Such a decompaction is just like proliferation that miners or road workers discover when the volume of extracted material is larger than the volume of the hole they have just dug. At the scale of the crust, proliferation in the brittle domain depends on the tectonic environment.

6.2.1 Fracturing in extensional environments

Extension, due to (sub)horizontal traction ($S_H = \sigma_3$) may occur locally, for example at the roof of a dome (core complex), or regionally as a consequence of rifting and back-arc or gravity spreading. Its signature, in term of structures, varies according to thickness and thermal structure of the crust (Buck, 1991). In extensional settings, fractures vary from vertical and purely tensional near the surface level to shallow dipping (listric faults, detachment faults) near the brittle–ductile transition where they become conformable with the horizontal plastic flow of the ductile crust.

From the source, located in the ductile crust, the driving pressure P_{sm} strongly increases up to a maximum at the brittle–ductile transition (Fig. 6.1). Note however that $P_{sm} > S_H$ is a necessary condition for fractures to initiate. In the absence of tectonic overpressure at the source P_s, a realistic condition in an environment under extension, P_{sm} exceeds S_H at a depth of about 7 km (according to Fig. 6.1) that is favourable for the development of dyke swarms. However, magma progression in dykes is made possible only if P_{sm} becomes larger than S_V, i.e. at a depth of about 3.5 km. In an extensive tectonic context, a large pluton may settle at a depth between 7 km and 3.5 km, for example by magma seeping into fractures taking place in-between lithological discontinuities. Finally, at a depth less than 3.5 km (again, under the conditions chosen in

Fig. 6.1) the pluton will be in position to thicken by pushing its roof upward, as in Fig. 5.8c, and eventually pressing its base downward. Thus, according to emplacement depths and local tectonics, several resulting pluton shapes are possible.

Cauldron subsidence, a situation typical of ring complexes, characterizes the most superficial level of alkaline magmatic complexes that comprise several levels (Bonin, 1982). It may occur in so-called 'anorogenic' tectonic conditions above mantle upwelling. Such complexes are responsible for crustal doming and regional scale radial extension (Fig. 6.2). The Monte Cinto Complex (Corsica, France) illustrates the superficial level of alkaline granites that are emplaced along with their associated volcanics, first mafic (basalts) then felsic (ignimbrites, pyroclastites), that are injected into a

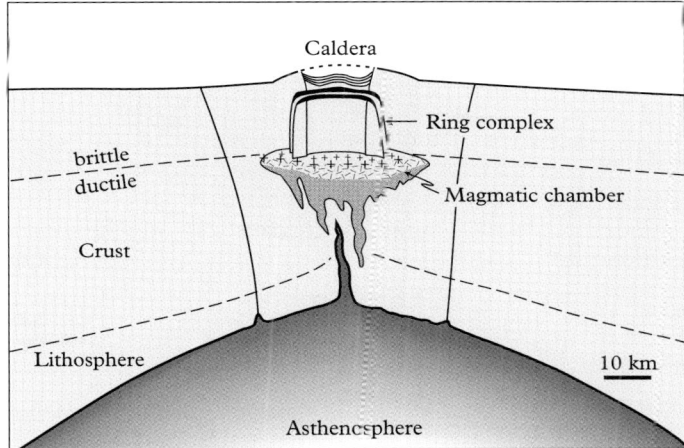

Figure 6.2 *Structural levels of anorogenic alkaline magmatism. A large doming of the crust, with radial extension, results from the isostatic equilibrium in response to asthenospheric upwelling and its subsequent dynamic uplift. Two emplacement levels occur in this system. One is the initial magma chamber, which is stagnant at the base of the brittle crust. It gathers the mantellic magma, more-or-less contaminated by crustal material, and results in bimodal plutonic complexes (mafic at the base and granitic at the top). The depth of brittle–ductile transition becomes shallower with time and with the increasing volume of the magma chamber. The level of ring complex and caldera formation constitute the second emplacement level. Concentric fractures over the whole thickness of the brittle crust due to doming are responsible for magma-infilled annular dykes that can form rings if viewed from above. Close to the surface, an important mainly rhyolitic volcanism takes place within a circular caldera. Ring diameters range from 4 to 30 km and even reach 65 km in Niger (Aïr). Diameters depend on the radius or curvature of the doming, hence on mantle plume activity but also on the depth of the erosion surface. After Bonin (1982).*

92 Emplacement and shape of granite plutons

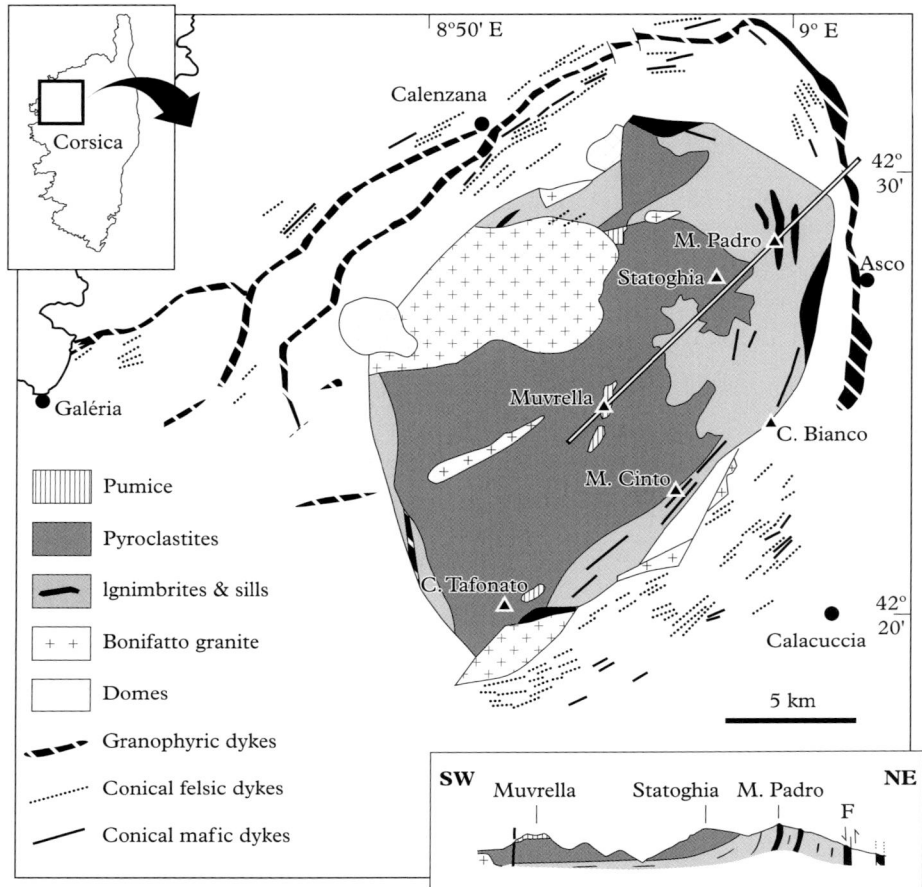

Figure 6.3 *Map and simplified geological sections of the Late Hercynian anorogenic alkaline complex of Monte Cinto (Corsica). After Vellutini (1977).*

progressively subsiding caldera. Concentric dykes, both mafic and felsic, illustrate the collapse, or subsidence, of the whole upper crust toward the central part of the complex, as the underlying magma chamber empties (Figs 6.2 and 6.3). Other ring complexes showing similar features are also well known in Nigeria (see Fig. 12.3).

Plutons that are emplaced very high in the crust in association with their volcanics, like in the Andes, are rimmed by vertical faults along which the pluton's roof was pushed upward like a piston (Fig. 6.4). In such a context, enclaves of country rocks may be detached from the roof zone and sink a few metres deep into the magma cauldron (Fig. 6.5). Since the thermal reservoir of a pluton is limited, it cannot assimilate large quantities of country rock. Therefore, this nibbling mechanism, also called 'stoping', cannot provide enough space for the magma to settle. The necessary room for magma settling,

Meeting dilatant sites 93

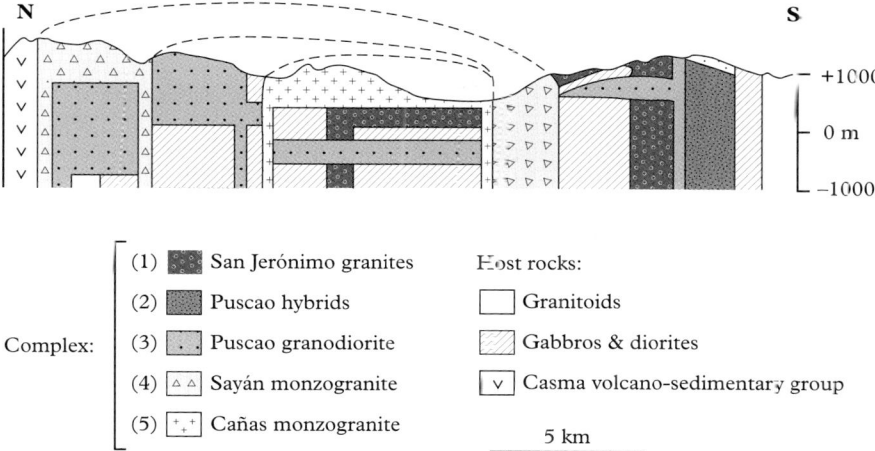

Figure 6.4 *Schematic geological section of the Huaura complex (~100 km north of Lima in the Peru coastal batholith: 11°10′ S, 77°15′ E). This complex is regarded as a repeated succession of cauldron subsidences that allowed the emplacement of plutons (1) to (5) in less than 2 million years (62–64 Ma) around the principal eruption centre. Note vertical scale exaggeration (×2) and the large uncertainty of the geometry at depth. After Bussell and Pitcher (1985).*

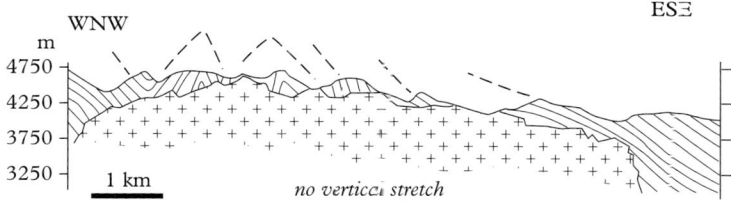

Figure 6.5 *Schematic section of the Chita pluton in the Andes (Argentina: 30°30′ S, 69°35′ W) showing the clear-cut contacts between pluton and host and suggesting that emplacement was assisted by stoping. After Yoshinobu et al. (2003).*

giving birth to a pluton, is provided through a combination of roof lifting and floor subsidence that compensated for the volumes of magma discharged from the underlying magma chamber.

When feeder dykes approach the surface, the magma may form a sill, as in Fig. 5.8c. The roof of the sill is the surface along which magma pressure was in equilibrium with the lithostatic pressure (at time of infilling). This is why the surface level of Monte Amiata (Tuscany, Italy), a locality hosting a hidden sill, has been uplifted by about one kilometre giving a large, more than 50 km-wide, doming (Fig. 6.6). In the Tuscany magmatic province, granitic magmatism took place from 14 Ma to less than 1 Ma, younging

94 *Emplacement and shape of granite plutons*

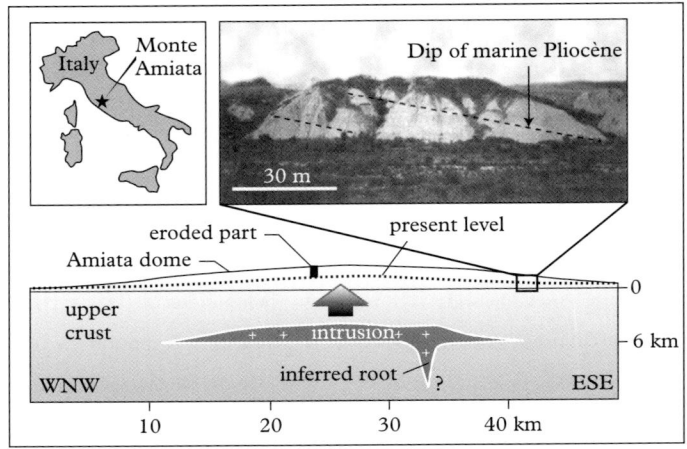

Figure 6.6 *The hidden granite dome of Amiata in Tuscany (Italy) reconstructed from gravity data. The geological evidence of roof uplift is given by the dip of the Pliocene formation (box). Intrusion age is estimated at 3.6 Ma. The whole area to the north of the Tyrrhenian Sea underwent a recent E-W crustal extension accompanied by emplacement of granitic plutons (Montecristo, Giglio, Elba, Campiglia, Gavorrano) between 7 and 5 Ma, and subsurface magma chambers (Amiata and Larderello). The latter is known for its geothermal production. After Acocella (2000).*

toward the east, related to the opening of the Tyrrhenian Sea. Westerman et al. (2004) have shown that these plutons are made of superimposed and more-or-less imbricated sills, from several hundreds of metres to 2 km in thickness.

The famous leucogranite lenses of the Himalayan chain (Fig. 6.7) also consist of sills that were emplaced at shallow depth during extension, and rapidly dissected by thinning of the overlying strata due to gravity spreading. Normal faults at the roof of the lenses were probably facilitated by the heat coming from the sills, allowing décollement surfaces to form. Hence the brittle–ductile transition, where listric faults are supposed to root and participate in tectonic denudation, may be limiting for the upward progression of plutons.

6.2.2 Fracturing in compressive environments

Submitted to horizontal compression ($S_H = 1$), the brittle crust potentially develops thrust faults ($S_V = 3$), or strike-slip faults ($S_V = 2$) if lateral escape of the rocks is possible. In compression, the critical stress needed for rupture (σ_{crit}) is larger than in extension, on the order of 150 MPa at the brittle–ductile transition (as compared with 100 MPa in tension: Fig. 6.1). In the absence of tectonic overpressure (P_s) the driving pressure for rupture will be weaker than in extension and its intersection with S_V (then

Meeting dilatant sites   95

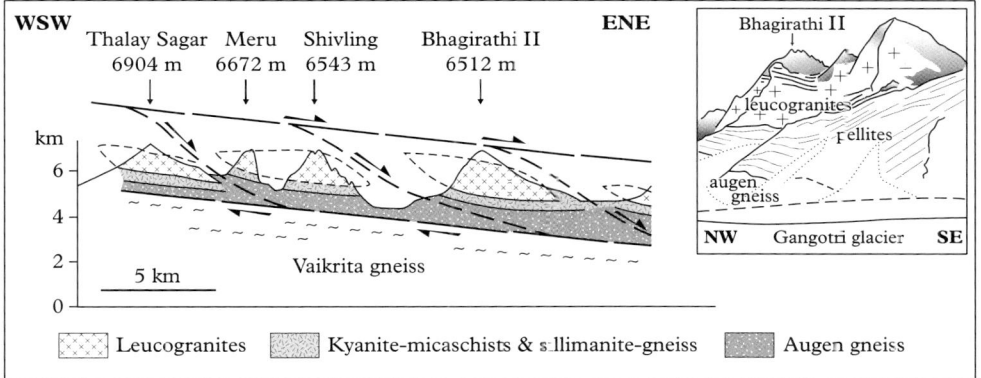

Figure 6.7 *Section along the Gangotri glacier (Himalayas) showing the geometry of the leucogranite bodies. Right: view in section normal to the length of these bodies. After Searle et al. (1993).*

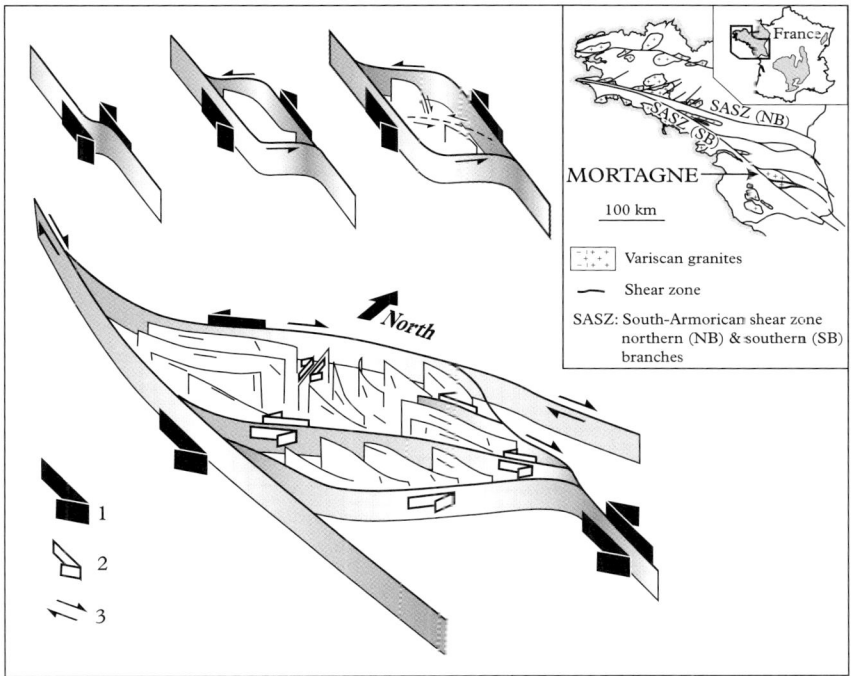

Figure 6.8 *Emplacement model of the Variscan Mortagne-sur-Sèvre granite pluton (Armorican massif). A sinistral extensional relay of the South Armorican Shear Zone has produced a dilatant jog (pull-apart) into which the Mortagne granite was emplaced. This model considers that the main shear zone (arrow 1) and its relays (arrow 2) had a sinistral sense before being reworked by the dextral shear (arrow 3) that characterizes the whole Armorican massif, reworking the northern and south-western borders of the pluton. The main trajectories of the magmatic foliation are represented inside pluton. After Guineberteau et al. (1987).*

96 *Emplacement and shape of granite plutons*

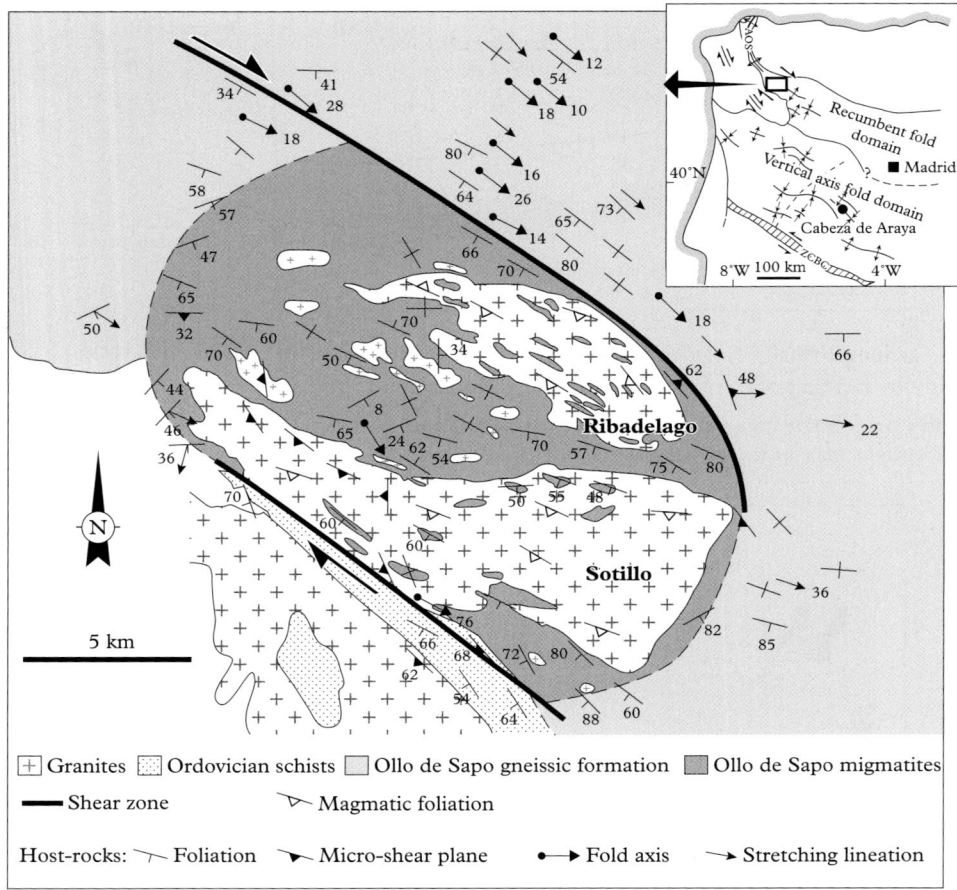

Figure 6.9 *Geological map of the region of Sanabria (Spain). Inside an extensive relay located between the two main branches of this dextral shear zone, two granite plutons (Ribadelago and Sotillo) were emplaced amongst the Ollo de Sapo gneissic formation (light grey) that forms a migmatitic dome at their roof (dark grey). Fold axes in the host rocks, almost parallel to the shear zone (and not in en-échelon configuration) suggest that shearing was compressive, hence that the dilatant jog was the result of a dextral transpression. From Vegas et al. (2001).*

with S_H) will occur quite high in the crust. This illustrates the major role that P_s should play for a pluton to make room for itself in a purely compressive environment. However, and even in compressive regimes, large dilatant sites may appear if strike-slip faulting is present (see Figs 6.8 and 6.9). Such sites may be infilled by magma and the granite thus formed may lift its roof if $P_{sm} > S_V$. By further lifting of its roof the magma may then form a laccolith, a kind of tabular pluton, up to a maximum thickness $e = P_m/\rho_{crust}.g$.

Note that as soon as P_{sm} becomes larger than S_H, magma progression is made possible up to the surface (volcanism) depending on the mechanical properties of the overlying rocks. More importantly, the addition of overpressure coming from the source (P_s), has the property of shifting P_{sm} to the right-hand side by the value P_s (Fig. 6.1) and this will favour an emplacement at still lower depths. Below the brittle–ductile transition, the mechanics of pluton emplacement is completely different due to the drastic decrease of P_{sm}. Variations naturally exist in the diagrams of Fig. 6.1, according to local tectonic environments, depth of magma source and brittle–ductile transition, and of course the intensity of P_s.

Strike-slip systems are responsible for horizontal displacements of crustal units, sometimes with bulk rotations around a vertical axis. Dilation sites due to changes in fault orientations during strike-slip, or to fault-relays, are responsible for open volumes into which a magma may find some room. Such tectonic objects, called pull-apart structures, are common along the large scale faults of the globe (San Andreas, Anatolia, Dead

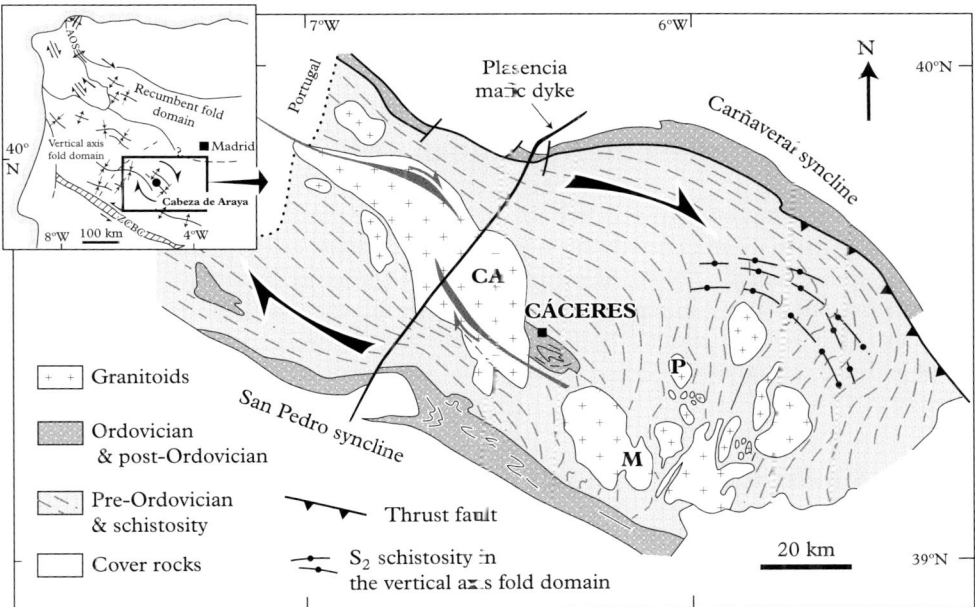

Figure 6.10 *Geological sketch map of the Hercynian batholith near Cáceres in Extremadura (Spain). To the south of the Ibero-Armorican arc (southern portion of the Iberian Central System), the Cáceres anticlinorium, made of pre-Ordovician slightly metamorphic schists, squeezed between the Carñaveral and San Pedro synclines, is considered as a dextral shear zone, up to 40 km in width. Shearing was responsible for the clockwise rotation (up to 90°!) of the S_1 regional schistosity (itself refolded around vertical axes in the hinge zone with development of a secondary schistosity). The resulting dilatant zone favoured the emplacement, to the east of Cáceres, of several plutons (Plasenzuela: P; Montanchez: M). To the west of Cáceres, the deeply rooted pluton of Cabeza de Araya (CA; see Fig. 6.16) occupies a dilatant relay in-between two branches of a dextral shear zone. After Castro (1985) and Audrain et al. (1989).*

Sea). Any magma present at depth on its way from the source is welcome in such sites, resulting in plutons such as the Mortagne-sur-Sèvres pluton, emplaced along one of the southern branches of the South Armorican Shear Zone (Fig. 6.8).

Toward the surface, strike-slip faults are observed to subdivide into branches, both in map view and in section. These arrangements lead to flower structures that are associated with pull-apart basins. At depth, strike-slip faults evolve into straighter and less branched shear zones, concentrating strain within a few hundred metre wide zones of deformation. Granitic magmas that are present along such zones of strain localization, driven by their buoyancy and eventually by pressure at the source, will constitute elongate plutons along shear zones, made of narrow strips or relay domains. Such a structure characterizes the main body of the Closepet complex in India (see Fig. 5.13).

In the large scale Variscan shear zone system of Central Iberia (Spain) the twin plutons of Ribadelago and Sotillo were emplaced within the Ollo de Sapo gneissic formation, at extensional sites connected by a dextral shear zone (Fig. 6.9). Further south in the Central Iberian System, the Cabeza de Araya granitic complex was emplaced in a similar way in a region subjected to bulk dilatancy related to a strike-slip fault system (Fig. 6.10).

6.3 Granite emplacement in the ductile crust

Although most granite magmas reach the brittle, upper part of the crust, this is not the fate for all of them. Some plutons are trapped in the ductile crust, possibly due to a lack of segregation efficiency, eventually due to a hard-to-fracture host material, or to a lack of transport efficiency ultimately aided by an insufficient overpressure at the source. Note also that several granitic bodies display a cumulative geochemical composition possibly due to an unknown liquid fraction that may have escaped higher into the crust (see Fig. 5.13).

6.3.1 Sills and sheet-like granites

A volume of magma that gathers within the ductile crust will form a stock whose shape will be primarily guided by the fabric of its gneissic host. Subhorizontal reflectors, attributed to impedance contrasts (impedance: seismic velocity multiplied by density) between formations, are well imaged by seismic reflection profiles. These contrasts, attributed to compositional variations, to the presence of fluid-rich layers or magmas and/or to shear fabrics, constitute the layered crust, typical of deep seismic imagery.

The subhorizontal, also called 'stratoid', granite sheets of Neoproterozoic age in Madagascar were emplaced into high-grade gneisses and migmatites (Fig. 6.11) and could be an example of such a layered crust although their impedance contrast with their host is not that high. They are made of a few decimetres to 500 metres-thick layers of quartz-syenite to alkali feldspar granites attributed to differentiation of deeper mafic magmas with the possible contribution of partial melting of granodiorites in the lower crust (Nédélec et al., 1995:

Granite emplacement in the ductile crust   99

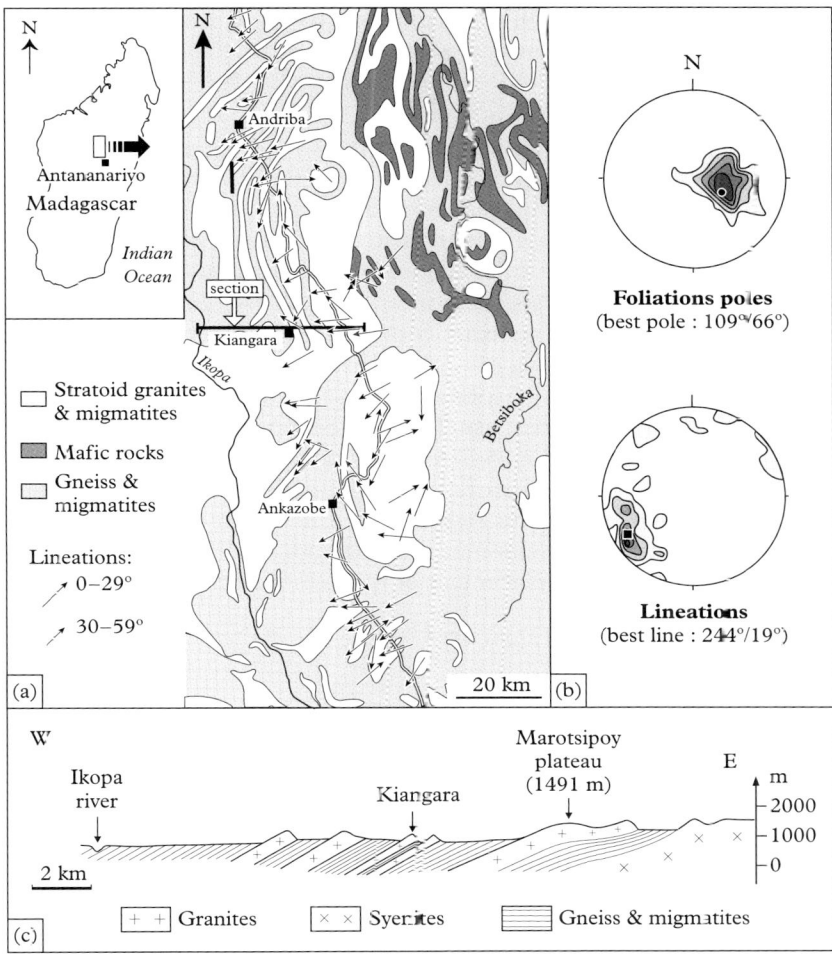

Figure 6.11 *Stratoid granites in the Pan-African basement of Madagascar, giving remarkable landscapes and forming cuestas as in a sedimentary basin, the alkaline granites being harder than their gneissic hosts. (a) Geological and structural map (magmatic lineations) of the area to the north-west of Antananarivo; the contorted aspect of the geological limits is due to the (westward) shallow dips of the granite layers. (b) Orientation diagrams of the magmatic foliation poles and lineations in the stratoid granite layers (Schmidt, lower hemisphere; 86 measurements). (c) E–W cross-section at the latitude of Kiangara (17°22′S). After Nédélec et al. (1994).*

Fig. 2.26). These granite sheets, as well as their magmatic fabric, are conformable with the foliation of the shallow west-dipping gneissic–migmatitic hosts. Along with their hosts, these stratoid granites display a magmatic lineation (measured by the anisotropy of magnetic susceptibility, or AMS, technique: see Chapter 10) whose azimuth is very constant over large areas and reveals the direction of magma flow during emplacement. These stratoid granites of Madagascar are therefore viewed as a succession of thin sills that were emplaced in a crustal pile which at the same time was ductilely flowing during extensional tectonics.

6.3.2 Migmatitic domes

Migmatitic domes, or core complexes (Fig. 6.12), are contained in hot thin crust where the isotherms are closely packed indicative of a high thermal gradient, possibly due to mantle upwelling. Rocks from migmatite domes have undergone strong partial melting without substantial segregation of the melt (see Fig. 3.15). The resulting poorly drained material is heterogeneous, commonly full of restites, and its mechanical behaviour, at least at the beginning of doming, is close to the behaviour of its gneissic source country-rocks. The arch-shaped surface separating migmatites from its host suggests that the partially molten material was beginning its gravity-driven ascent when it stopped ascending (see Fig. 5.5). As a consequence, the temperature and pressure conditions, as recorded by thermo-barometric methods, may vary on the sides of the gneiss–migmatite interface. This interface is also a shear zone acting as a normal fault and allowing partial exhumation of the migmatitic dome. Indeed crustal extension is often invoked for the emplacement of migmatitic domes. However, substantial

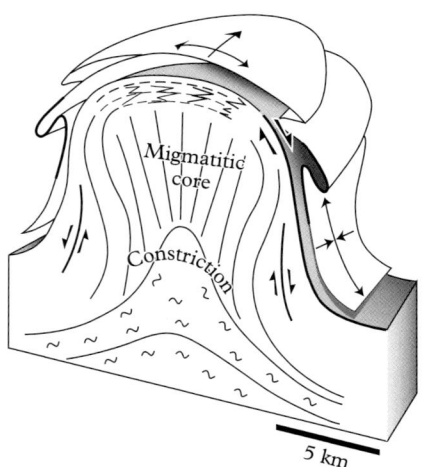

Figure 6.12 *Schematic section of a gneiss dome illustrating its main structures: migmatitic core and high grade metamorphic cover highly flattened on top, stretched and showing series of folds on its limbs; detachments often initiate at the gneiss-migmatite interface. After Whitney et al. (2004).*

extension is not required since, in a ductile crust, upwelling of material in one location may be balanced by sinking of material elsewhere. In the Variscan Pyrenees a recent study has concluded that the migmatite domes at Aston and Hospitalet (France) were squeezed upwards along shear zones in a large-scale compressive tectonic environment (Denèle et al., 2007).

6.4 Emplacement depth

Observation of granite plutons at the surface results from the interplay between erosion and tectonic uplift, in turn responsible for several kilometres of upper crust stripping. How can we know the depth at which a magma crystallized? There are two possible approaches. One is based on metamorphic conditions recorded in the immediate host (see Chapter 7). The other, based on pluton mineralogy, is developed in this section.

Few minerals provide precise barometric information, i.e. information about the pressure at time of crystallization. A particular case in peraluminous granites is magmatic andalusite whose stability domain is narrow, its pressure of crystallization being less than 300 MPa (Fig. 6.13), corresponding to a depth of less than 12 km. In these granites, and contrary to common belief, garnet is not a marker of high pressure since it contains a high fraction of spessartine (Mn-bearing garnet) that makes it stable at low pressure.

In I-type granitoids, hornblende (a calcic amphibole) is used as a barometer. Hammarstrom and Zen (1986) have shown that the number of Al-atoms in hornblende linearly increases with depth. This Al-in-hornblende barometer was calibrated experimentally by Schmidt (1992) and an example is provided in Fig. 12.7:

$$P \text{ in kb} (\pm 0.6) = -3.01 + 4.76 \text{ Al (total atoms)}$$

As for every barometer, strict conditions must be respected for its use. It works for common mineral assemblages in I-type granites, comprising plagioclase + quartz + K-feldspar + biotite + Fe–Ti oxides (ilmenite and/or magnetite) + sphene, in equilibrium with temperatures around 750 °C. In addition, since the crystal cores were formed at depths larger than that of final magma emplacement, hornblende compositions must be determined close to the mineral margins, i.e. to the mineral portion that formed in condition close to the solidus, thus avoiding Anderson's (1996) objection concerning the sensitivity of this barometer to temperature. The barometer helps to determine the emplacement depth of calc-alkaline plutons with a precision of about ± 60 MPa (more than 3 km of uncertainty!) for emplacement depths usually ranging from 200 MPa to 1 GPa (2 to 10 kb). Finally, the (rare) presence of magmatic epidote is often considered as pointing to crystallization at high pressure (Zen and Hammarstrom, 1984). However, there is no consensus concerning epidote since its stability strongly depends on the fugacity of oxygen at the time of crystallization.

102 Emplacement and shape of granite plutons

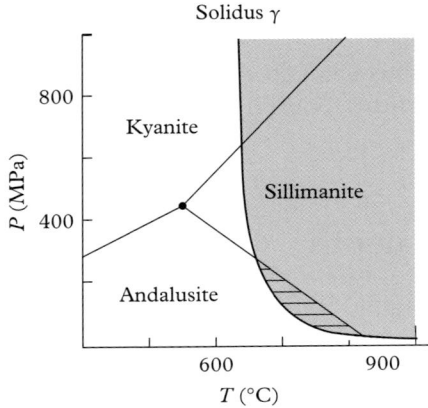

Figure 6.13 *Aluminium silicate (Al_2SiO_5) stability diagram (Robie and Hemingway, 1984) and hydrated solidus of granite (bold line) indicating that the presence of magmatic andalusite (hatched) corresponds to low-pressure (<300 MPa) production and emplacement of magma.*

6.5 Three-dimensional shapes of plutons

The shape of a pluton is easy to identify if the granite body is isolated in its country-rocks. Inside a batholith, formed by a set of side-by-side or imbricated plutons, the shape of an individual body is often hard to identify. The geological mapping of the pluton's margins and of its various exposed rock-types will not suffice for the understanding of the three-dimensional (3D) shape of each entity. More detailed structural mapping will be needed (see Chapters 9 and 10). However, the 3D shapes of isolated plutons are considered here.

Unless an excellent natural vertical section is available, as in the Andes or Himalayas (Figs 5.12a and 6.7), the geologist will not be able to observe the roots of the plutons. By measuring the dips at the pluton contacts, it may be concluded that a pluton is enlarging ('extravasated' shape) or narrowing with depth. On the basis of preconceived ideas on pluton shapes, for example resembling hot-air balloons, the geologist may conclude that the observed section is close to the top or the base of the pluton. This analysis quickly finds its limits. By chance, the density contrast that most granites have with their country rocks is responsible for a gravity anomaly, which is used to derive 3D distributions of rock masses and hence shapes of plutons with a better degree of certainty.

6.5.1 Bouguer anomaly map

To make a Bouguer anomaly map, one needs to obtain gravity and level measurements with precisions better than 0.1 mGal and 50 cm, respectively, arranged along one-dimensional profiles or preferably onto two-dimensional grids in map view. Profile length (or extension of the grid) must be several times larger than the maximum depth of the pluton.

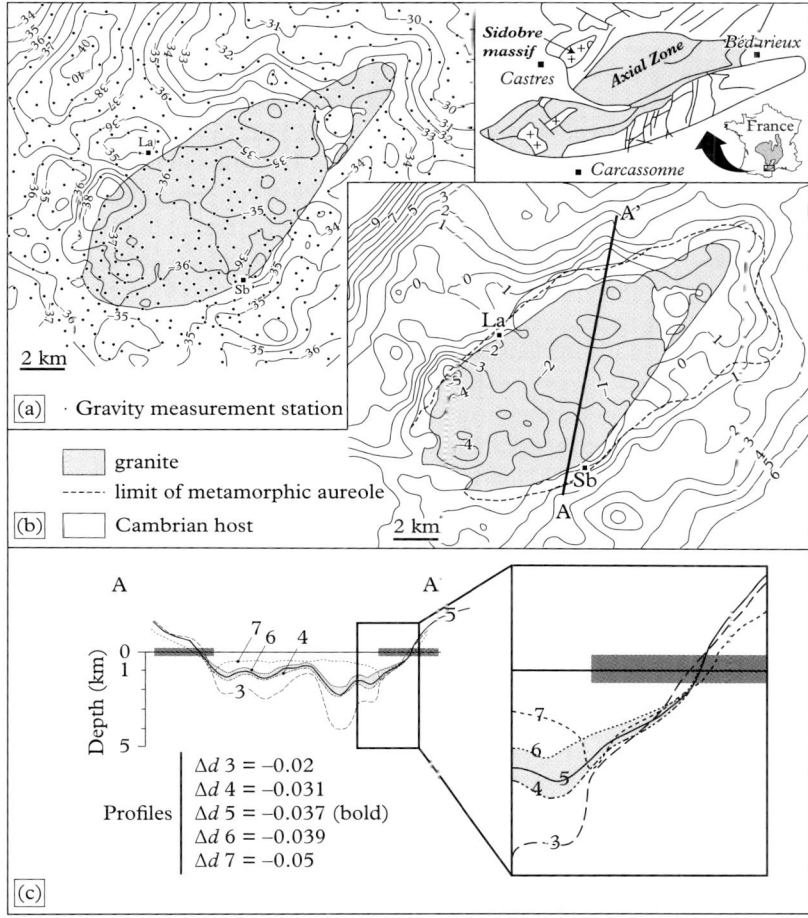

Figure 6.14 *Gravity study of the Sidobre granite pluton (Tarn, France) hosted in the Cambrian of south-west Massif Central. (a) Bouguer anomaly map (mgal) built from more than 600 measurements; Sb: St Salvy-de-la-Balme; La: Lacrouzette. (b) Residual anomaly map (contours in mgal). Dashed line: limit of contact metamorphic aureole. (c) Floor depth profile across AA' (no vertical exaggeration) extracted from the 3D model by using various density contrasts between granite and host. In grey, the domain which best fits granite–host contact (in map view) corresponds to the density contrast measured on rock samples ($\Delta d = -0.031$ to -0.039). After Améglio et al. (1994).*

The minimum pluton size that can be detected depends on the spacing of the data points. In practice, plutons and their surrounding country-rocks are well covered by a map grid spacing of one measurement per km^2. After the usual complete Bouguer corrections (latitude, free-air and topography), the Bouguer anomaly map is obtained (Fig. 6.14a), usually calculated for a reference density of 2.7 (see Dubois and Diament, 1997, and Lowrie, 2007).

6.5.2 Residual anomaly and density contrast

The Bouguer anomaly reflects the heterogeneity of mass distributions at depth. The amplitude of this anomaly shows variations at large wavelengths and is called the 'regional anomaly'. It reflects large deep-seated heterogeneities. This regional anomaly often looks like a shallow dipping and slightly undulated surface. Its equation is determined in order to subtract its values from those of the Bouguer anomaly. The resulting, high frequency, anomalies are reported in map view and determine the residual anomaly map (Fig. 6.14b). This map is rich in information concerning the pluton. At a given depth the residual anomaly is proportional to the volume of the pluton, to the density contrast between granite and host-rocks, and decreases according to the inverse of the square of the depth. In practice, for a given density contrast, the amplitude of the anomaly depends on the volume of the underlying granite; for a given granite volume, the amplitude of the anomaly depends on the density contrast.

The quality of a gravity model, hence the quality of the model pluton shape, will depend on the magnitude of the density contrast. This is why modelling the shape of a granite pluton hosted in rocks having similar densities, like other plutons, is a very hard task. Precise density determinations of the rocks in question (using a pycnometer for volume measurements, the double-weighting method for mass measurements, and taking into account the eventual porosity due to surface decompression) would ensure the ideal conditions for gravity modelling. Densities of granitic rocks vary according to their content in iron, from 2.63 for leucogranites to 2.83 for tonalites. Densities of country rocks are much more variable. Mica schists ($d \geq 2.65$) and gneissic rocks ($d \geq 2.80$) frequently display higher densities than granites. This is at the origin of the popular belief that granites always give negative anomalies. However, lower country-rock densities are possible although less common, such as those for limestones ($d \sim 2.53$) and quartzites ($d \sim 2.60$).

6.5.3 Modelling pluton depth and floor shape

Gravity modelling is based on direct or inverse techniques. The direct technique consists in choosing a priori a mass distribution at depth (corresponding to the inferred geometry of the pluton–country rock interface), the associated densities, and in calculating the resulting gravity anomaly. This anomaly is then compared to the observed gravity profiles that constitute the map of the residual anomaly. Through several iterations, the pluton shape at depth and/or density contrasts are progressively refined minimizing the differences between the observed and calculated anomalies along selected profiles. Implementation of this technique is rather straightforward. However it can be used only along profiles, or series of profiles, and not in two dimensions. In addition, this technique provides non-unique solutions, hence introducing potential ambiguities.

The inverse technique ignores, in principle, all a priori information. Model parameters are directly calculated either from the measured values, or from the values interpolated for the intersections of the residual anomaly grid map. Calculations are based on the subdivision of the area under study into vertical, elementary and contiguous prisms, each one having a constant density, its height being calculated according to the anomaly value ascribed to the prism. The resultant, calculated anomaly map is then compared with the observed one, and the differences between measured and calculated values are iteratively minimized by adjusting heights and densities of the prisms. Although more realistic, this technique does not provide a unique solution. However, its stability can be easily tested, particularly with respect to the sensitive parameter of density contrast (Fig. 6.14c).

6.5.4 Pluton shapes

A map representing the depth of a pluton's floor gives an idea of its shape, except that the thickness of the eroded portion remains unknown. In the Sidobre granite (Fig. 6.15), the topography of the pluton's floor is observed to be irregular with a few limited sectors deeper than 3 km, the mean depth not exceeding 2 km. Although local heterogeneities may modify the calculated depths, it is tempting to consider that this pluton has the shape of a sill displaying several roots. A detailed examination of its internal

Figure 6.15 *Calculated floor depth (in km) of the Sidobre pluton, showing a weakly rooted, flat-floored pluton characteristic of a sill. The poor adjustment of the calculated zero contour (with respect to the real one) is attributed to the simplistic model used for the regional anomaly as well as to local density variations due to the presence of several granite dykes in the country rocks, and to the presence of late, NW-SE directed faults (dotted). After Améglio et al. (1994). See also the map of the magmatic structures (Fig. 10.19).*

106 *Emplacement and shape of granite plutons*

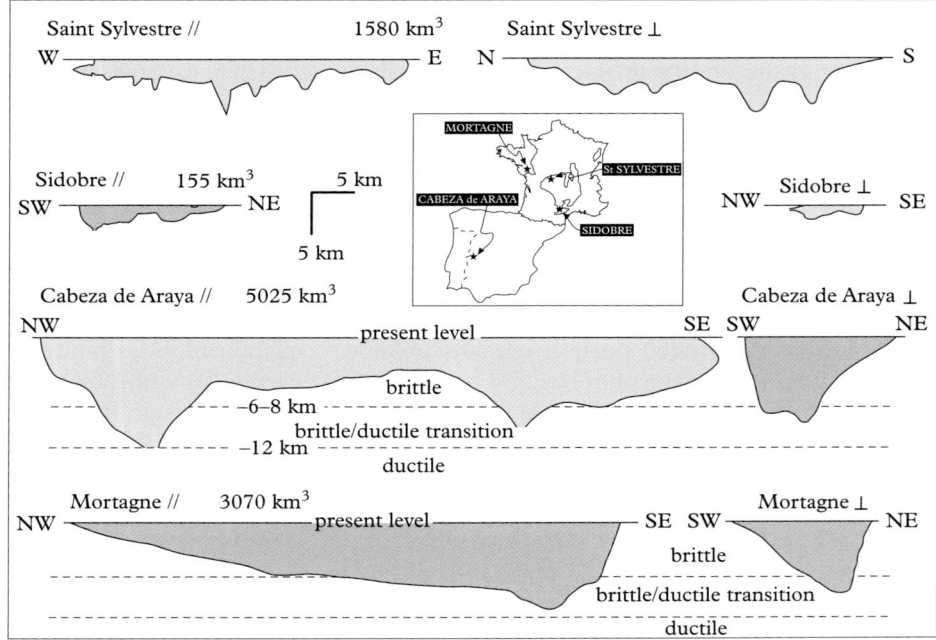

Figure 6.16 *Longitudinal (//) and transversal (⊥) sections through four granite massifs from the French and Spanish Hercynian chain. Although there remains some uncertainty about the eroded thickness, these sections help to distinguish small-volume plutons (such as sills) from voluminous, edge-shaped and deeply-rooted plutons which have a limited number of feeder zones. The roots are located below the level where the dips of pluton margins change, ascribed to the brittle–ductile transition (at time of pluton emplacement). These sections come from gravity models (see also Fig. 9.16 for Saint-Sylvestre; Fig. 6.15 for Sidobre; Fig. 6.10 for Cabeza de Araya; and Fig. 6.8 for Mortagne). From Améglio et al. (1997).*

structure (magmatic foliations and lineations) may help to clarify the ambiguity (see Fig. 10.19).

By contrast, other plutons appear as deeply rooted. This is the case of the Mortagne-sur-Sèvres pluton (Fig. 6.16) already examined for its emplacement in a large-scale fault zone of the upper crust related to an extensional fault relay. From the gravity model, the floor of this pluton is imaged as a ship's hull whose median line contains, from northwest to southeast, the deepest parts of the pluton down to 10 km. Another example is the Spanish Cabeza de Araya pluton, already examined in Fig. 6.10 for its emplacement in a large extensional jog related to a dextral shear zone. It is imaged as if it was fed by magma from two relayed tension faults, each one being at a depth of 6–10 km.

Whatever the pluton type (sill, laccolith or edge-shaped, as in the last example) the vertical thickness (H) of a pluton hardly exceeds its width (L). On the basis of studies combining gravity and structural analyses (Chapter 10), the floor of the pluton can be tentatively defined as flat or edge-shaped, a classification more-or-less related to the number of roots supposed (and in some cases ascertained) to be the magma feeding channels for the plutons (Fig. 6.16). Flat-floored plutons such as the Sidobre (Fig. 6.15) are considered as rather thin ($H/L < 0.5$) and have several root-zones. These characteristics could indicate that a lithological interface served as a magma trap and that the pluton was fed by several fractures, possibly via several magma batches. By contrast, edge-shaped plutons may be quite thick ($H/L \gg 0.5$) and bow-shaped in cross-section, the dip of their floor increasing with depth. Such plutons are elongate perpendicular to their bow-shaped section and seem to be fed at their base through a small number of deep and narrow roots.

6.6 Passive versus forced emplacement

This presentation comes from an old debate concerning diapirs that were considered either to forcefully intrude their country rocks or to progress passively upwards by disaggregation of their roof, a mechanism also called 'stoping', pieces of country rocks accumulating inside the pluton or even at its floor (Fig. 6.5). As a consequence, clear-cut contacts between pluton and virtually undeformed country rocks were considered to support a passive emplacement. By contrast, should the schistosity or foliation of the country rock be conformable with the pluton's contact, the pluton was therefore concluded to result from a forceful emplacement.

In any event, it is accepted that a pluton progressively enlarges at its emplacement site through a limited number of feeding zones, by the way of either a single batch that lasts no more than a few 10^5 years, or through a variable number of batches (Michel et al., 2008). As discussed previously, the assembling site of a magma is preferably a dilatant volume located in a more-or-less brittle crust and controlled by extensional or wrench tectonics.

As soon as a granitic magma begins to collect, various situations are possible whose end-members are as follows. In one case, granite emplacement is controlled by magma pressure (forceful emplacement) which is all the more important that overpressure at the source is large, and that pressure does not fail as the magma column rises. Pressure may help the magma to push up the overlying strata (Figs 6.1 and 6.6) or to inflate the pluton in its emplacement site. Doming, which accompanies the emplacement of a sill or laccolith (Fig. 6.6), may help dome-denudation by gravity collapse of the overlying rock-pile. Alternatively, magma emplacement is controlled only by tectonics (passive emplacement), for example when a pull-apart opening, related to a strike-slip system, favours the immediate suction of a magma as would happen in a syringe. As already mentioned, tectonic denudation may also increase the rate at which a magma may collect in a reservoir, allowing the pluton to further expand due to the reduced weight of the overlying rocks. In nature, passive and forceful emplacements can act together in various proportions but one mode may become dominant over the other.

6.7 Conclusions

Upward magma transport, which takes place before magma assembly as a pluton, will stop its progression either because of an increase in magma viscosity when approaching its solidus, due to a drop in temperature, or to magma pressure equilibration where the feeding fractures reach a plastic unit. In a few cases, the granitic magma reaches the surface giving as rhyolitic or ignimbritic volcanism. Emplacement of granite plutons is therefore conditioned by the physical properties of both the magma and the country rocks, and depends on the geotectonic context, both for magma production and for its dynamic aspects. Extensional and strike-slip tectonic environments play a key role in the production of magma volumes. The variety of these environments is responsible for the variety of shapes of plutons.

Finally, it is important to note that pluton shapes are closer to sills or laccoliths than to hot-air balloons or inverted water drops. Hence they are larger in map view than thicker in cross-section. Three-dimensional shapes of plutons are difficult to determine using classical geological tools, mainly because vertical sections are lacking. In favourable cases (isolated plutons, density contrasts) the shape can be modelled by gravity studies. In Chapters 9 and 10, the emplacement scenario of a pluton can be specifically refined by combining geological and geophysical data with information extracted from microstructures and fabrics.

7

Thermomechanical aspects in the country rocks around granite plutons

The thermal effect induced by a pluton is found in the aureole of contact metamorphism surrounding the granite. The principal aim of this chapter is to determine the rate at which the thermal 'front' propagates through the country rocks, and to determine the extent of its propagation. The simplest but less realistic case is when the thermal pulse from hot magma intruding the crust is purely conductive, i.e. does not imply any movement of fluids into the country rocks. However, fluids are very efficient in carrying heat and chemical constituents. They equilibrate the temperatures, facilitate metamorphic reactions and help the country rocks to fracture, developing veinlets, veins and dykes, confirming the presence of hydrothermal activity.

Considering intrusions close to the surface (less than 12 km deep for 'normal' geotherms), rock deformation due to emplacement is only recorded temporarily because metamorphic and rheological conditions rapidly change with time and distance from the cooling pluton. The warmed crust may trigger shearing, which can be normal, inverse or strike-slip depending on the tectonic situation. For large volumes of granite magma a 'regional' zone of contact metamorphism will be recorded in the surrounding crustal rocks when emplacement takes place at shallow depth. At deeper levels of emplacement in the crust, the contact metamorphic aureole cannot be distinguished from the effects of earlier widespread regional metamorphism.

7.1 Conductive heat transfer

Heat transfer (Q, in W m^{-2}), from the pluton margin to host country rock with no fluid movement, obeys the first Fourier law:

$$Q = -k \, gradient(T)$$

Box 7.1 Conductive heat transfer: the virtue of non-dimensional variables

Figure 7.1 shows the decrease with time of a thermal perturbation. It illustrates the generality of non-dimensional relations: temperatures and distances are represented as ratios,

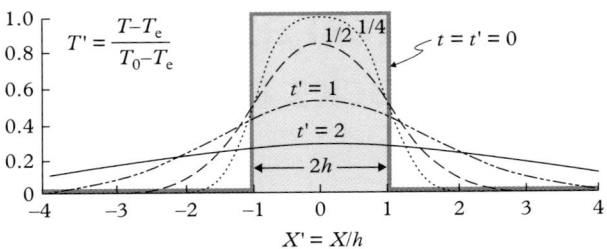

Figure 7.1 *Purely conductive cooling of a dyke. Temperatures in kelvins. Numerical application: along the vertical axis, taking $T_e = 573\ K\ (300\ °C)$ and $T_0 = 1073\ K\ (800\ °C)$, T values become: $T = 673\ °C$ for $T' = 0.2$, $T = 773\ °C$ for $T' = 0.4$ and $T = 1073\ °C$ for $T' = 1$; along the horizontal axis and for the $T' = 1$ curve (used here as a reference for thermal decay) for which $t = h^2/\kappa$, by taking $h = 1\ m, x' = 1, 2, 3\ldots$ become $1\ m, 2\ m, 3\ m\ldots$ and $t' = 1$ will correspond to 106 seconds (one and a half weeks), taking $h = 1\ km$, $t' = 1$ will correspond to 3.17 million years, and taking $h = 10\ km$, $t' = 1$ will correspond to 31.7 million years. After Spear (1993).*

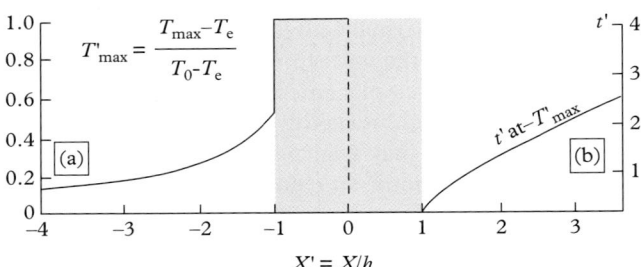

Figure 7.2 *Representation of T'_{max} (left-hand side (a)) and t' at temperature T'_{max} (right-hand side (b)) as a function of $x' = x/h$. The thermal maximum readily appears close to the contact, then slowly migrates outwards together with a slow decrease of the maximum temperature. Numerical application: to the left (using the same T_e and T_0 values as in Fig. 7.1), $x' = 1, 2\ldots$ gives $x = 1\ m, 2\ m\ldots$ for $h = 1\ m, x = 1\ km, 2\ km\ldots$ for $h = 1\ km$; to the right, for a 2 m wide dyke ($x' = 1$), the maximum temperature is reached within two weeks at 1 m from contact ($\approx 430\ °C$); it takes 24 days at 2 m ($\approx 390\ °C$) and 32 days at 3 m from the contact ($\approx 365\ °C$); for a 2 km wide body the same temperatures are reached within 43,000, 66,000 years and 88,000 years at, respectively, 1 km, 2 km and 3 km from the contact. After Spear (1993).*

Box 7.1 *continued*

respectively $T' = (T - T_e/T_0 - T_e)$ and $x' = x/h$, and the different curves representing time are labelled as $t' = (\kappa t/h^2)^{1/2}$. Numerical applications are 'readily' obtained by calculating T', x' and t' with the proper values of T_0, T_e, x, h, t and κ. At $t' = 2$, the thermal perturbation has almost disappeared. Above all, this figure illustrates that the cooling rate greatly depends on the available amount of heat, since for a 2 m-wide dyke (h =1 m), t' =1 is reached in less than 2 weeks while for a width of 2 km (h =1 km) it takes slightly less than 37.5 years.

The maximum temperature reached, and hopefully recorded by mineral parageneses resulting from contact metamorphism, is given by T'_{max} (Fig. 7.2a). Arrival time of the heat front at the maximum temperature and with increasing distance from the pluton (hence with decreasing T) is given by t' at T'_{max} (Fig. 7.2b).

---

where k is the thermal conductivity of the host rock (J s^{-1} m^{-1} K^{-1}). The negative sign before k means that the thermal flux moves down the thermal gradient, i.e. toward lower temperatures (in one dimension: $Tx + dx < Tx$; gradient$(T) = dT/dx < 0$ for $Q > 0$).

In order to respect the principle of energy conservation, the variation of temperature with time within a unit 'volume' of material ρdx (in one dimension) requires this variation to be compensated by a variation of the thermal flux ($dQ = Qx - (Qx + dx)$) between x and $x + dx$ in the 'volume':

$$dT/dt = (1/\rho c_p).(-dQ/dx),$$

where c_p, the heat capacity, represents the number of calories needed to increase by 1 °C the unit 'volume' ρdx. This helps to establish the second Fourier law for heat conduction:

$$dT/dt = \kappa(d^2T/dx^2),$$

where $\kappa = 1/\rho c_p$ is the thermal diffusivity of the material which usually is close to 10^{-6} m^2 s^{-1}. In three dimensions, $dT/dt = \kappa \nabla^2 T$, where ∇^2 represents $(d^2T/dx^2 + d^2T/dy^2 + d^2T/dz^2)$.

The above should be enough to understand the evolution of a thermal profile (in one dimension) with time, and with distance from a hot contact (i.e. with x increasing). The simplest example is given by a basaltic dyke of thickness $2h$ and at temperature T_0 instantaneously intruding its host at temperature T_e (see Fig. 7.1 in Box 7.1). The analytical solution of the heat equation is the following (in one dimension):

$$T' = (T - T_e/T_0 - T_e) = 0.5[\text{erf}(h - x/2(kt)1/2) + \text{erf}(h + x/2(kt)1/2)],$$

where erf is the error function found in every profile where diffusion only (conduction) is concerned.

As far as contact metamorphism is concerned, the curves describing the maximum temperature T'_{max} and its arrival time t' as a function of $x' = x/h$ provide important information, assuming that cooling is purely conductive (Fig. 7.2).

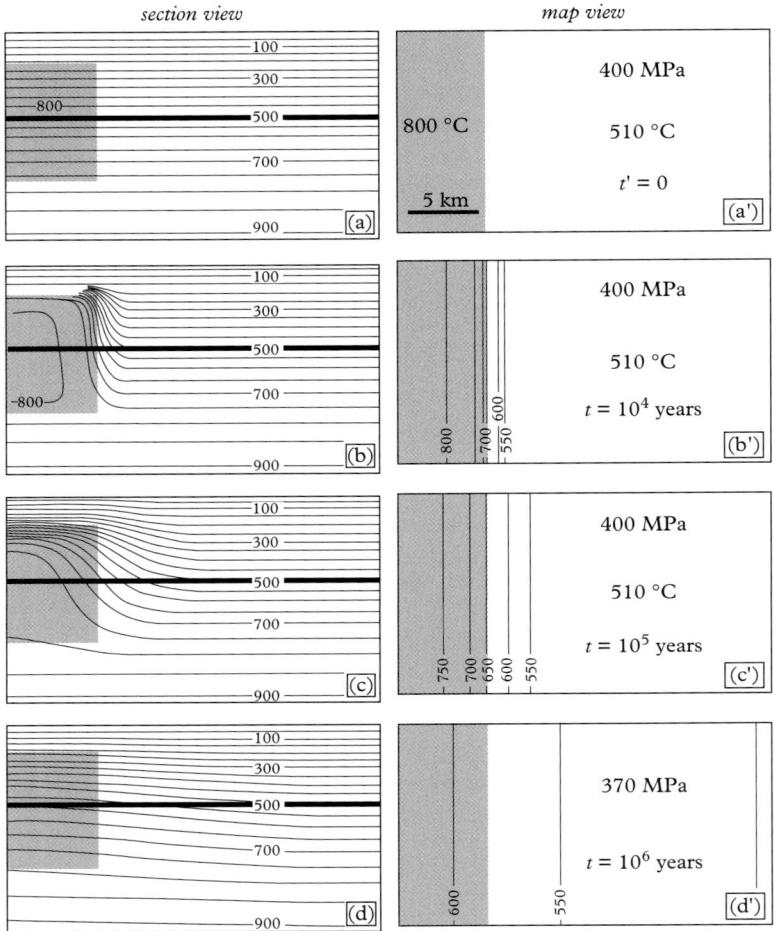

Figure 7.3 *Two-dimensional model of conductive cooling applied to the Quotoon pluton (British Columbia). (a–d) Sections at $t = 0, 10^4, 10^5$ and 10^6 years, respectively, assuming an erosion rate of 1 mm/year. (a'–d') Progression of the isotherms with time (map view). After Harrison and Clarke (1979) and Spear (1993).*

To better explain the implication of thermal modelling in terms of metamorphic parageneses, cross-sections and maps can be examined to illustrate the evolution of isotherms that develop around an intrusion (Fig. 7.3). Clearly, the maximum temperature is readily reached close to the contact in a few thousand years while more than one million years are needed at approximately 3 km from the contact, corresponding to a temperature barely exceeding the initial temperature of the country rocks. An erosion rate of 1 mm/year is taken into account in this example, making the isotherms to tighten, and hence the thermal gradient to increase.

These models remain rather simple. They do not take into account the three dimensions of space, and particularly the variation of T_e with depth. In addition the latent heat of crystallization should be included in the model since the restored heat is at least two orders of magnitudes larger than the heat produced by radioactive decay. Taking an extreme situation, a magma at its liquidus that would instantaneously crystallize, i.e. without thermal transfer (two conditions that never happen) would increase its temperature by more than 350 °C. In fact, models take into account the latent heat of crystallization by adding a few tens of degrees to T_0, or by using a modified thermal capacity $c_p{}^* = c_p + \Delta H/\Delta T$, where ΔH represents the latent heat and ΔT the temperature interval of crystallization.

7.2 Convective heat transfer

In the general situation, a pluton that settles at a shallow depth, about 10 km deep ($P <$ 300 MPa) in a fractured crust, fluids, that may come from different sources, will move inexorably through the fracture network. This phenomenon is called advection. The pluton itself may deliver a fluid phase at the time of its crystallization. The country rocks, if made of slightly metamorphosed sediments, will produce fluids too, either trapped in the porosity of the sediment or due to the metamorphic reactions implying dehydration of minerals. In addition the country rocks may contain meteoric fluids coming from the surface. For P,T conditions exceeding those of the critical point ($T = 376$ °C, $P = 22.1$ MPa), water (the most common fluid) forms a single phase, called supercritical, highly compressible with a low density (0.5 to 1 g m^{-3}), much less viscous (75 Pa s) than water in surface conditions (10^3 Pa s) and chemically highly reactive. Consequently, when introduced in convective loops linking the surface to the pluton, aqueous fluids and their chemical contents suffer several phase changes that may be responsible for the formation of ore deposits (see Chapter 14).

Convective heat transfer is more difficult to model than conductive transfer. In high permeability country rocks (permeability $> 10^{-18}$ m^2 or $\sim 10^{-6}$ darcy), a very common situation at shallow depth, the movement of fluid, once it overcomes the viscous resistance, is entirely due to the low density contrast of the fluid with respect to the country rocks (Archimedes' principle). Convection becomes established as soon as a critical value of the Rayleigh number (around $R_c = 40$) is reached, and concerns a large volume around the pluton. The vigour and regime of the convection (steady state or periodic) that takes place for $R > R_c$ will depend (1) on the geometry of the hot boundary (the heat source); (2) on the temperature contrast between this boundary and the porous host; and (3) on the petrophysical properties of the host (permeability, fracture density, heterogeneity, anisotropy, . . .) and on the fluid properties at specific temperatures (liquid, vapour or supercritical). By contrast, in low permeability host rocks the fluid is rather easy to pressurize, hence more able to fracture the host rocks. The fluid and its heat will tend to leave the host, eventually after having deposited ore minerals in veins and dykes because of a pressure decrease. Numerical simulations of such phenomena are based on highly nonlinear coupled equations, making each situation a specific case.

114   *Thermomechanical aspects in the country rocks around granite plutons*

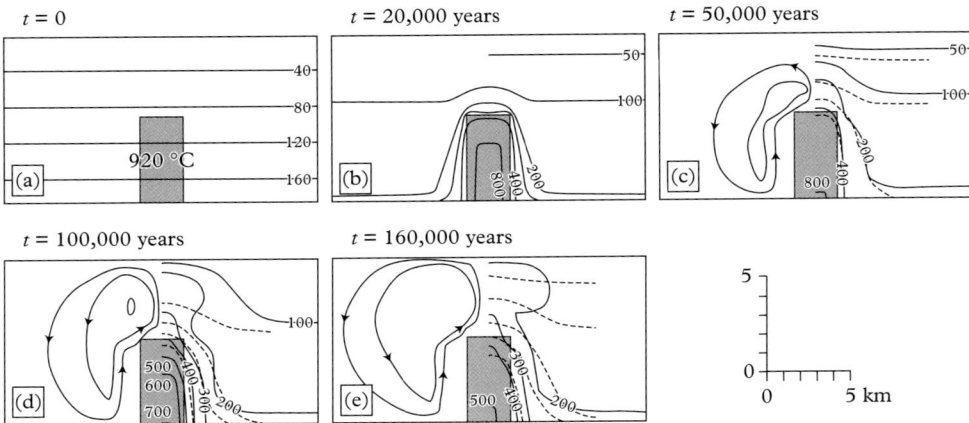

Figure 7.4 *Numerical model describing the cooling of a pluton through convective flow in the country rocks (advection). In this model, a 3 km wide pluton settles at a depth of 4 km (top part) in a crust subjected to an ordinary geothermal degree (~25 °C/km). (a–e) Five stages of cooling between t = 0 and t = 160,000 years. On the left, the anticlockwise convection cells are represented for (c), (d) and (e); on the right side of the pluton the cells (not represented) rotate clockwise; the isotherms of the convective model are indicated with solid lines; for comparison, the dotted isotherms are those for pure conductive cooling. After Norton and Knight (1977).*

Remember that this cooling mode by fluid circulation, after a delay needed for convection to begin, i.e. about 20,000 years for the chosen model (Fig. 7.4), tends to tighten the isotherms against the sides of the pluton and to enlarge the thickness of the metamorphic and hydrothermal aureole situated above the pluton. Within sectors submitted to fluid circulation for a long period of time, outside as well as inside the pluton and particularly at its roof, hydrothermal parageneses, greisens, quartz-rich veins and dykes, eventually accompanied by economic mineralization will abundantly develop (see Chapter 14).

7.3 Diachronic metamorphism and rheological changes at the contact

Time delay between pluton emplacement and the arrival of T_{max} at a given distance from the contact zone can be observed in the field, and more particularly under the microscope by comparing deformation structures with mineral parageneses due to contact metamorphism. At the immediate proximity of the pluton quickly subjected to T_{max}, metamorphic minerals may develop before being themselves deformed due to the stress of pluton emplacement. By contrast, at a distance from the pluton, the newly formed minerals (which record a lower metamorphic degree than against pluton) may seal a deformation event that took place and cease before the arrival of the heat front, and thus appear as post-kinematic minerals (Fig. 7.5).

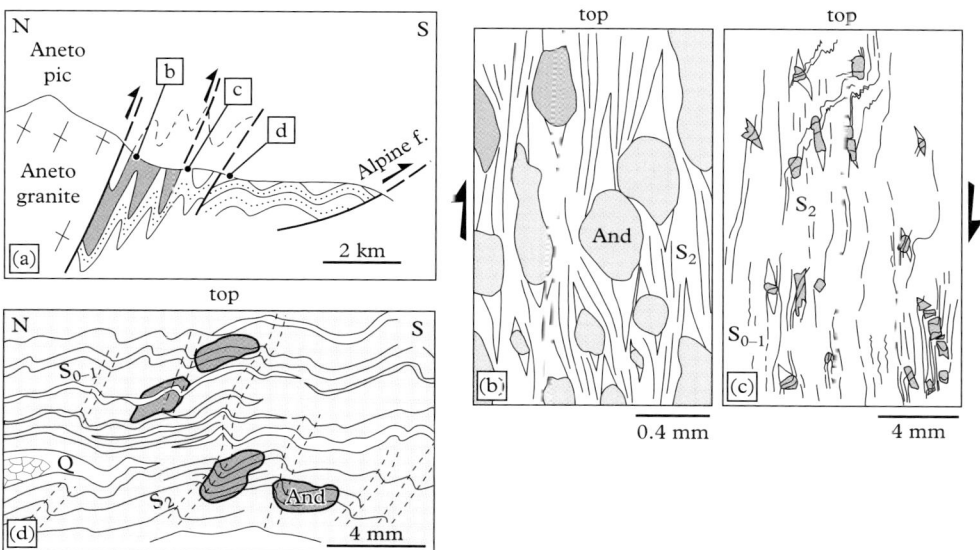

Figure 7.5 *Structures and microstructures observed in the metamorphic aureole of the Maladeta granite massif (Pyrenees; westen Aneto unit; see Fig. 7.6 for location), illustrating the relationships existing between contact metamorphism and deformation (D_2) induced by emplacement of a granite pluton. (a) Cross-section of the southern country rocks (Silurian and Devonian) strongly deformed over more than 2 km, and showing an Alpine décollement to the south. (b) 300 m from contact: andalusite-bearing schist (spotted schist) as observed in section normal to schistosity and parallel to the L_2 line which has a steep plunge to the north; andalusite is affected by D_2 (stretched, displaying pressure shadows and sheared along reverse, top-to-south, faults). (c) 1300 m from contact: moderately deformed (stretched parallel to L_2) scapolite crystals in a carbonate (scapolite is a tetragonal mineral corresponding to a mixture of marialite, sodic, and meionite, calcic, end-members). (d) 1700 m from contact: S_2 appears as a simple crenulation cleavage, axial plane of F_2 micro-folds healed by andalusite crystals witnessing that the arrival of the heat front post-dated the slight D_2 deformation due to emplacement. After Evans et al. (1998).*

Contact metamorphism also leads to dehydration of the host rock, if this host rock is of a sedimentary nature, hence modifying its rheological properties, mainly by impeding the slow deformation mechanism of pressure solution. Therefore, temperature may not be sufficient (except very locally) for plastic deformation to be efficient. In the Bassiès granite massif (French Pyrenees) for example, the country rocks located close to the contact zone are observed to be 'hardened' hornfels by the effect of contact metamorphism that stiffened the rocks after a relatively slight amount of deformation. Farther from the contact zone, regional deformation continued to develop after the emplacement of the pluton, meanwhile the contact aureole was moving outward (Fig. 7.6).

116 *Thermomechanical aspects in the country rocks around granite plutons*

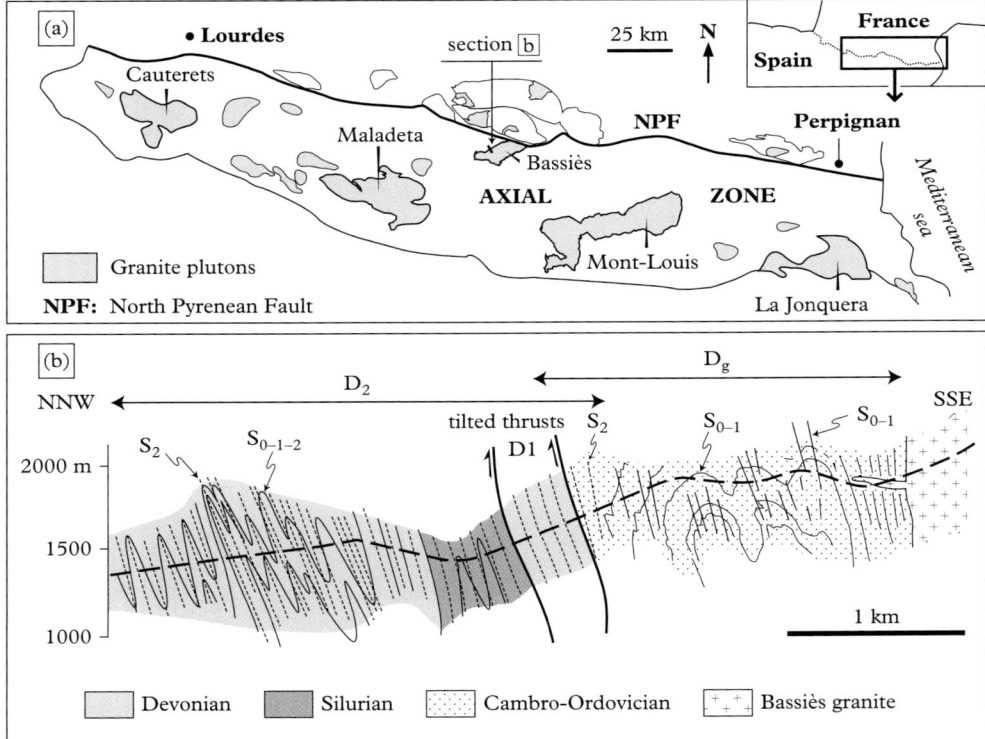

Figure 7.6 *(a) Location of cited granitic plutons in the Axial Zone of the Central Pyrenees; NPF: north-Pyrenean fault. (b) NNW–SSE section in the Bassiès deformed country rocks. The D_g deformation, coeval with the granite emplacement and the transient contact metamorphism, can be distinguished from the regional deformation D_2. In the D_g domain, S_g deforms S_{0-1} with a pervasive foliation and a strong stretching lineation parallel to the NE–SW pluton border. Farther from contact (~1 km) S_{0-1} and S_2 planar structures are recognized. The pluton was emplaced before D_2, by the end of D_1, with its carapace hardened by contact metamorphism having largely avoided an imprint of D_2 deformation. After Evans et al. (1997).*

7.4 Parageneses of contact metamorphism

The paragenesis represents the mineral assemblage reflecting the pressure and temperature conditions suffered by the rock during metamorphism. In addition, the newly formed minerals depend on the chemical nature of the country rock being metamorphosed in the contact aureole.

7.4.1 Pelitic rocks

The contact aureole of the Flamanville granite massif (Normandy, France; Fig. 7.7) is a good example of contact metamorphism showing a succession of parageneses

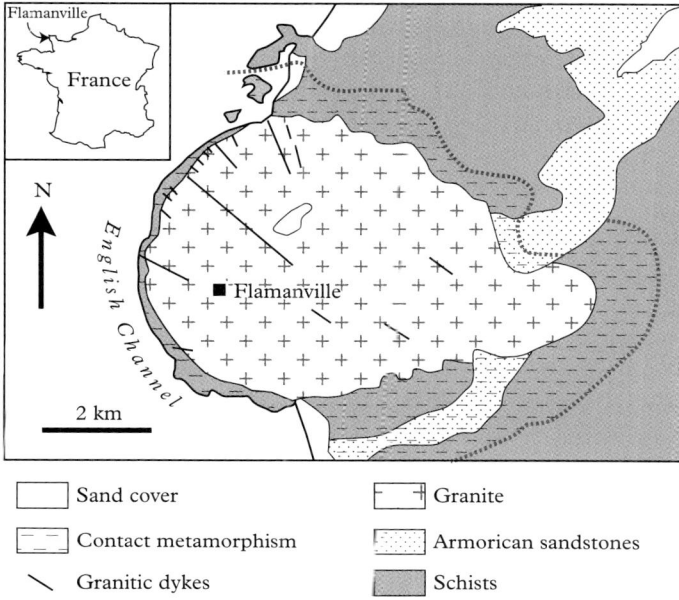

Figure 7.7 *The late-Variscan Flamanville granite massif intrudes sedimentary units. Along the eastern side of the pluton the metamorphic aureole is wider in the schists than in the Armorican sandstones whose highly silicic (quartz-rich) composition is not favourable for the development of metamorphic parageneses. After the geological map of Cherbourg area (BRGM, 1/50,000).*

due to increasing temperatures as one approaches the magma emplacement zone (Fig. 7.7). Pelites are transformed into hornfels, i.e. dark and compact rocks with no visible foliation. Andalusite, Al_2SiO_5, is a typical component of metapelitic hornfels. The (P,T) stability domain for andalusite along with other contact metamorphic conditions are given in Fig. 7.8. Indeed, contact metamorphism is not limited to the low pressure domain (≤ 200 MPa) characterized by andalusite. According to the depth of pluton emplacement, contact metamorphism under greenschist or amphibolite facies may be observed. Against the pluton, or very close to its contact, andalusite may be replaced by sillimanite. However, kyanite-bearing aureoles are exceptional since, with increasing lithostatic pressure, the thermal contrast between the magmatic intrusion and its country rocks lessens to such a point that contact metamorphism cannot be distinguished from regional metamorphism. Finally, contrary to mafic plutons that correspond to the cooling of very hot magmas, intrusion temperatures of granitic magmas are seldom high enough to reach anatexis at their contact.

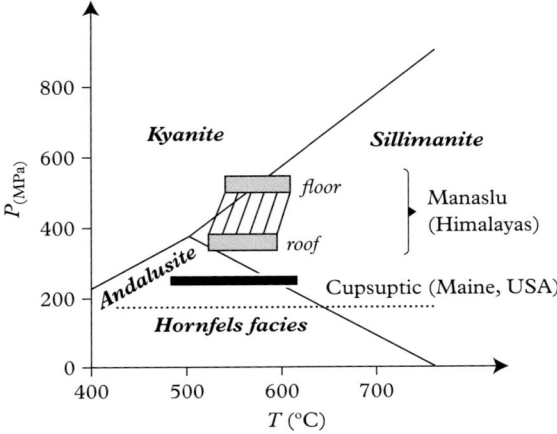

Figure 7.8 *Examples of P,T conditions due to contact metamorphism. Contrary to common belief, the metamorphic conditions determined from contact aureoles are not restricted to hornfels. Hornfels parageneses are defined for pressures lower than 200 MPa, hence are observed around rather shallow intrusions. The Cupsuptic pluton (Maine, USA) displays a metamorphic aureole superimposed on the regional greenschist facies metamorphism. The aureole successively traverses biotite, andalusite and sillimanite isograds, corresponding to conditions of the amphibolite facies. Another example is given by the top and base of the Manaslu leucogranitic pluton (Himalayas) where kyanite and sillimanite correspond to an emplacement depth of 13–18 km. The tectonic setting of the Himalayan leucogranites is detailed in Chapter 12. After Pattison and Tracy (1991) and Guillot et al. (1995).*

7.4.2 Carbonate rocks

The case of carbonates in the contact aureole is more complex than for pelites since the metamorphic reactions of carbonates involve decarbonation. Hence, a carbonic fluid is released that usually mixes with the aqueous fluid evolving from the granite (Fig. 7.9a and b). In addition to pressure and temperature, the composition of the resulting hydrocarbonic fluid is another key parameter that controls the stability fields of the metamorphic minerals. This is why no P,T diagram is proposed for the metamorphic parageneses in carbonates or marbles. Figure 7.9c illustrates the representative case of wollastonite resulting from the following reaction:

$$\text{calcite}(CaCO_3) + \text{quartz}(SiO_2) = \text{wollastonite}(CaSiO_3) + CO_2$$

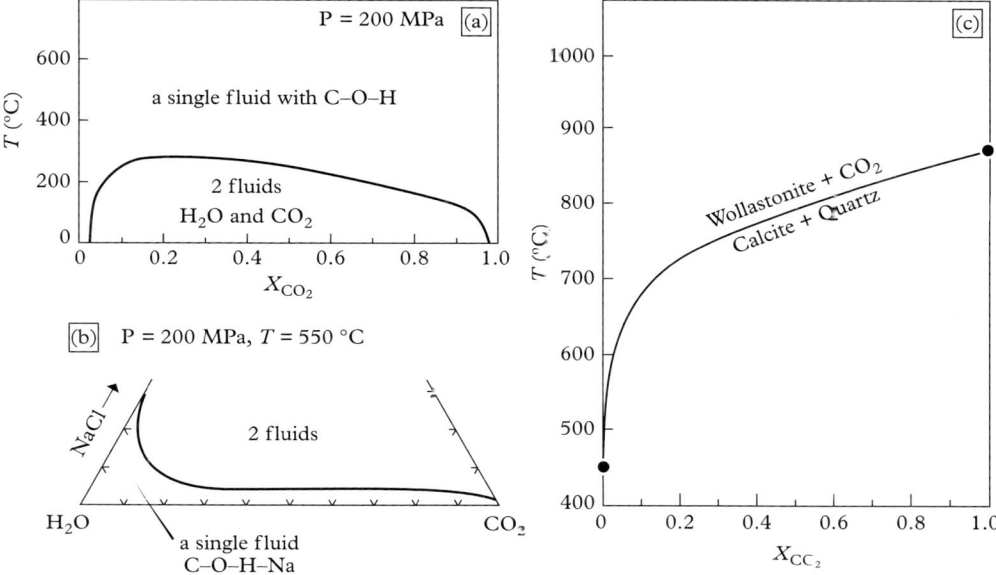

Figure 7.9 (a, b) Miscibility domain of H_2O and CO_2 fluids: the domain is very large at $T>200$ °C and is considerably reduced in the presence of NaCl; after Bowers and Helgeson (1983). (c) (T, X_{CO_2}) diagram for the bivariate reaction of wollastonite formation: note that formation of wollastonite at low temperature (<650 °C) is possible only if $X_{CO_2} < 0.1$; after Spear (1993).

The presence of specific elements in the fluid may play a decisive role. For instance, a small quantity of NaCl increases the immiscibility domain of aqueous and carbonic fluids. NaCl will remain in the aqueous phase. Such questions of eventual unmixing, element partitioning between two fluid phases, and fluid-rock interactions are very important for developing metallogenic deposits (Chapter 14).

In open systems, i.e. when the fluids import some elements and dissolve others, carbonate-bearing country rocks are transformed into skarns, more precisely into exoskarns (Figs 7.10 and 7.11). Note that endoskarns result from hydrocarbonic fluid retroactions within the granite. Skarns often show successive zones in which a few different mineral species are present, a feature characterizing open thermodynamic systems. Some of the mineral species may be concentrated to give economic deposits. This is the case of the tungsten (or wolfram)–bearing skarn at Salau (French Pyrenees) that contains scheelite ($CaWO_4$) which was mined continuously until 1996 (Fig. 7.12). This ore mineralization developed against the Hercynian 'de la Fourque' granodiorite at the contact with Palaeozoic limestones that were metamorphosed into diopside-bearing layered marbles. Pyrrhotite (iron sulfide)- and scheelite-bearing skarns are found in the contact zone with granodiorite.

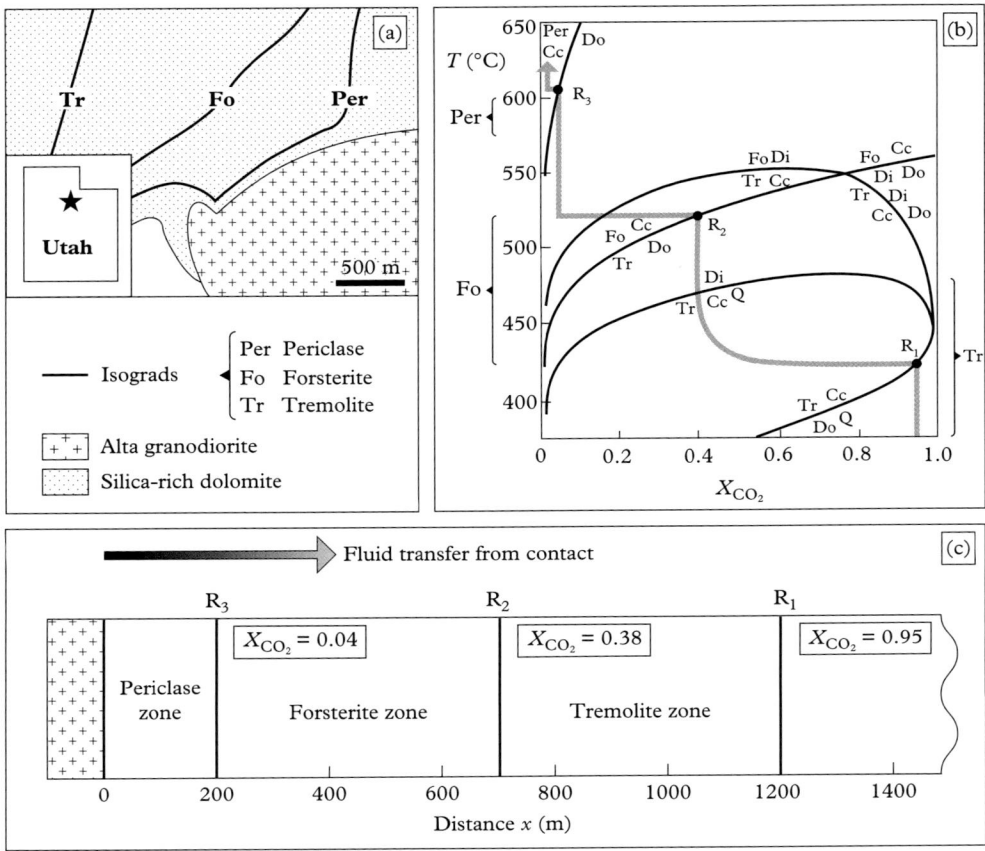

Figure 7.10 *The Alta granodiorite (Utah, USA) was emplaced into marbles, causing the formation of skarns. (a) The isograds correspond to reactions R_1, R_2 and R_3 that successively operate with increasing temperature but also with compositional change due to the release of water during magma crystallization (b, c). After Cook and Bowman (2000).*

7.5 Thermomechanical aspects at the crustal scale

7.5.1 Regional contact metamorphism

Around a large intrusion emplaced in the lower crust, contact metamorphism cannot be distinguished from regional metamorphism. Thermal modelling of a 10 km-thick granodiorite laccolith instantaneously emplaced at a depth between 30 and 40 km illustrates this situation (Fig. 7.13a). The thermal peak due to emplacement is eliminated on cooling with time and gives rise, 10 million years later, to a geotherm warmer than the previous one (Fig. 7.13b). The model shows that everywhere, except in the intrusion

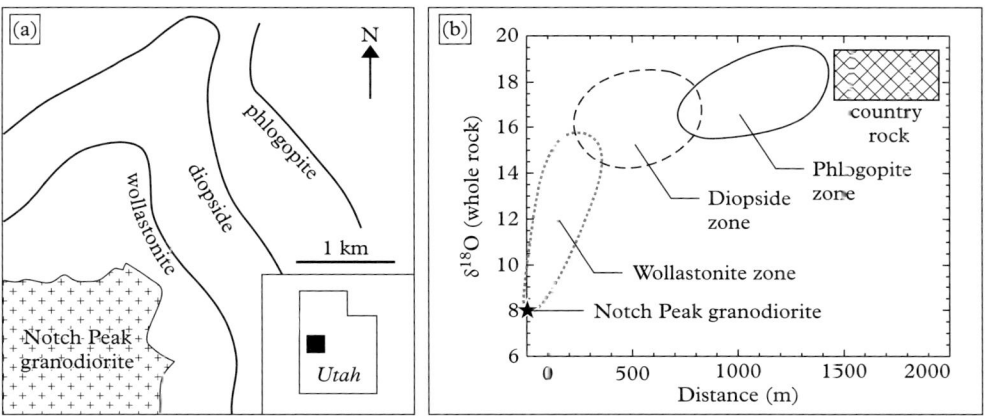

Figure 7.11 *(a) Metamorphic isograds in the skarns at contact with the Notch Peak granodiorite (Utah, USA). (b) Analyses of the oxygen isotopes demonstrate that a magmatic fluid ($\delta^{18}O = 8‰$) seeping through the country rocks was mixed with the fluid coming from the carbonates ($\delta^{18}O = 12‰$) as shown by the intermediate $\delta^{18}O$ values corresponding to the resulting C–O–H fluid. After Nabelek (2002).*

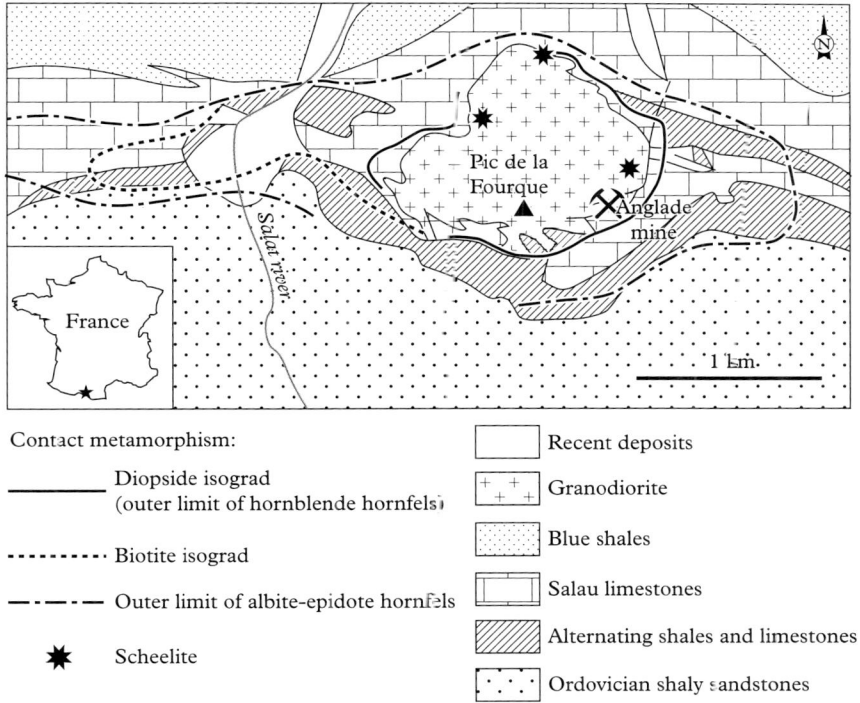

Figure 7.12 *Pic de la Fourque granite body: contact metamorphism and W-ore mineralization (Salau, Pyrenees). According to the geological map of Aulus-les-Bains (BRGM, 1/50,000).*

122 *Thermomechanical aspects in the country rocks around granite plutons*

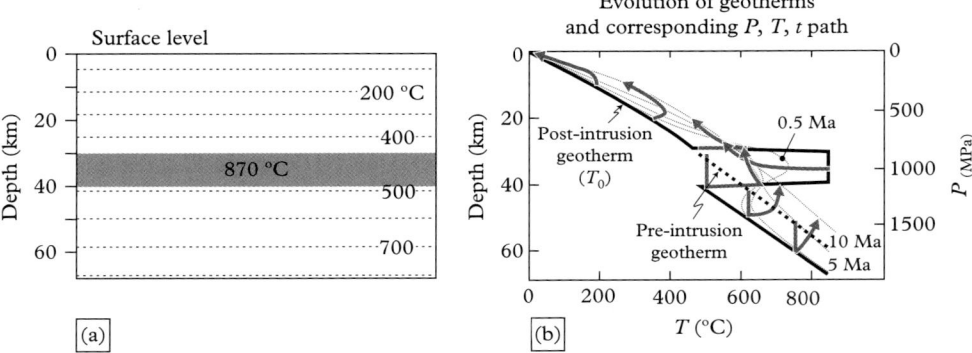

Figure 7.13 *Thermal modelling of a laccolithic intrusion. (a) A granodioritic magma with T = 870 °C settles at depths between 30 and 40 km, in the ductile crust having an average geothermal gradient of 14 °C/km. (b) Calculated geotherms obtained by thermal relaxation of 0.5, 1, 5 and 10 Myr after magma emplacement, and corresponding P-T-t paths (in grey). The calculation takes into account: an erosion rate of 1 mm/year, a (classical) thermal flux from the mantle of $Q^\star = 30$ mW m^{-2}, and a crustal heat production of 2 μW m^{-3} (which is about twice the value for a 'normal' crust because the magma concentrated radioactive elements U, Th and K). After Spear (1993).*

itself, the *P-T-t* paths follow the same trends. Sediments located up to 20 km above the intrusion will develop high temperature metamorphic parageneses. By contrast, below the intrusion the rocks are less sensitive to the thermal effect since their (regional) metamorphic parageneses are already of high grade, and they record a pressure increase rather than a temperature rise. Note that, for a given volume of magma in the laccolith, the waning of the thermal peak is more sluggish if magma injection is pulsed. Such repeated magma inputs are common in dykes and sills as well as in many zoned plutons (Chapter 11).

The Hercynian crust of southern Calabria (Italy) features an example of regional contact metamorphism in a crust that was thinner and hotter than in the previous example, corresponding to a late-orogenic context (see Section 12.5.2). In this example, thanks to a large post-Hercynian tilting, a crustal section of about 25 km thick can be observed. A 10 km-thick layer of crystallized dioritic magma at the base grading to tonalitic, granodioritic then to granitic at the top, was sequentially emplaced at depths between 10 and 20 km (Fig. 7.14a). Emplacement took place during regional deformation that gave a pervasive south-dipping gneissic foliation to this magmatic pile of rocks. To the south, i.e. at the top of the intrusive layer in the upper crust, the isograds of metamorphism can be traced. They increase from south to north, from chlorite, biotite, garnet, staurolite-andalusite and finally sillimanite-muscovite against the granite corresponding to P,T conditions of approximately 250 MPa and 620 °C. The P,T data obtained from these successive parageneses lead to reconstructing a palaeo-thermal gradient of about 60 °C/km. To the north, under the intrusion in the lower crust, a thermal gradient of 30–35 °C/km can be determined. Immediately below the granite,

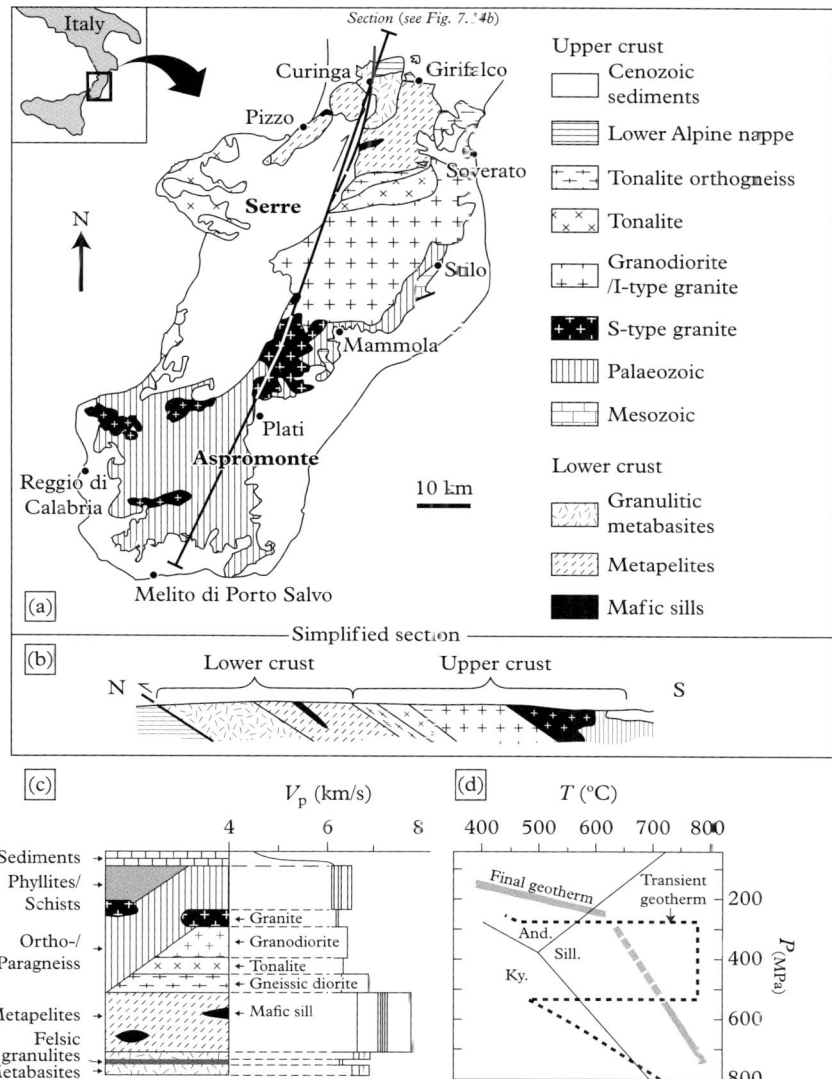

Figure 7.14 *High temperature and low pressure metamorphism of southern Calabria (Italy), an example of 'regional' contact metamorphism. (a) Geological map: the units representing the lower and upper crust are located, respectively to the north and south of the plutonic complex. (b) Geological cross-section of the Calabrian crust. (c) Schematic profile and corresponding seismic velocities. (d) P-T diagram of the Calabrian crust showing contrasted geotherms above and under the granitic intrusion. After Graessner and Schenk (1999).*

the rocks were equilibrated at temperatures close to those of the upper contact of the magmatic layer (~690 °C) but at a much higher pressure (~550 MPa) due to the overburden (Fig. 7.14b).

7.5.2 Crustal decoupling and HT-LP metamorphism

In numerous orogenies, multiple intrusions of granitoids carry large quantities of heat toward upper crustal levels and are directly responsible for the high temperature, low pressure (HT-LP) metamorphism and related parageneses. Indeed such a heat transfer tightens the isotherms in the upper crust and, as a consequence, is at the origin of thinning and softening of the crust which, in turn, becomes easier to deform. Locally, at the roof of such granitic masses of magma, commonly dome-shaped due to gravity traction, low-angle detachments, or sub-horizontal decoupling sites or shear zones, allow the cover rocks to slide away from the centre of the intrusion. Such a subtractive contact due to tectonics amplifies the thermal gradients recorded by the mineral parageneses and is a characteristic feature of thermal domes (see Section 6.3.2). At the regional scale, granitic masses intruding the crust at certain depths, forming temporary low viscosity levels, may form thin layers spreading over large areas as sills or laccoliths. A spectacular example of such a scenario is provided by the so-called 'stratoid' granites that abound in the Neoproterozoic basement of Madagascar (Fig. 6.11) as a set of sills, whose total thickness is about a few kilometres. The subhorizontal shear recorded by these rocks through their typical mineral fabrics (see Chapter 9) indicates that the magmatic sheets have suffered a substantial crustal extension along an approximately E–W direction, as indicated by the mineral lineation. This magmatic and tectonic event is responsible for temperatures of 700–750 °C at depths of 15–20 km, thus recording a very high geothermal gradient, which is likely not only due to heat conduction in the crust, but also to heat advection by magmas.

Crustal-scale shear zones, several hundreds of kilometres long and several kilometres wide in map view, are rooted in the mantle, as shown by thermal (Fig. 7.15) and gravity models, as well as by the presence of CO_2-enriched fluids of deep origin (Pili et al., 1997). Peridotites located below the Moho are more sensitive to shear heating, or friction heating, than the overlying crustal rocks (Fig. 7.15a). The lower crust deforms in a ductile way at Moho temperatures (~800 °C), preventing stress from increasing. The subsequent low heat production also limits the rise of temperature (Fig. 7.15b and c). By contrast, a peridotite at Moho temperatures is rigid, hence is able to sustain high stress values that increase friction heating. Through shear heating at the Moho, the peridotite may therefore gain a few hundreds of degrees in temperature, and the heat transferred upwards may generate felsic magmas by partial melting at the base of the crust. These low density melts will tend to rise along the shear zone, increasing *de facto* the thermal transfer as shown by the observed HT-LP parageneses.

Crustal strike-slip zones thus combine thermal, mechanical and petrological aspects, as in the large N–S directed Angavo Pan-African shear zone of Madagascar in which thermal transfer occurred by heat advection inside the shear zone as well as by lateral conduction as far as 60 km from the shear zone (Nédélec et al., 2000;

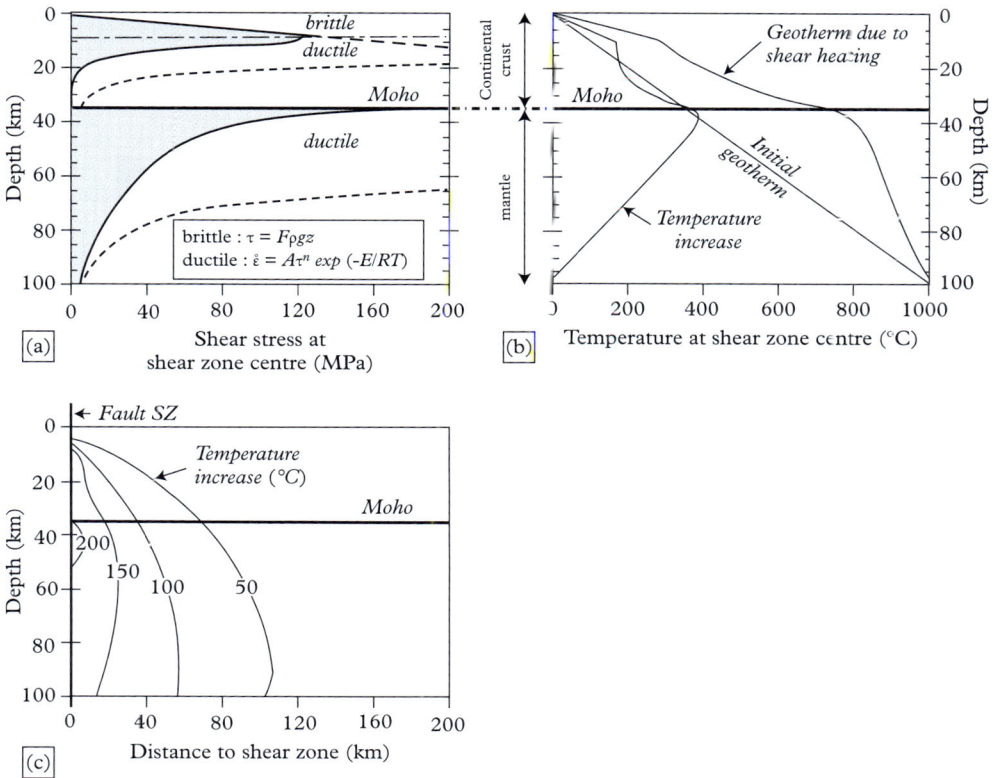

Figure 7.15 *Numerical modelling of shear heating along a crustal scale shear zone (a) Strength profile in the middle of the shear zone as a function of depth; broken line: without shear heating; solid line: steady-state shear heating. (b) Temperature profile within the shear zone and corresponding geotherm due to shear heating, compared to the initial geotherm. (c) Profiles of temperature increase, as a function of distance from the shear zone and depth, due to shear heating. Parameters are displacement rate: 3.3 cm/year; thermal conductivity: 2.5 W m^{-1} K^{-1}; friction: $F = 0.6$. Rheology of the crust is illustrated by the Westerly granite (Hansen and Carter, 1983), and rheology of mantle by the Aheim dunite (Chopra and Paterson, 1981). After Leloup et al. (1999).*

Grégoire et al., 2009). Another example is provided by the Ailao Shan shear zone (Fig. 7.16) that belongs to the Red River fault system and, as a consequence of the collision between India and Eurasia, accommodated the south-eastward displacement of Indochina with respect to South China. Modest shear zones, for instance less than 50 km long and much less than a kilometre in width, are not necessarily rooted in the mantle. Therefore, since the viscosity of the crust decreases as soon as its temperature increases, as indicated elsewhere, shear heating is much less efficient in such shear zones.

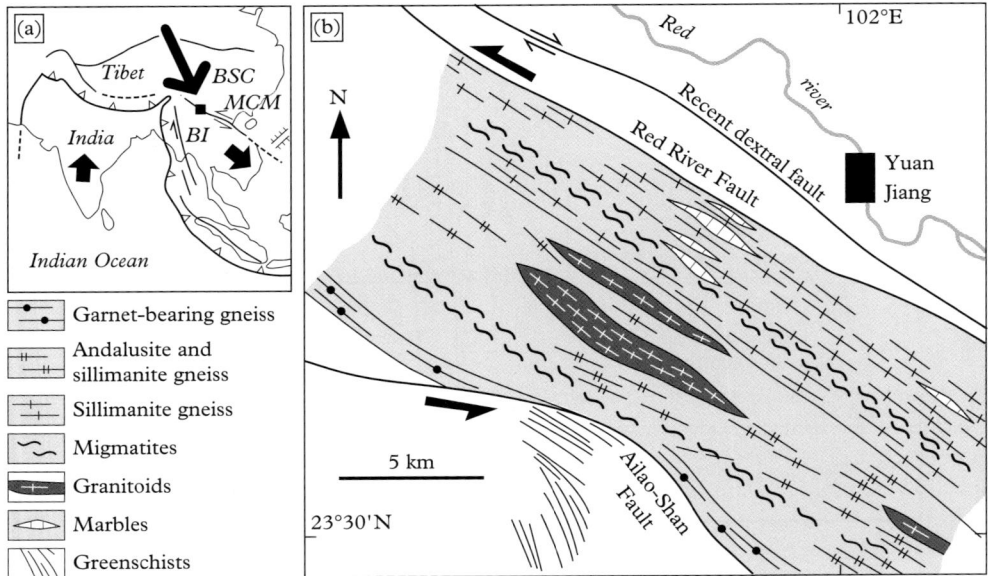

Figure 7.16 *Geological sketch of the Ailao Shan–Red River shear zone in the region of Yuan Jiang (Yunnan, China). (a) This Tertiary sinistral shear zone has accommodated the south-eastward expulsion of the Indochina crustal block (IB) with respect to the South-Chinese block (SCB) over a distance greater than 500 km, allowing the opening of the southern China Sea (SCS). (b) The ~10 km-wide highly deformed and metamorphic domain is shown in grey. The P,T conditions during prograde metamorphism are 400–500 MPa and 650–700 °C. The granitoids are Miocene in age (23 Ma; monazite and xenotime U/Pb data). The end of deformation is recorded in greenschist facies conditions (~380 MPa, 500 °C). After Leloup and Kienast (1993).*

7.6 Conclusions

Magma heat dissipation is achieved both by conduction through the country rocks and by convection of the induced motion of the fluids bathing the fractured environment. Heat dissipation by convection is by far the most efficient mode. It is responsible for dyke systems that eventually contain ore minerals classically represented around, and particularly in the roof zone of granite plutons. The parageneses due to contact metamorphism around an intrusion are effective indicators of the maximum temperatures reached by the country rocks. The same is not valid for pressures. Thermomechanical models displaying the distribution of isotherms in the crust, due either to crustal thickening followed by thermal relaxation, or to transformation of mechanical work into heat along shear zones, show that crustal melts can be generated in these conditions and contribute to heat transfer. It can be concluded that granitic magmatism, with its induced major localized and transient rheological perturbations, is a key factor in the evolution of an orogenic belt.

8
Crystallization of granitic magmas

In this chapter, we examine important aspects of magma crystallization, from the formation of the first nuclei to the fate of the late-magmatic fluids. Although the processes described below can happen in magmas of different composition, we shall focus on granitic magmas in the broadest sense, such as magmas of intermediate to felsic compositions. The structures of magmatic rocks can be described with respect to sizes and shapes of the mineral grains, which are important features controlled by nucleation and crystal growth, depending on the conditions of crystallization. A third aspect, the crystal preferred orientation will be examined in Chapter 9. The study of magma crystallization under natural conditions is only possible in very special cases, such as the Makaopuhi lava lake in the crater of the Kilauea volcano (Hawaii). In addition, the crystallization of an experimental silicate liquid, a magma analogue, can be reproduced in industrial processes. However, such experiments have a limited duration and are restricted to simple systems, with only one or two phases; hence they are not appropriate as simulations of the crystallization of granitic magmas rising through the crust.

8.1 General considerations on nucleation and crystal growth

8.1.1 Nucleation

Nucleation is the appearance of the first crystalline germs or nuclei, consisting of 10 to 10^3 atoms. Although very small (10^{-3} to 10^{-2} mm), a nucleus has the same characteristics as a crystal, namely an ordered spatial distribution of its constituent atoms. Theoretically, the first nuclei should appear at the equilibrium temperature T_L (the liquidus temperature) when the liquid free energy (G_l) and the crystal free energy (G_c) are equal. For these very small crystallites, characterized by a high surface/volume ratio, the surface free energy, γ, must also be considered and, thus, the first stable nuclei only appear at $T < T_L$ (Fig. 8.1). Moreover, the nucleation of minerals with a simple crystalline structure is easier than the nucleation of other minerals, such as the three-dimensional (3D) framework silicates. Nucleation is easier when the entropy difference is larger; hence

128 *Crystallization of granitic magmas*

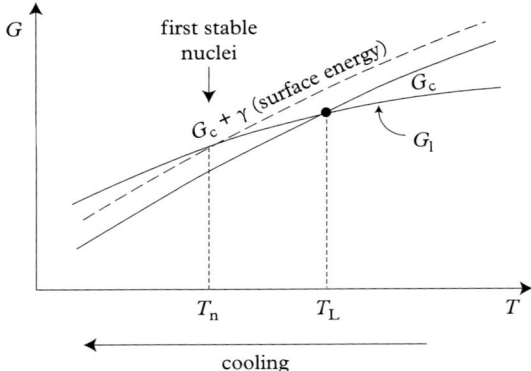

Figure 8.1 *Evolution of free energies (G) versus temperature (T) during cooling of a silicate liquid (single phase); if $T < T_L$, the crystal free energy is lower than the liquid free energy, but the first stable nuclei only appear at T_n.*

there is a natural succession of minerals with an increasingly difficult nucleation, in the following order: iron oxides, pyroxenes, plagioclases, alkali feldspars, for instance. The nucleation rate also depends on the temperature difference with the equilibrium temperature $\Delta T (= T_L - T)$, the so-called liquid undercooling (Fig. 8.2). As regards plutonic rocks, the cooling rate $\Delta T/\Delta t$ has almost no influence on the nucleation rate, i.e. on the number of nuclei formed during a specific time unit. In other words, the characteristic times associated with nucleation and crystal growth processes are much shorter than the cooling duration of plutons. This is the reason why most plutonic granitic rocks are always holocrystalline, i.e. fully crystallized.

The considerations above apply to homogeneous nucleation, when nuclei appear randomly in the liquid. However, nucleation in magmas is more often heterogeneous, i.e. activated at the contact of a surface, often a previously crystallized mineral or even a source restite, whatever their composition. A conspicuous illustration of the heterogeneous nucleation is the crystallization of biotite surrounding zircons in peraluminous granitic magmas. Nevertheless, the heterogeneous nucleation of plagioclase appears especially favoured on another plagioclase crystal, or at least on another tectosilicate crystal.

8.1.2 Crystal growth

Crystal growth is much better studied than crystallite nucleation, because it is more easy to reproduce and to follow experimentally. The crystal growth rate depends on the degree of undercooling ΔT, defined by an asymmetrical bell-shaped curve (Fig. 8.2). The maximum growth rate corresponds to a ΔT value determined by the mineral species and by the composition of the enclosing liquid.

Crystal growth is controlled by the most sluggish of the following processes:

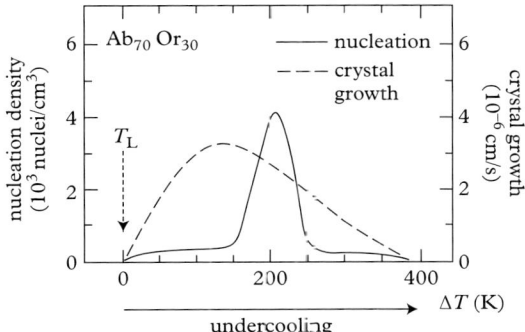

Figure 8.2 *Evolution of nucleation and crystal growth rates versus undercooling for a silicate liquid of composition $Ab_{70}Or_{30}$ with 4.3% dissolved water. Note than the nucleation peak always follows the maximum crystal growth, i.e. occurs at a greater undercooling. After Fenn (1977).*

- the reaction of atom fixation at the liquid–crystal boundary;
- the diffusion of atoms in the liquid, either atoms that will constitute the growing crystal and that must migrate to its boundary or atoms that do not enter in the crystal lattice and that must migrate in the opposite direction;
- the dissipation of the latent heat of crystallization formed at the liquid–crystal boundary.

This last process has no consequence at the crystal scale, because heat diffusion in a liquid is much faster than atomic diffusion (see Section 4.4.1). On the other hand, it can be a limiting factor at the intrusion scale. Let us consider that the atom fixation reaction is the most important process in near-liquidus conditions. The atomic diffusion processes play a significant role only for a significant undercooling. In this case, the atomic transfer is delayed by the increased viscosity of the silicate liquid. By contrast, the viscosity is decreased by dissolved water in the liquid, which is responsible for melt depolymerization; the atomic diffusion is thus favoured and by consequence the crystal growth rate is increased (Fig. 8.3). This is the case for the giant pegmatite crystals grown in a hydrous silicate liquid (London, 2009).

The crystal growth rate depends on the crystal lattice directions, hence the typical habits of idiomorphic minerals (squat plagioclases, flattened biotites, elongate amphiboles . . .). The differences in crystal growth rates are enhanced by the undercooling. A large degree of undercooling, for instance if $\Delta T > 40\ °C$ on both sides of an interface, will produce acicular or skeletal crystal shapes. Such shapes are rarely observed in plutonic rocks, with noticeable exceptions. For instance, the acicular apatite crystals in hybrid rocks testifies to the fast cooling of a mafic magma in contact with a felsic magma (see Chapter 4).

130 *Crystallization of granitic magmas*

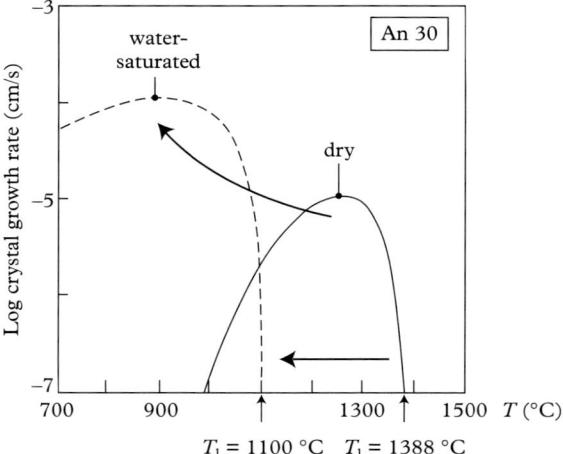

Figure 8.3 *Crystal growth rate of a plagioclase of composition An_{30} versus temperature and water content. The solid curve corresponds to anhydrous conditions at atmospheric pressure and the dashed curve corresponds to a water-saturated liquid at a pressure of 200 MPa. The dissolved water decreases the liquidus and favours crystal growth despite the lower temperature. T_L: liquidus temperature. After Lasaga (1998).*

Rocks crystallized at shallow levels display peculiar structures due to the fast cooling of the magma. They contain large (sub)automorphic crystals grown at depth (the phenocrysts) and much smaller crystals rapidly grown in near surface conditions: this is the microgranular texture. In the granophyric texture (or graphic micropegmatite), quartz and feldspar represent quickly grown crystals leading to imbricated domains of one crystal into the other, with shapes resembling those of ancient cuneiform writing. When looking down the microscope for extinction under crossed polars, it is possible to recognize all domains belonging to the same period of crystal growth as well as to identify the initial core that served as a nucleation site on which the crystal grew (Fig. 8.4). This texture is common in subvolcanic A-type granites that crystallized quickly at very shallow levels.

Rocks crystallized at deeper levels display fine, medium or coarse grains, respectively, corresponding to millimetre, several millimetre and centimetre sizes. The texture can be iso- or anisogranular. In the first case, it is called porphyric, when large (sub-)automorphic crystals, or megacrysts, up to several centimetres in length, are present. This is the case of the so-called 'horse's teeth' granites found in Cornwall, that contain orthoclase megacrysts, whose magmatic origin was initially disputed, but is now widely accepted (Vernon and Paterson, 2008). A texture with a large range of crystal sizes is called 'seriate', and 'equant' when the crystals are all the same size and shape.

General considerations on nucleation and crystal growth **131**

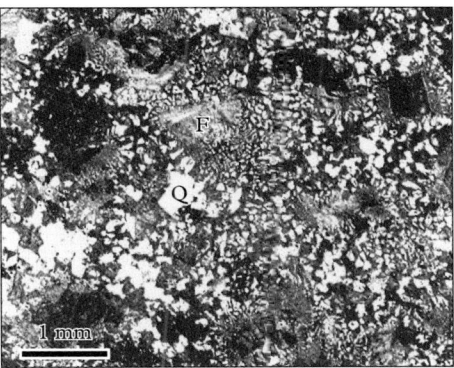

Figure 8.4 *Microphotograph (crossed polars) of a granophyric texture (alkali-feldspar granite from Isle of Skye, Scotland); F: alkali feldspar, Q: quartz.*

8.1.3 Crystal size distributions (CSDs)

Rocks with a microgranular texture testify to an obvious relationship between the size and number of nucleated crystals. For example, fast crystallization results in the formation of numerous small grains. Marsh (1988) demonstrated that the size L (length) and the number of grains N (per volume unit) are two characteristic variables related thermodynamically, which can be determined by a quantitative study of the crystal size distribution (CSD). The important parameters that play a key role are the nucleation rate J and the crystal growth rate G represented in the following equations:

$$L = C_L (G/J)^{0.25} \text{ and } N = C_N (J/G)^{0.75},$$

where C_L and C_N are constant values. The narrow range of grain sizes is explained by the small value of the exponent in the equation that gives L. For instance, a nucleation rate 5000 times faster will yield grains only 4 times smaller at the end of the crystallization period.

The CSD is obtained by studying the population density $n(L)$, i.e. the number of crystals at all successive size intervals (ΔL) per volume unit (Fig. 8.5a). It is difficult to measure $n(L)$ for small ΔL intervals. Therefore, $n(L)$ is determined from the cumulative curve of grain numbers $N(L)$, which represents the total number of grains of sizes lower or equal to L, at different L values (Fig. 8.5b). The slope of the tangent to the curve provides $n(L)$ by the following relation:

$$n(L) = dN(L)/dL,$$

where $n(L)$ is a number per unit length and per unit volume, expressed in mm^{-4}. As a rule, L is determined as the diameter of the circle of equal area to the studied grain in

132 *Crystallization of granitic magmas*

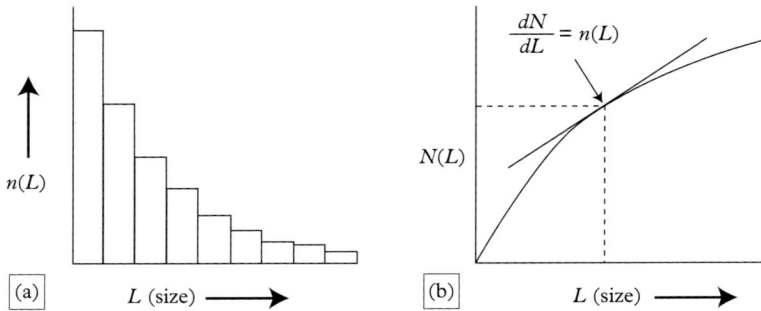

Figure 8.5 *Crystal size distribution. (a) Histogram of grain density per each grain size intervals. (b) Definition of the population density n(L) using the cumulative curve of grain sizes.*

order to remove any peculiarities of the grain shapes. Mutually orthogonal rock sections (corresponding to the three structural reference planes) are considered to avoid the effect of any mineral preferred orientation. The results, which have been acquired from a surface (two dimensions), must be raised to the power 3/2 to represent a 3D volume. Image analysis techniques and digital processing are required for easier and faster determination of CSDs (see Section 9.4.1).

The usual representation of a CSD is a semi-logarithmic diagram, of the form $\log(n)$ versus L. In the case of batch crystallization with a constant crystal growth rate and a nucleation rate increasing exponentially with time, the CSD is represented by a straight line with a negative slope equal to $-1/Gt$, whose y-intercept is $\log(n°)$ (Fig. 8.6a); t is the residence time of the studied mineral and $n°$ is the final nucleation density, i.e. the nucleation density corresponding to the smallest grain size, not to be confused with n_0, the nucleation at the beginning of crystallization, which is much smaller (Fig. 8.6b). Natural CSDs are often akin this theoretical model. Any phenomenon modifying the crystallization of silicate liquids will also modify the straight line of the reference CSD, as can be observed in industrial crystallizers, which serve to interpret natural cases. Thus, a gap corresponds to an interruption of crystallization (e.g. during a magma mingling event). A bend in the largest size range corresponds either to an accumulation or to a fractionation process, hence concave CSD lines that are respectively curved upward or downward (Fig. 8.7a). A lack of small grains at the left end of the CSD line results from a process called 'Ostwald ripening': by an annealing phenomenon occurring at the end of crystallization, the smaller grains are engulfed by the larger ones due to grain boundary migration (Fig. 8.7b).

Studying the CSD of a granite sample is therefore a quantitative method that is complementary to qualitative texture analysis and that sometimes can even replace traditional geochemical methods. Only a few similar studies have been performed so far. Two cases will be presented here. The first one deals with a syenite sample and

General considerations on nucleation and crystal growth **133**

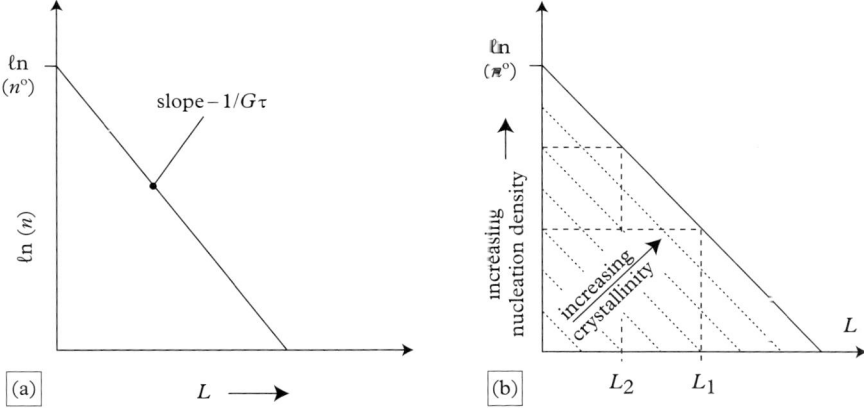

Figure 8.6 *CSD diagrams typical of batch crystallization with constant crystal growth rate (G = 1) and exponentially increasing nucleation rate. (a) Reference line with negative slope. (b) Evolution of the CSD during crystallization: the largest grains are the oldest, but also the fewest ones; for instance compare grains appeared at t_1 with a nucleation rate n_1, reaching the size L_1 at the end of the crystallization, and grains appeared at t_2, with a nucleation rate n_2, which show the size L_2. After Marsh (1988).*

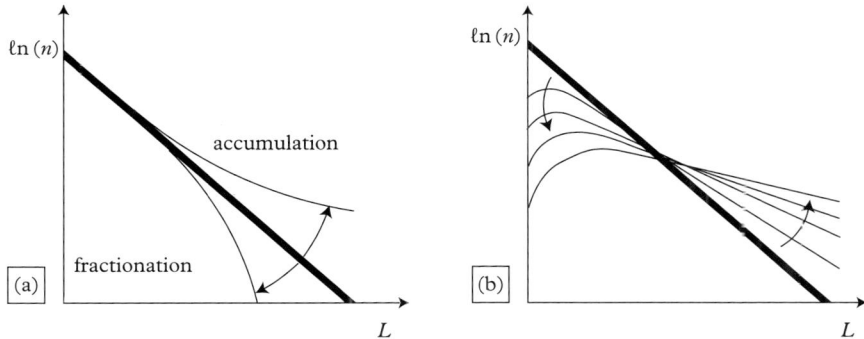

Figure 8.7 *CSD diagrams typical of fractionation or accumulation processes (a) and of an annealing process (b), after Marsh (1998). In the latter case, the rotation with respect to the reference line is in agreement with the model of Dehoff (1991).*

compares two interstitial phases: clinopyroxene and magnetite (Fig 8.8). The second example provides an explanation for the origin of the K-feldspar megacrysts in the Cathedral Peak granodiorite from the Sierra Nevada batholith (Fig. 8.9; see location of the batholith in Fig. 12.8).

134 *Crystallization of granitic magmas*

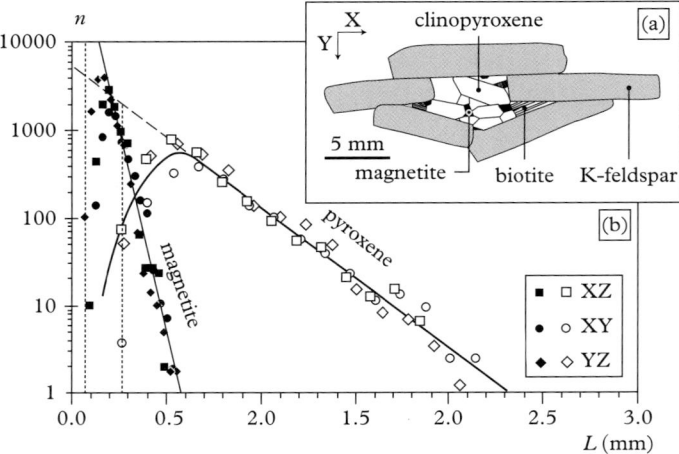

Figure 8.8 *Textural study of a syenite sample from Lebel (Canada). (a) In thin section, clinopyroxene and magnetite crystallized as interstitial phases after K-feldspar (but before biotite). (b) The magnetite CSD displays a much steeper slope than the pyroxene CSD, suggesting a shorter residence time in the magma, provided that both phases have nearly similar crystal growth rates. The pyroxene CSD is characterized by a lack of small grains possibly resulting from a grain boundary migration process at the end of crystallization; this annealing process may have been responsible for the change of the CSD slope (see Fig. 8.7b), apparently suggesting a pyroxene residence time longer than the magnetite residence time. Magnetite, crystallized later, was not subjected to annealing; the low population densities at the smallest grain sizes are but an artefact due to the resolution limit (pixel) of the digital image. Owing to the CSD study, it is suggested that pyroxene crystallized before magnetite, a conclusion that could not have been derived from thin section observations alone. After Launeau and Cruden (1998).*

8.1.4 Evolution of crystallinity with time and its rheological consequences

The magma crystallinity, i.e. the percentage of crystals in the magma or solid fraction, S, changes with time following a more or less steep sigmoidal curve (Fig. 8.10). The S fraction can be calculated using the Johnson–Mehl–Avrami equation, for a constant nucleation rate $\mathcal{J}$ at a constant crystal growth rate G:

$$S = 1 - \exp[-(p/3)\mathcal{J}Gt].$$

Even with an exponentially increasing nucleation rate, the curve keeps a sigmoidal shape, because the type of equation is not changed.

General considerations on nucleation and crystal growth 135

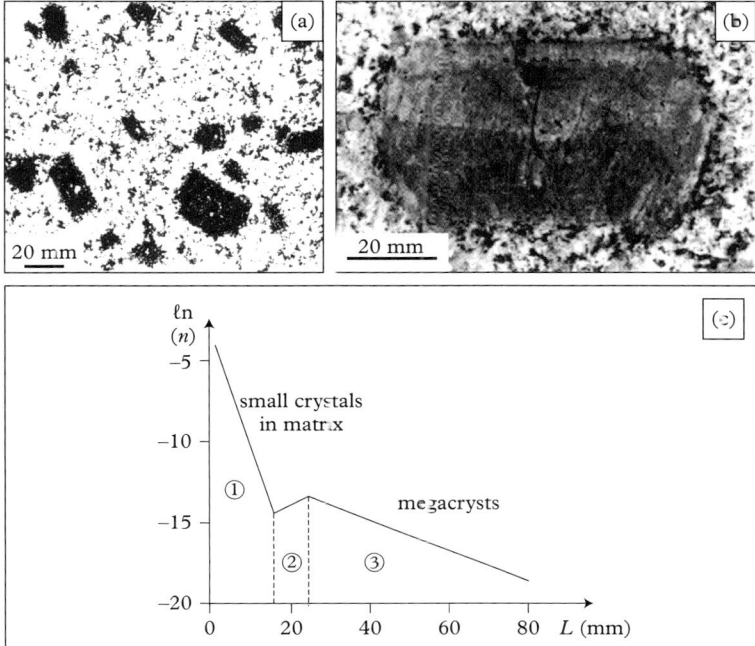

Figure 8.9 *Microcline megacrysts in Cathedral Peak granodiorite (Sierra Nevada, California). (a) Digital image of a rock section stained with sodium cobaltinitrite selectively revealing K-feldspar (here in black; other minerals in white); a preferred orientation of the K-feldspar megacrysts can be noticed. (b) Microphotograph of a megacryst displaying a growth zoning indicated by small amphibole inclusions in planes parallel to crystal faces, as a proof of magmatic growth. (c) The composite CSD is made of a segment (1) with a steeply negative slope corresponding to the small crystals of the matrix and a segment (3) with a less steep negative slope corresponding to the megacrysts, which can result from a longer residence time in the magma and/or from a faster growth. Segment (2), characterized by a positive slope, shows the loss of the smallest megacrysts: the largest ones grew at their expense without any nucleation. This event may occur if the magma is emplaced in warm enough country rocks to remain quite a long time at near-liquidus conditions. In such a situation, the crystal growth will be fast, but the nucleation will be nearly stopped. The small crystals will form afterward, when cooling begins at last. After Higgins (1999).*

Crystallinity directly influences the viscosity, hence the rheology of the magma. The influence of crystallinity on the development of mineral fabric will be studied in Chapter 9. Figure 4.2 shows the evolution of magma viscosity as a function of crystallinity (or crystal load). The viscosity evolution is represented by a sigmoidal curve characterized by a substantial increase in viscosity up to levels typical of a solid rock, even before full crystallization of the magma. Two important thresholds must be considered

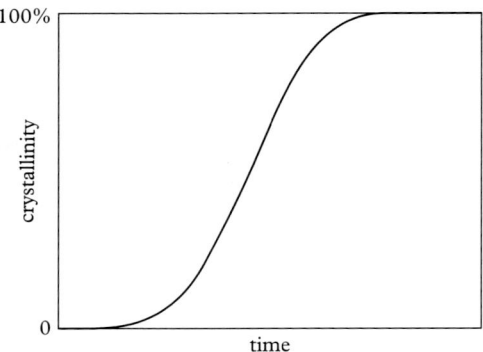

Figure 8.10 *Theoretical evolution of crystallinity with time, following the Avrami equation (Avrami, 1939).*

(see Fig. 4.2). The rigid percolation threshold ($S \geq 55\%$) is reached when crystals come in contact with each other, building a rigid framework where the residual melt can still be mobile. This crystalline framework is able to sustain deviatoric stress, but can still be deformed, for instance in a magmatic shear zone. From the next threshold ($S \geq 75\%$) or locking threshold, the interlocked crystals cannot move, because they constitute a compact stack. Fractures may still appear at the crystal scale and may even be recognized at the outcrop scale (protodykes or late-magmatic dykes: Fig. 8.11).

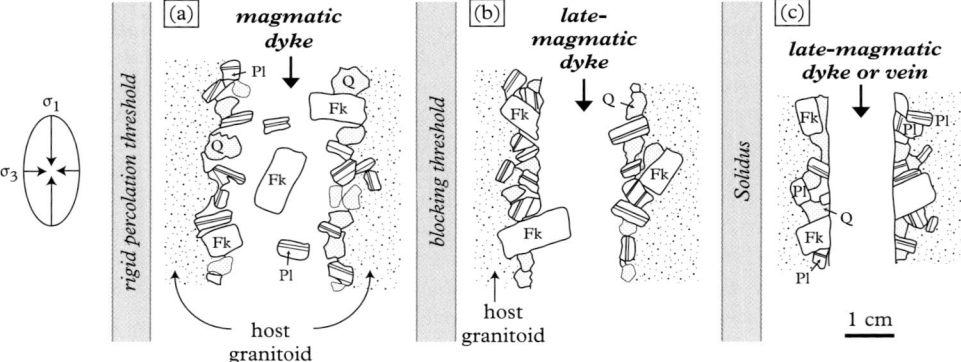

Figure 8.11 *Different types of magmatic dykes s.l. (a) Magmatic proto-dyke with no broken crystals; the proto-dyke has tapped the residual leucogranitic melt. (b) Late-magmatic dyke displaying a few broken crystals along its walls, whereas other crystals, unbroken, straddle the contact between the dyke and its host; the aplitic or pegmatitic infilling confirms that this dyke formed later than the previous one. (c) Post-magmatic dyke with straight walls cutting all crystals belonging to the host. After Hibbard and Watters (1985).*

8.2 Order of crystallization of minerals

8.2.1 Textural observations

In hand specimens and under the microscope, it is easy to distinguish euhedral, idiomorphic or automorphic crystals (from the Greek *morph*, shape, and *auto*, itself), which grew parallel to their crystal faces and thus developed their own typical habit, from anhedral, interstitial or xenomorphic crystals (from the Greek *xenos*, foreign) (Fig. 8.12). The first ones crystallized earlier and freely in the melt, whereas the others grew in the remaining spaces. A few (sub)automorphic crystals may reach large sizes; for instance, K-feldspar megacrysts up to several centimetres in length can be observed in some porphyritic textures. However, large crystals are not necessarily the earliest formed. They often contain trails of euhedral inclusions of other minerals (e.g. biotite or amphibole), probably derived from early nucleation, but presently included in the feldspar because of its faster crystal growth rate (Fig. 8.9b). Petrographic observations help to establish the crystallization order of minerals in granitic rocks. Some mineral successions are commonly observed and constitute the so-called empirical Bowen series (for instance the crystallization order of ferromagnesian minerals: pyroxene—amphibole—biotite). However, these Bowen series do not have any compelling character and it will be noted later that the crystallization order of minerals is dominantly influenced by the composition of the magma.

Moreover, crystallization of different mineral phases is not necessarily a sequential process. A mineral phase can have a long period of crystallization, superimposed on part or the whole of the crystallization interval of another mineral phase. Some minerals, for

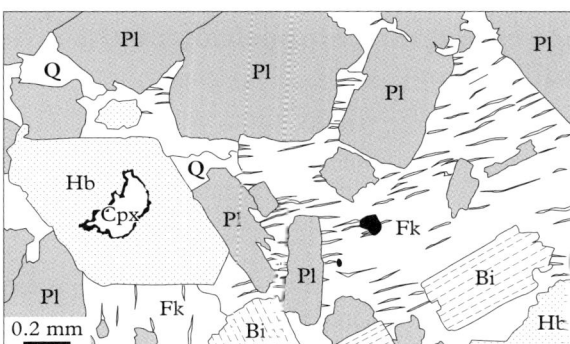

Figure 8.12 *Typical magmatic texture with automorphic crystals (plagioclase: Pl and hornblende: Hb), interstitial xenomorphic crystals (perthitic orthoclase: FK, and quartz: Q). Relict clinopyroxene (Cpx) in hornblende was saved from full consumption by an early (peritectic) reaction in the monzodioritic magma. After Weiss and Troll (1988).*

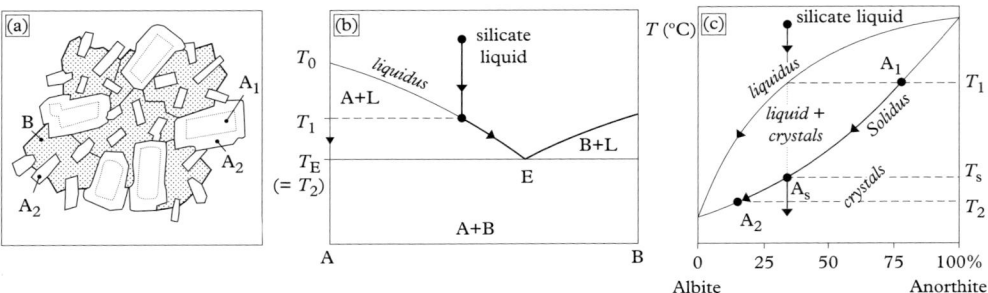

Figure 8.13 *Plagioclase crystals in a granitoid often display a range of compositions and can be also distinguished by their sizes, shapes or textural connections with the other minerals. This provides the way to reconstruct the crystallization stages of a magma. (a) Simplified example (after Bard, 1980): the anorthite-rich core (A_1) of a zoned plagioclase crystallized at the liquidus. (b) At the eutectic point, mineral B crystallized together with plagioclase A_2, whose composition is more albite-rich. (c) When temperature decreases below T_1, plagioclase composition follows the solidus, becoming more and more sodic; theoretically, plagioclase should always remain in equilibrium with the melt in order to reach composition A_S at the end. Actually, the plagioclase–melt equilibrium is so sluggish that excess anorthite remains in the plagioclase core, and hence excess Na in the liquid, which will finally crystallize as a plagioclase with composition A_2 (more sodic than A_S).*

instance plagioclase, display an evolving composition throughout magma crystallization. Microscopic observation, eventually supplemented by a few *in situ* mineral analyses with the electronic microprobe if necessary, is usually sufficient to determine the whole crystallization history (Fig. 8.13).

8.2.2 Influence of magma composition on the order of crystallization

Granitic magmas usually contain a few per cents of water dissolved in the silicate melt. Water content plays a role in the crystallization order of minerals. This point was explored with the help of many experimental studies, including the works of Whitney (1975) and of Naney and Swanson (1980), both summarized later. All their experiments were performed at 800 MPa. The system considered by Whitney (1975) only consisted of $NaAlSi_3O_8$–$KAlSi_3O_8$–$CaAl_2Si_2O_8$–SiO_2 and H_2O, but enabled the author to reproduce the crystallization order of feldspars and quartz in granitic and granodioritic magmas (Fig. 8.14). If plagioclase is always the first liquidus mineral in granodioritic magmas, it is not always the same in various granitic magmas. For example, when there is more than 6 wt % water dissolved in the melt, K-feldspar becomes the first phase to crystallize. Quartz always appears to crystallize after plagioclase in granodioritic magma whereas it nucleates before or after alkali feldspar in granitic magmas depending on their water content (Fig. 8.14).

Naney and Swanson (1980) reproduced similar experiments after addition of Fe and Mg to the $NaAlSi_3O_8$–$KAlSi_3O_8$–$CaAl_2Si_2O_8$–SiO_2–H_2O system, which

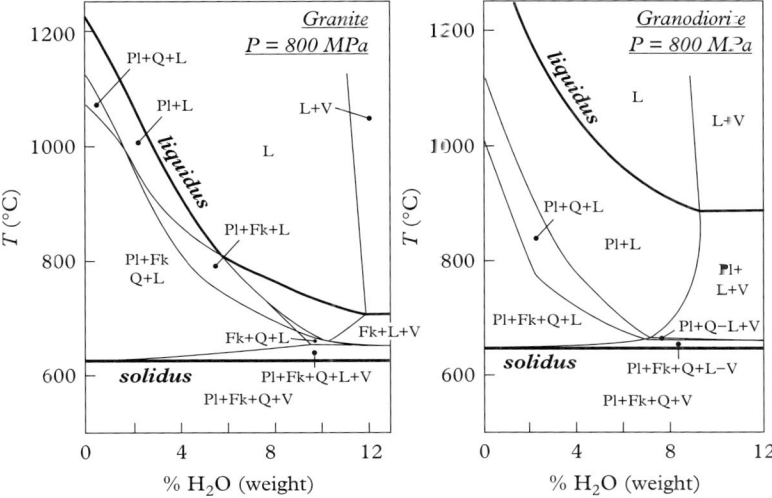

Figure 8.14 Phase diagrams of Whitney (1975). In both cases, the strong influence of the water content on the liquidus temperature can be observed. A hydrous fluid phase (V, for vapour) coexists with the melt (or silicate liquid, L) at $T > T_{liquidus}$ only for minimum water contents of 10 to 12 wt %. In other cases, as crystallization is going on, the dissolved water becomes more and more concentrated in the residual melt, which finally reaches water saturation even for rather low water contents in the whole system. This is the reason why a hydrous fluid phase can appear toward the end of crystallization. Kf: K-feldspar; Pl: plagioclase; Q: quartz.

enabled the authors to investigate the formation of ferromagnesian silicates (Fig. 8.15). Hornblende crystallizes only in granodioritic magmas provided H_2O content is ≥3 wt %. It is the first liquidus phase if H_2O ≥4 wt %. Below this value, plagioclase crystallizes first. Orthopyroxene can appear in granitic and granodioritic magmas provided that water contents are low or moderate, but this mineral is not especially stable below 800 or 900 °C, as it may react with the melt at lower temperatures. Orthopyroxene-bearing granites (charnockites) are therefore derived from dry and hot melts. Crystals involved in peritectic reactions may sometimes be observed as protected relicts included in the new phase. In granitic magmas, biotite is always observed below 900 °C, but its position in the mineral succession mainly depends on the water content. It is the first phase to crystallize at liquidus conditions if H_2O ≥ 8 wt %, whereas it appears only in fifth position after orthopyroxene, K-feldspar, clinopyroxene and quartz, if the water content is close to 3 wt %. This can be deciphered from textural studies: early automorphic biotite indicates a water-rich magma, whereas late interstitial biotite indicates a relatively dry magma. Petrographic observations can therefore provide an estimation of the initial water content of magmas.

140 Crystallization of granitic magmas

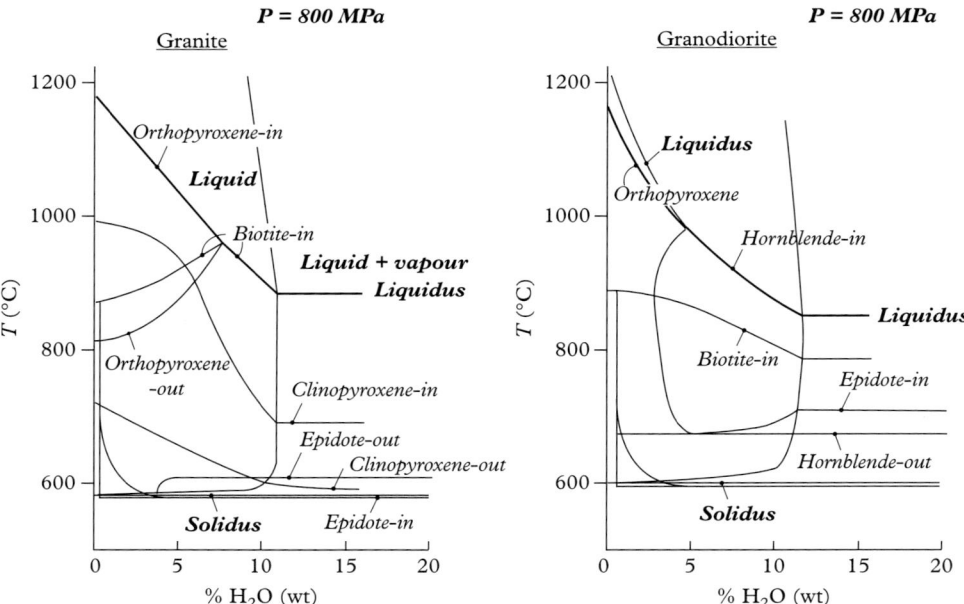

Figure 8.15 *Phase diagrams from Naney and Swanson (1980) with stability fields of ferromagnesian silicates (quartz and feldspars omitted for sake of clarity); note the formation of magmatic epidote just above the solidus, because the experiments were performed at rather high pressures and under appropriate oxygen fugacity conditions. Dashed line: water saturation curve.*

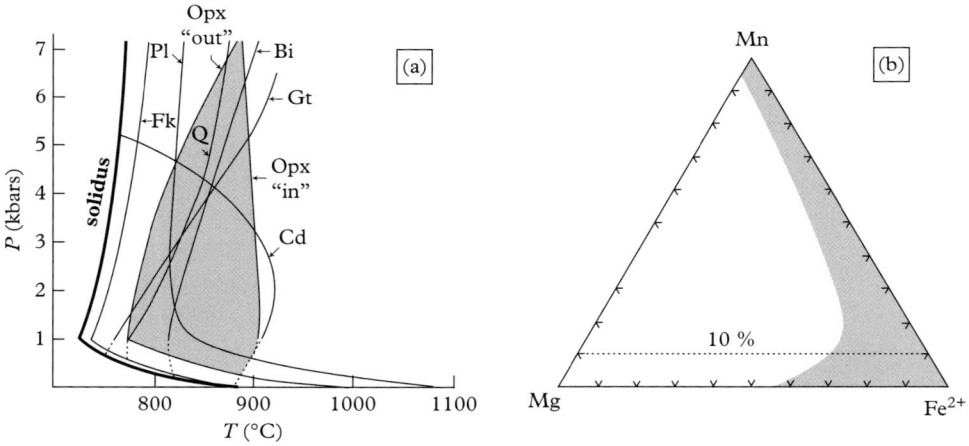

Figure 8.16 *(a) Experimental phase diagram of Clemens and Wall (1981) for an S-type granitic magma containing 4 wt % dissolved water; the orthopyroxene (Opx) stability field is grey-shaded; Gt: garnet; Cd: cordierite. (b) Composition domain of magmatic garnets, in grey. After Dahlquist et al. (2007).*

These observations are of consequence for I-type granitoids. Clemens and Wall (1981) experimentally studied the case for S-type granitoids. Quartz, a mineral supposed to crystallize late, is actually a rather early mineral in S-type granites, where it is often observed as automorphic crystals, with faces at a characteristic angle of 60°, included in later-crystallized K-feldspar. Biotite is always present at temperatures lower than 850 °C. The small stability field of orthopyroxene is still observed at high temperatures (Fig. 8.16a). Orthopyroxene should disappear with cooling due to a peritectic reaction; indeed, it is at least partly replaced by a symplectic association of biotite and quartz surrounding orthopyroxene crystals. Clinopyroxene and hornblende never appear in S-type magmas; conversely, garnet and cordierite are often observed in such magmas and can yield a pressure estimate, because cordierite is only stable at pressures lower than 500 MPa (Fig. 8.16a). Magmatic garnet is usually Mn-rich, containing substantial amount of spessartine (Fig. 8.16b). In strongly peraluminous granitic magmas ($\geq$ 2% normative corundum), primary muscovite crystallizes from the melt at near-solidus temperatures and under a large range of pressures.

8.3 Fractional crystallization and magmatic differentiation

8.3.1 History of magmatic differentiation

Any process separating the first crystallized minerals from the residual melt will modify the magma composition: it is the so-called magmatic differentiation due to crystal fractionation. It is not the only mode of magmatic differentiation. Indeed, it has been shown in Chapter 4 that magma mingling and mixing can also generate new magma compositions, hence producing different magmatic rocks. However, fractional crystallization is the main process responsible for the formation of a large diversity of cogenetic rocks inside the same pluton or in a group of intrusions, corresponding to magmatic suites. This concept was first elaborated more than a hundred years ago in the pioneering work of Harker (1909) and then by Bowen (1928), and was strengthened by the studies of mafic layered igneous complexes such as the Skaergaard complex in Greenland (Wager and Deer, 1939). At their footwall, these layered intrusions display rocks called 'cumulates', which resulted from the accumulation of early crystals (olivine, for instance). Cumulates have their counterparts at higher levels as the so-called 'evolved' or 'differentiated' suites, resulting from crystallization of a magma enriched in residual melt. Several mechanisms can be responsible such as gravitational settling of crystals (following Stokes law), dynamic crystal sorting by convection currents along the walls of the magma chamber, and, finally, compaction of the crystal mush and upward expulsion of the interstitial liquid remaining among the crystals of the cumulate. All these mechanisms require that the melt has a low enough viscosity, a condition easily fulfilled in a mafic (basaltic) magma, but not necessarily in a felsic (granitic) magma, whose melt has a viscosity higher by about four orders of magnitude (even without any crystals). Nevertheless, although felsic magmas do not always form intrusions as spectacular as mafic layered complexes, the diversity of rocks observed in a granitic pluton is sufficient to admit that fractional crystallization can also operate in such rocks and should not be questioned (see Chapter 11).

8.3.2 Evidence of fractional crystallization

The occurrence of a magmatic differentiation process to explain a suite of cogenetic rocks can be verified using binary diagrams, or Harker diagrams, that represent the evolution of an oxide (or of a trace element) versus SiO_2. In the case of fractional crystallization, broken straight lines are observed and can be ascribed to the fractionation of a given mineral (Fig. 8.17). A negative correlation (corresponding to a negative slope) against silica means that the studied oxide (or trace element) enters into the composition of a fractionating mineral.

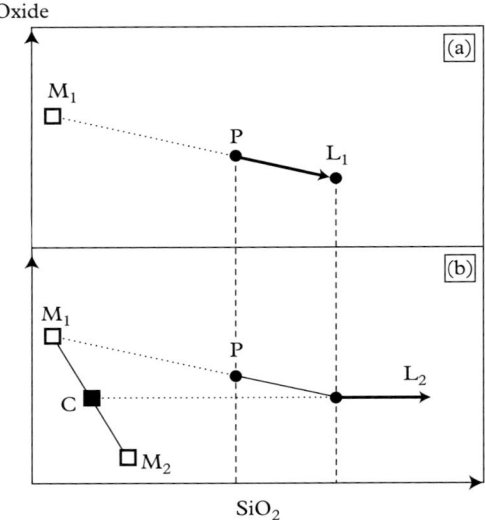

Figure 8.17 *Harker diagrams of the type (oxide) = f(SiO_2) resulting from a fractional crystallization process. (a) A mineral M_1 crystallizes from a parental magma P and, if this mineral is removed (or fractionated), the magma evolves following the arrow toward L_1. (b) Then, a second mineral M_2 begins to crystallize and will also fractionate; the cumulate C now contains M_1 and M_2, and the residual liquid evolves toward L_2. Notice the change of slope of the line representing the evolution of the liquid (or liquid line of descent) as soon as the second crystallizing mineral begins to fractionate.*

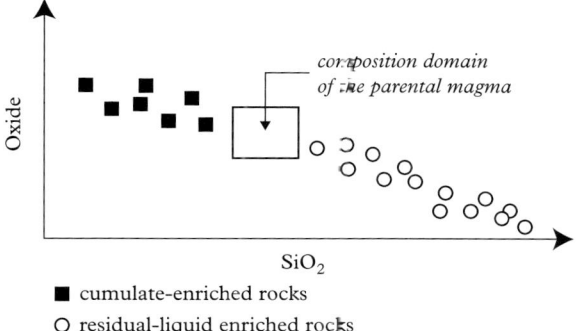

Figure 8.18 *Graphic determination of the composition of the parental magma in a suite of cogenetic rocks resulting from fractional crystallization with incomplete extraction of the cumulate.*

It is possible to make the distinction with magma mixing if at least one change of slope is observed, which can be ascribed to the initiation of the fractionation of a newly crystallizing mineral. By contrast, magma mixing only results in straight unbroken lines (mixing lines). The converse proposition is not true: a unique straight line may correspond to a case of magma mixing, or to a restricted fractionation process involving only one mineral in the parental magma, or finally to variable degrees of partial melting of the protolith (the magma source). It must be also emphasized that, in felsic to intermediate magmas, whose viscosities are higher than in mafic magmas, fractionation of the cumulate minerals often remains incomplete. A suite of rocks, either more or less enriched in cumulus crystals or in residual liquid, is obtained. Between the plots of the two groups, a compositional gap can be observed and corresponds to the likely composition of the parental magma (Fig. 8.18).

Precise determination of the modal composition of the fractionating mineral assemblage requires the use of numerical modelling based on the Harker approach, such as the XLFRAC model of Stormer and Nicholls (1978). An example of such a numerical calculation is proposed in Table 8.1. These mass balance calculations using major elements require a precise knowledge of the composition of the minerals likely to fractionate.

It is also possible to use trace elements (i.e. minor elements analysed in parts per million or ppm). Compatible elements (fully expressed as elements compatible within the crystal lattice of rock-forming minerals) are distinguished from incompatible elements. The first ones are elements easily substituted in a crystal lattice, because of their small ionic radius or because of their convenient electric charge, for instance the transition elements (elements that are adjacent to iron in the classification: Ni, Cr, Co, V), which form either divalent cations entering the same sites as Fe^{2+} or Mg^{2+}, or even smaller trivalent cations. By contrast, incompatible elements correspond to cations with a large ionic radius (the so-called 'large ion lithophile elements' or LILE, such as

Table 8.1 Example of calculations regarding the Kobé pluton in Cameroon using a fractional crystallization model (Njanko et al., 2006); calculation results in italics. The granite magma (C_l) may have been derived from the tonalite magma (C_p) by fractional crystallization of a cumulate whose composition was calculated using the analysed compositions of tonalite minerals. In this process, the granite represents a residual liquid fraction of 42.67%. This numerical modelling is acceptable because the residual sum of squares is lower than 1.

	Rocks		Mineral compositions					Fractionated cumulate
	Parental magma	Residual liquid	analyzed			types		
	C_p (tonalite)	C_l (granite)	Hornblende	Biotite	Plagio (An29)	Orthoclase	Apatite	Magnetite	C_s
SiO$_2$ (%)	62.99	72.92	44.67	38.43	60.83	65.00	0.00	0.00	50.97
Al$_2$O$_3$	15.52	13.92	8.67	14.88	24.42	18.00	0.00	0.00	17.96
Fe$_2$O$_3$*	5.68	1.59	18.69	21.04	0.00	0.00	0.00	100	11.45
MnO	0.07	0.00	0.00	0.00	0.00	0.00	0.00	0.00	0.00
MgO	1.38	0.27	11.75	13.38	0.00	0.00	0.00	0.00	3.46
CaO	2.91	0.67	11.97	0.00	6.16	0.00	56.7	0.00	5.97
Na$_2$O	3.27	3.10	1.66	0.00	8.39	0.00	0.00	0.00	4.32
K$_2$O	5.30	6.39	1.20	9.83	0.19	17.00	0.00	0.00	4.25
TiO$_2$	1.17	0.17	1.39	2.44	0.00	0.00	0.00	0.00	0.53
P$_2$O$_5$	0.39	0.10	0.00	0.00	0.00	0.00	43.3	0.00	1.08

Modal composition of the fractionated cumulate =
- Hornblende 12.87%
- Plagioclase (An29) 48.95%
- Biotite 14.57%
- Orthoclase 15.14%
- Apatite 2.5%
- Magnetite 5.98%

% of residual liquid = 42.67%

Sum of squared residuals = 0.93

Box 8.1 Compatible and incompatible elements

Compatible elements are easily retained in crystals, because they are easily substituted in the atomic sites due to their relatively small ionic radius and charge. This is the case for the transition elements (Ni, Cr, Co, V), which give either bivalent ions easily entering the same sites as Fe^{2+} or Mg^{2+}, or smaller trivalent cations. By contrast, incompatible elements correspond to cations with a large ionic radius (LILE: large ion lithophile elements), for instance: Rb and Ba, and/or with a high charge (HFSE: high field strength elements), for instance: Th and U, thus complicating their substitution into the crystal lattices. Consequently, they remain as long as possible in the melt or in the hydrous fluid: they are therefore hydromagmatic. Cations are plotted with respect to their ionic radii and charges in the Goldschmidt diagram (Fig. 8.19), where different types of elements are easily distinguished.

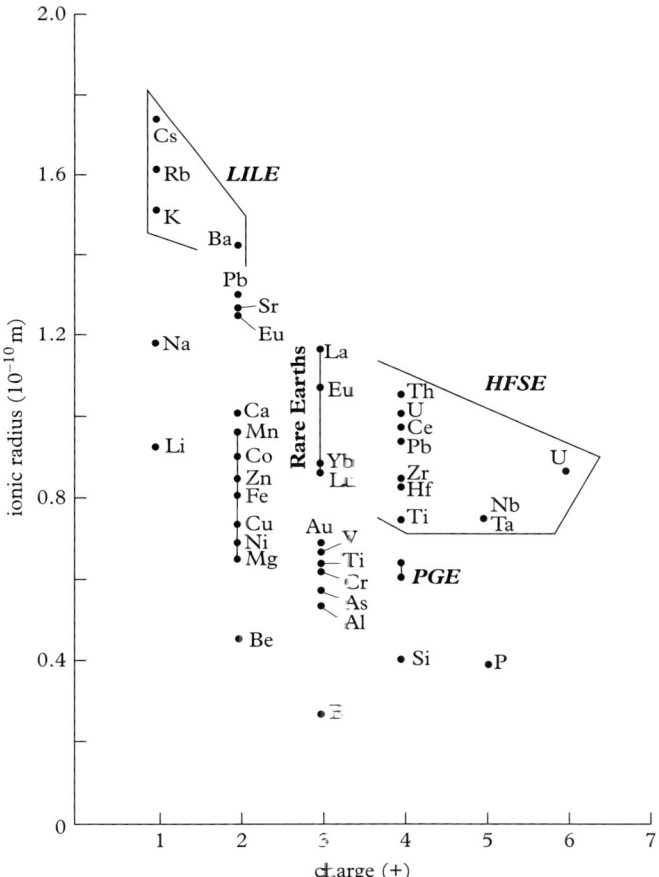

Figure 8.19 *The different types of elements depending on their charge and ionic radius. PGE: platinum-group elements.*

146 *Crystallization of granitic magmas*

Box 8.1 *continued*

Notice that the compatible or incompatible behaviour of an element is also strongly influenced by the melt composition and by the nature of the mineral assemblage crystallized in the melt. For instance, Yb (ytterbium), a heavy rare earth, behaves compatibly when garnet is present (see Fig. 13.9b), but incompatibly when this mineral is lacking.

Rb and Ba) and/or with a high electric charge (the so-called 'high field strength elements' or HFSE, such as Th and U), hence their difficulty to enter into a crystal lattice (see Fig. 8.19 in Box 8.1). Harker diagrams of the type: trace element = f(SiO$_2$) can be used to identify compatible elements (negatively correlated with silica) and incompatible elements (positively correlated with silica). The behaviour of a trace element depends on its distribution coefficient D, which is the ratio of two concentrations:

$$D = C_{mineral} / C_{liquid}.$$

The distribution coefficients of compatible elements are larger than 1; those of incompatible elements are smaller than 1. The values of these coefficients have not all been determined precisely and may also change when the melt composition evolves, an additional complication in numerical modelling. The distribution coefficient for an element also relies on the nature of the crystallized mineral in equilibrium with the melt. For the whole rock, the formulation of the bulk distribution coefficient (D_{rock}) is the sum of the distribution coefficients of each mineral multiplied by the weight fraction X of each mineral in the rock modal composition. For instance:

$$D_{rock} = X_1 D_{(mineral\ 1)} + X_2 D_{(mineral\ 2)} + X_3 D_{(mineral\ 3)},$$

with $X_1 + X_2 + X_3 = 1$ for a rock containing three minerals.

In a magmatic system undergoing Rayleigh fractional crystallization (progressive precipitation of its cumulate), the concentration C_l of a trace element in the residual liquid depends on the initial concentration C_p in the parental melt, on the remaining liquid fraction F and on the bulk distribution coefficient D of this element:

$$C_l / C_p = F^{(D-1)}.$$

The resulting curves are presented in Fig. 8.20a. For comparison, the curves corresponding to partial melting characterized by a melt remaining in equilibrium with the residual solid until its removal as a batch (the so-called 'batch' melting) are presented in Fig. 8.20b. They correspond to the equation:

$$C_l / C_p = 1/[F + D(1-F)],$$

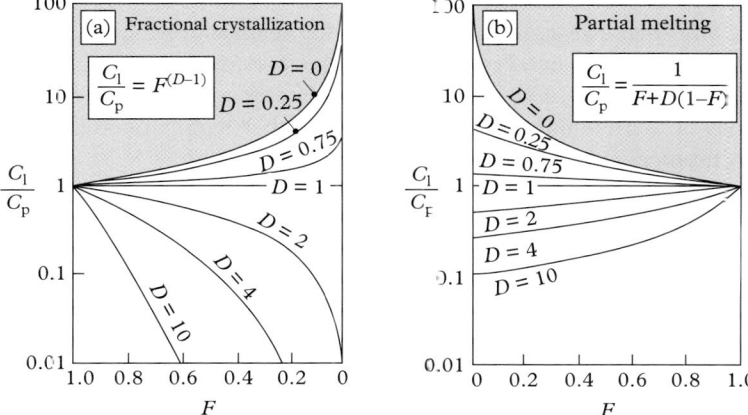

Figure 8.20 *(a) Rayleigh fractional crystallization: evolution of concentration ratios of a trace element in the residual melt (C_l) and in the parental magma (C_p) for different values of the bulk distribution coefficient D as a function of the residual liquid fraction F. It can be observed that a highly compatible element (D > 10) is very rapidly depleted in the crystallizing magma, whereas the concentration of a highly incompatible element (D < 0.1) does not change significantly until the end of crystallization, when it begins to increase very fast. (b) Batch partial melting: the concentrations of compatible elements (at low F values, i.e. at low fractions of melting) are not as low as in (a) for the same F values; see also Fig. 8.21. After Arth (1976).*

where F is the fraction of melting (liquid fraction) and C_p the initial concentration of the element in the solid. At identical F liquid fractions and for large D values, compatible elements are much more depleted in the case of Rayleigh fractional crystallization than in the case of batch partial melting.

This property is used to reveal a crystallization process with the help of log–log diagrams (logarithmic variation diagrams), or more precisely diagrams plotting concentrations of compatible versus incompatible elements with logarithmic coordinates, theoretical support of which was presented by Cocherie (1986). The penultimate equation is written in the logarithmic form for the compatible element:

$$\log C_l(\text{compatible}) = \log C_p(\text{compatible}) + [D(\text{compatible}) - 1]\log F,$$

and similarly for the incompatible element, as $\log C_l$ (incompatible). Both equations are combined to derive only one equation without F, which corresponds to the equation of a straight line:

$$\log C_l(\text{compatible}) = a \log C_l(\text{incompatible}) + b$$

whose slope (a) is the ratio $[D(\text{compatible}) - 1]/[D(\text{incompatible}) - 1]$.

Compositions of cogenetic rocks derived from a fractional crystallization process are distributed along such a straight line in the log–log diagram of Fig. 8.21. As a matter of fact, this straight line is obtained by plotting the trace-element concentrations in the logarithmic variation diagram. The selected trace elements must have the most contrasted behaviour: one of them highly compatible and the other highly incompatible in order to get the best graphical results.

In the log–log diagram, the crystallized solid fraction (the cumulate fraction) plots along a straight line (S) parallel to the line (L) of the residual liquid, but located to the left (i.e. at lower concentrations of the incompatible element). Rocks corresponding to mixtures of cumulate and residual liquid plot in between. Actually, if mixtures contain less than 50% of cumulate, their representative points will plot close to the straight line of the residual melts. Owing to the logarithmic variation diagram, it is not only possible to recognize a plutonic suite derived from a fractional crystallization process, but also to identify the cumulative products as well as the rocks that most resemble the parental magma. An application of this method is presented for the La Jonquera granite in the eastern Pyrenees (Fig. 8.22).

8.3.3 Mechanisms of fractional crystallization

Several mechanisms are able to separate the early crystallized minerals from the residual melt and not only one process will operate in the same pluton.

The gravity settling of crystals follows Stokes' law, which predicts that the rate of settling (v) depends on the crystal radius r, on the density contrast ($\Delta\rho$) between the crystal and the liquid, and on the liquid viscosity h:

$$v = 2r^2 g \Delta\rho / 9h.$$

Gravity settling is inefficient in granitic melts because of their relatively high viscosities. Conversely, it is easier to fractionate a cumulate in a basaltic magma chamber than in a granitic one.

Fractional crystallization may also rely on other processes: descending convection currents may transport and deposit crystals along the walls and the floor of a magma chamber. Whereas convection starts easily in a large enough basaltic chamber, this is not the case in a granitic chamber. Indeed, the expression for the Rayleigh number (which must be higher than 1000 for the convection to occur) has the melt viscosity in its denominator. Therefore, convection is hindered by a high viscosity. However, it can occur in granitic magma chambers as long as the crystal fraction is not too high. This is the case of shallow magma chambers, filled by hot, near liquidus, magmas (it is worth recalling from Chapter 5 that crystals resorb in quickly ascending magmas: see Fig. 5.11). Nevertheless, shallow intrusions will experience fast cooling, hence restricting the lifespan of any magmatic differentiation (unless the magma chamber is periodically replenished by new magma).

In the lower continental crust, the occurrence of a magma chamber as an isolated volume of magma is very unlikely. Under these conditions, magma crystallization is slow

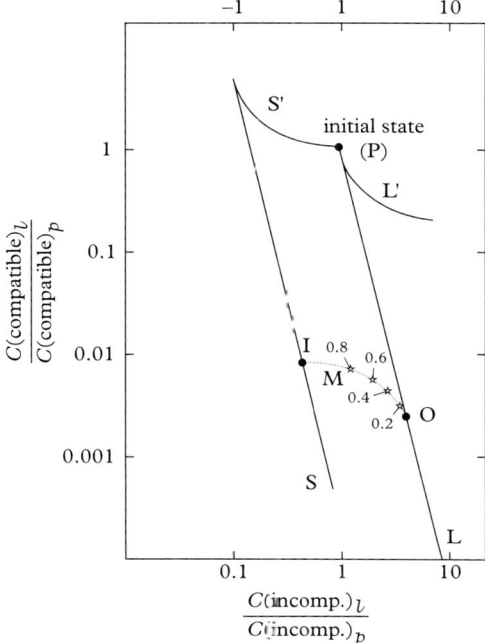

Figure 8.21 *Log–log diagram for a pair of trace elements (compatible vs. incompatible). The theoretical diagram is built for concentration ratios, i.e. concentrations divided by the initial concentration in the parent P. Point P is therefore plotted at (1; 1) coordinates. For practical use, P is undetermined and only the analyzed concentrations in compatible and incompatible elements are plotted and not their ratios with respect to an unknown parental composition, hence a resulting similar diagram despite the necessary translation. The straight lines L and S illustrate the evolution of the residual liquid (L) and of the cumulate (S) during a fractional crystallization process, whereas curves L' and S' illustrate the partial melting case. Curve M represents the mixture of residual liquid and cumulate (numbers refer to different cumulate fractions in the mixture). After Cocherie (1986).*

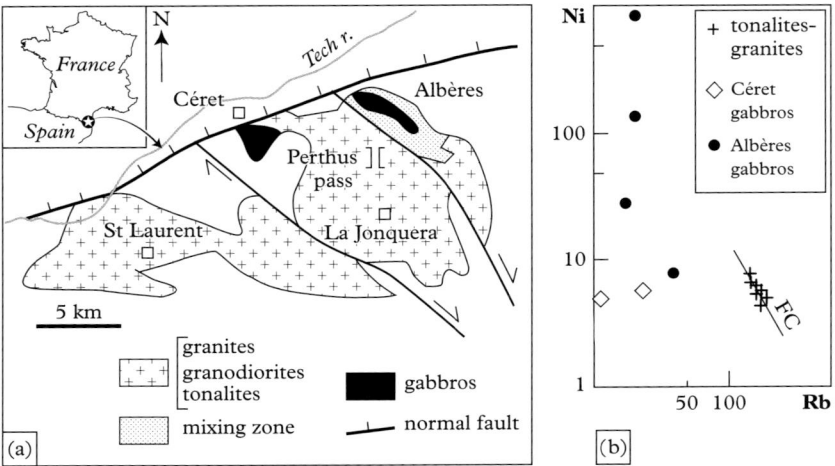

Figure 8.22 *The Hercynian massif of Saint-Laurent–la Jonquera (eastern Pyrenees). (a) Geological sketch map. (b) Diagram of (log Ni) vs. (log Rb), showing that the tonalitic to granitic suite is cogenetic and results from fractional crystallization (FC); the Céret gabbros can be regarded as the corresponding cumulates, whereas the Albères gabbros represent coeval but not cogenetic gabbros variably hybridized with the granitoids. After Cocherie (1985).*

and the residual melt is able to migrate out of the crystal framework as soon as the crystal mush is deformed, at least when the solid fraction is more abundant than the rigid percolation threshold: this is the filter-press mechanism. Migration of the residual liquid can be spatially restricted, for instance contributing to the filling of a nearby magmatic dykelet. In other cases, the residual liquid can contribute to the formation of a large stock. In the Closepet batholith discussed in Chapter 5 (Fig. 5.13), the residual melt from the main plutonic mass first migrated into magmatic shear zones, then into vertical dykes to feed surficial granitic intrusions.

8.3.4 Assimilation and fractional crystallization

Isotopic signatures sometimes provide evidence of interactions between the magma and its country rocks, corresponding to an assimilation process. In this case, magmatic differentiation can occur together with assimilation of the country rocks, hence the so-called AFC (assimilation–fractional crystallization) process. The diagram Sr_i versus SiO_2 can be used to disclose an AFC process (Fig. 8.23). The consequences of the AFC on the trace-element distribution can be formulated mathematically. Actually, it is difficult to demonstrate an AFC process with the help of trace elements, unless there is a strong compositional contrast between the country rocks and

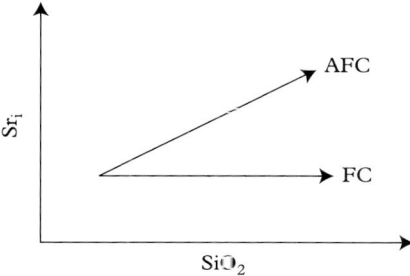

Figure 8.23 *Diagram of Sr_i vs. SiO_2, showing how to discriminate fractional crystallization alone (FC), where no modification of the isotopic signature occurs, from assimilation and fractional crystallization (AFC), where Sr_i is modified.*

the magma. In addition, assimilation is necessarily restricted according to energy balance considerations, because it requires a huge heat transfer at the expense of the magma, hence increasing the crystallization rate. Nevertheless, it is possible to consider that basaltic magma underplating or intraplating in the lower continental crust may interact with it. This situation was experimentally reproduced by Patiño Douce (1995), resulting in a magma characterized by an isotopic composition intermediate between crustal and mantle signatures. A residual granitic melt can be extracted after fractional crystallization of this magma (Fig. 8.24). A-type granites having geochemical characteristics pointing to an important mantle contribution may result from such an AFC process, rather than from fractional crystallization of a sole basaltic magma.

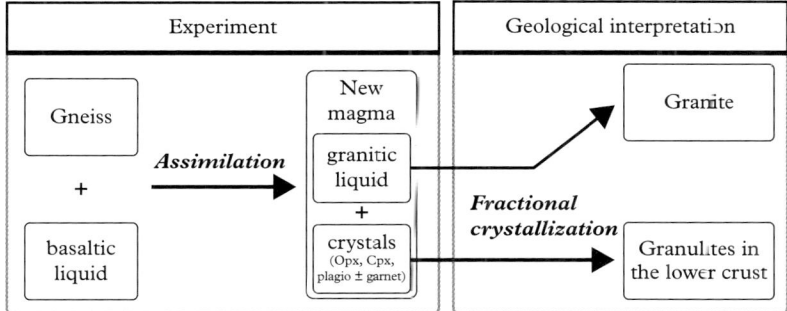

Figure 8.24 *Result of an assimilation experiment of gneiss in a basaltic liquid and possible geological consequences. After Patiño Douce (1995).*

8.4 Late-magmatic processes

8.4.1 Composition of residual granitic melts

A residual granitic melt is close to the eutectic composition. It is water-enriched as a consequence of the crystallization of a large volume of anhydrous minerals (feldspars and quartz). It is also enriched in minor or trace incompatible elements. Among them, fluorine and boron play a crucial role, because they decrease the solidus temperature and the melt viscosity (Fig. 8.25). Hence, these elements will be responsible for an enhanced magmatic differentiation due to a larger temperature interval between liquidus and solidus and, therefore, a longer crystallization lifespan. Besides, by decreasing the melt viscosity, they will also make melt segregation easier.

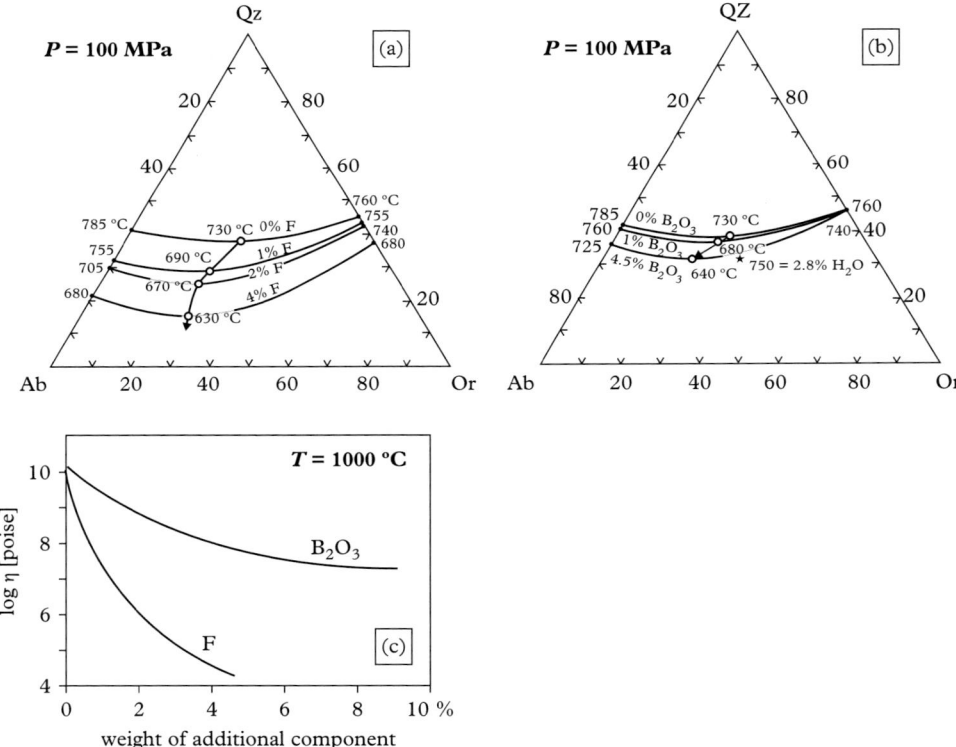

Figure 8.25 *Influence of fluorine (a) and boron (b) on an experimental haplogranitic melt: the cotectic curve moves and the solidus temperatures decreases due to F (after Manning, 1981) and B (after Pichavant, 1987). (c) Effect of F and B on viscosity. After Bagdassarov et al. (1993).*

8.4.2 Magma water saturation and its consequences

It has been shown above (Section 8.2.2) that the water content of a magma has an influence on the mineral crystallization order. Figures 8.14 and 8.15 display the water-saturation curve. Left of this curve the magma is water-undersaturated. On the other side of this curve, a hydrous fluid phase (V) is present together with the melt, or silicate liquid (L), and the eventually crystallized minerals. On the outcrop, miaroles or vugs, as irregularly-shaped cavities a few centimetres in size, are pieces of evidence for the appearance of a fluid phase. These cavities are filled with centripetally grown crystals that formed from the fluid phase. Indeed, the fluid phase contains many dissolved elements and especially abundant silica (Fig. 8.26). Some authors regard this phase as a water-rich silicate melt. This terminological dispute has something to do with the debate on the origin of pegmatites. In this chapter and in Chapter 14, we agree to call 'hydrous fluid' any phase containing more than 50 wt % water.

Pressure is the main factor controlling water saturation in a silicate melt. It has been previously emphasized (Fig. 2.8a) that a granitic melt can contain a lot of dissolved water at high pressure, i.e. at depth, whereas a melt ascending into a shallow crustal level will rapidly reach water saturation. Magmas crystallizing at depth can remain water-undersaturated until the end of crystallization. Any dissolved water will enter into crystallizing hydrous minerals. Conversely, magmas crystallizing at shallow levels can give rise to large volumes of water as a free hydrous phase coexisting in equilibrium with the silicate melt.

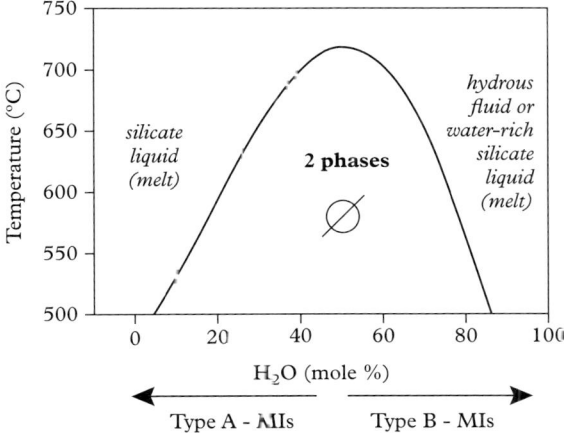

Figure 8.26 *Binary diagram for silicate melt and H_2O (actually, there are other constituents) showing the domain where two phases coexist. This diagram is based on melt inclusion (MI) study. Two different types (A and B) of melt inclusions can be identified depending on their water content: see Chapter 14. After Thomas et al. (2006).*

154 *Crystallization of granitic magmas*

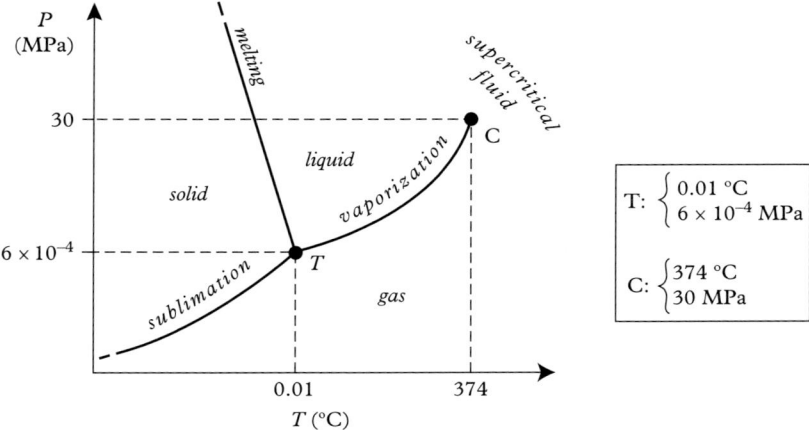

Figure 8.27 *Water state diagram (axes graduation not to scale): the vaporization curve (separating liquid and gas domains) ends at the critical point beyond which there is only one phase of intermediate density, the supercritical fluid.*

The hydrous fluid phase is nearly always a supercritical fluid, i.e. a fluid whose physical properties (e.g. density) are intermediate between the properties of liquid water and those of gaseous water (vapour), because the water critical point is located at conditions much less severe than the stability conditions of magmas (Fig. 8.27). This fluid phase appears before melt crystallization comes to its end and can be called a 'late magmatic' fluid. However, it will still persist after complete crystallization and is then called 'post-magmatic' fluid. This hydrous fluid contains dissolved silica as well as several incompatible elements derived from the residual melt. The concentrations of incompatible elements depend on their distribution coefficient between the silicate melt and the

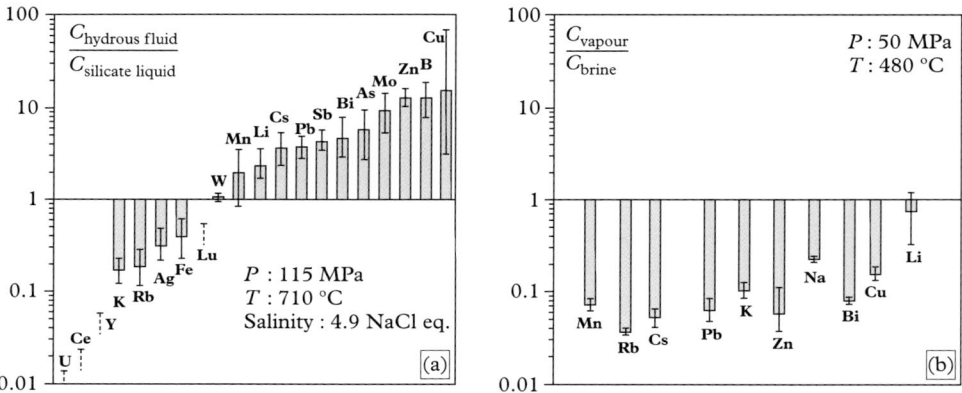

Figure 8.28 *(a) Distribution coefficients between hydrous fluid and silicate liquid (or melt), and (b) between vapour and brine, in two Tertiary granites of New Mexico. After Audétat and Pettke (2003).*

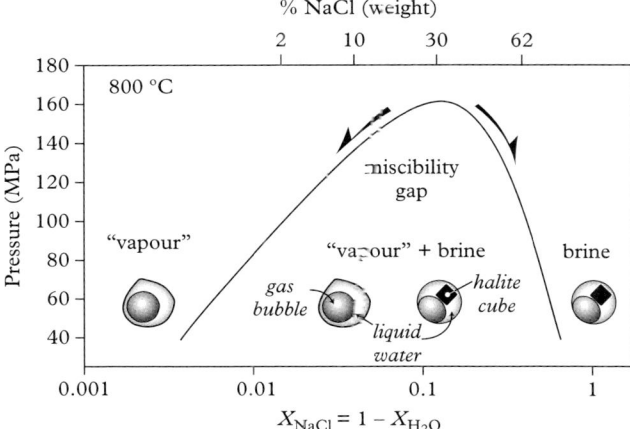

Figure 8.29 *Location of the solvus for the NaCl–H$_2$O system in an isothermal diagram at 800°C and typical features of fluid inclusions at room temperature: notice that the hydrous supercritical fluid has decomposed in liquid water and gas, and that salt previously dissolved in the brine crystallized as a halite cube. After simon et al. (2005).*

hydrous fluid (Fig. 8.28a). Finally, the fluid phase itself can give rise to two different phases by exsolution (Fig. 8.29): this is the so-called second boiling event, implicitly regarding the water saturation of the melt as the first event. The two newly formed phases are, on one hand, a brine highly concentrated in dissolved elements and, on the other hand, a hydrous phase with a lesser content in dissolved elements, that is here called 'vapour', although it can still be a supercritical fluid (Fig. 8.28b).

8.4.3 Pegmatites and aplites

The question of the origin of pegmatites and aplites naturally ensues from the above considerations dealing with water saturation in granitic magma. Pegmatites are rocks of granitic to leucogranitic compositions, characterized by very large grain sizes (from one centimetre up to one metre in size!). By contrast, aplites are quartzo-feldspathic rocks displaying a fine grain size (from a few millimetres to less than one millimetre). Both rock types are often observed together as dykes in granites. Pegmatites are also observed as dykes either in the roof zone of their source granitic pluton or at some distance away from this pluton, or even without any apparent relation with a granitic pluton, possibly because of the lack of any outcrop at the current level of erosion.

Pegmatites often contain minerals of economic interest (see Chapter 14), but there is no consensus on their mode of formation. Nevertheless, their large grain sizes are generally regarded as the consequence of a high water content that would have favoured

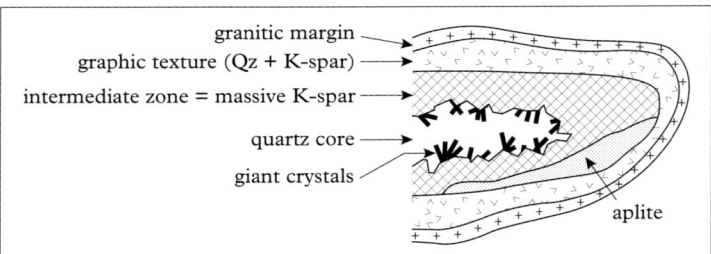

Figure 8.30 *Zoned pegmatite (sketch view in transverse section). After Cerny (1991).*

crystal growth by increasing the rate of diffusion of the chemical elements necessary to build the crystal lattices. Moreover, pegmatites often display a mineralogical and textural zoning (Fig. 8.30). This zoning results from the order of crystallization of minerals, from primary magmatic phases at the granitic wall (outer zone) to only quartz in the pegmatite core. Finally, quartz-infilled fractures are sometimes observed originating in the core and cutting the intermediate and outer zones, suggesting the effect of a late fluid. In the classical hypothesis of Jahns and Burnham (1969), pegmatites appear at the time when a granitic magma becomes water-saturated. The associated aplites would represent the last residual melt coexisting with the hydrous fluid that is expected to give a pegmatite or at least the intermediate and inner zones of a pegmatite. This genetic model was questioned by London (1986, 2008), who prefers the idea that pegmatites represent very water-rich magmas that possibly did not yet reach water saturation. In this hypothesis, aplites do not necessarily represent the last silicate melt and pegmatites could occur alone, as indeed they have often been observed.

Thomas and Davidson (2012) compiled water concentration data from melt inclusions in minerals from granites and pegmatites: granite melt inclusions have water concentration lower than 10% (wt %), whereas the water content of melt inclusions from pegmatites generally ranges from 10 to 40%.

8.5 Subsolidus mineral transformations

8.5.1 Hydrothermal alteration

The hydrous fluid is warm and acidic (because it contains H^+) and thus it will react with the early crystallized minerals. Alkali feldspars react to give secondary muscovite (not to be mistaken with the primary muscovite that may crystallize early in peraluminous magmas). This transformation may lead to a rock made of quartz and muscovite that is called a 'greisen'. At a somewhat lower temperature ($T < 500\,°C$), biotite will also begin to react and then becomes more or less chloritized. This reaction also produces oxides of iron and/or titanium, because Ti cannot enter in the chlorite lattice. These observations are common pieces of evidence for the action of a hydrous fluid phase. However, it is worth observing that this fluid phase may not have been of uniquely magmatic origin, since other fluids may percolate and react with the granite and its surrounding country rocks well after crystallization of the granitic magma.

8.5.2 Exsolution

Alkali feldspars do not form a continuous series in any situation (Fig. 8.31). At moderate to low pressures and for appropriate magma compositions, a sodic-potassic alkali feldspar may crystallize, but it will be not be stable on cooling. An exsolution or unmixing process

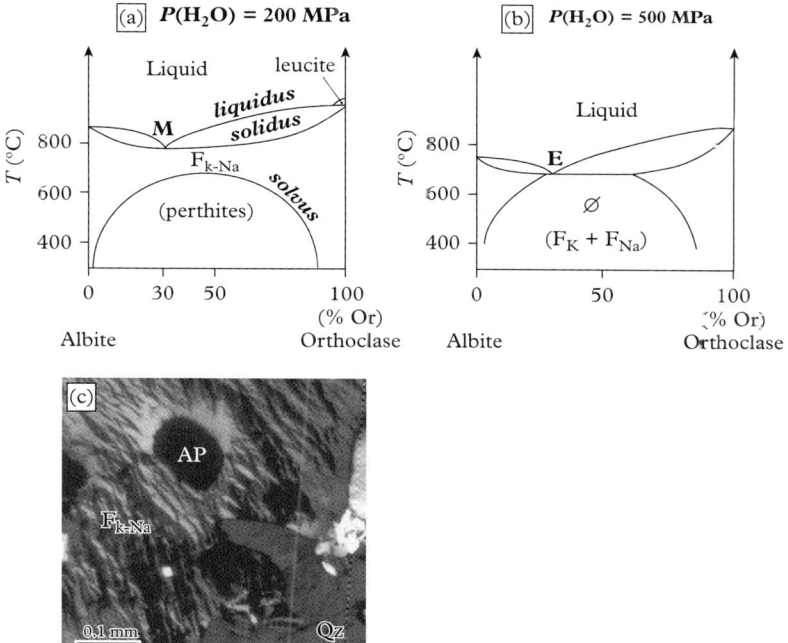

Figure 8.31 *(a) Binary phase diagram of the alkali feldspars at $P(H_2O) = 200$ MPa; liquid crystallization results in only one feldspar with an intermediate composition (between orthoclase and albite), which will exsolve perthites below the solvus temperature; compositions of the sodic and potassic domains evolve on both sides of the solvus; the exsolution process will stop as soon as the temperature is not high enough for an efficient diffusion of K and Na atoms. (b) Same binary diagram at 500 MPa; liquid crystallization results in directly forming two different feldspars, one potassic and the other sodic. (c) Microphotograph (crossed polars) of a perthitic alkali feldspar (F_{K-Na}) in a stratoid hypersolvus A-type granite from Madagascar (see Figs 2.24 and 6.11); potassic domains (microcline) can be distinguished from sodic ones (here characterized by a higher birefringence, hence a lighter grey colour); Ap: apatite, Qz: quartz.*

will occur at the solvus temperature, i.e. sodic domains will form in the host feldspar grain that will consequently become more potassic (Fig. 8.31c). The whole crystal is recognized as a single grain called 'perthitic' alkali feldspar (e.g. perthitic orthoclase). Mesoperthites correspond to feldspar grains with equivalent volumes of sodic and potassic domains. Hypersolvus granites contain only one alkali feldspar. At high pressures, the solidus is lowered until it comes in contact with the solvus; hence, crystallization of an intermediate (sodic-potassic) feldspar is no longer possible. Two different feldspars (a sodic plagioclase and an alkali feldspar of mainly potassic composition) will crystallize immediately to form a subsolvus granite. Notice that this potassic feldspar may still undergo restricted exsolution.

Myrmekites, i.e. quartz–plagioclase vermicular associations (or symplectites), are also regarded as products of a subsolidus reaction. They are commonly observed in granodiorites at the contact between potassic alkali feldspar and plagioclase. However, there is no consensus at the present time on their formation process.

8.6 Conclusions

Textures observed in thin sections reveal the crystallization order of minerals that can be explained by comparison with the crystallization of experimental silicate melts, as well as by using a quantitative approach, such as the study of crystal size distributions (CSDs).

It was already noted that water plays a major role in partial melting processes (Chapter 2). Here, its importance in the crystallization process is also confirmed. Indeed, granitic magmas often contain dissolved water. The water content determines the eventual crystallization of hydrous silicates (amphiboles, micas) and their position, early or late, in the mineral crystallization sequence. Dissolved water decreases the magma viscosity and thus may favour fractionation of early crystals. Finally, when the dissolved water reaches saturation, a free hydrous fluid phase appears in equilibrium with the melt. This fluid phase may react with the already crystallized pluton as well as with the country rocks, as explained in Chapter 7. Its key role in the genesis of economic ores deposits will be discussed in Chapter 14.

Fractionation of early crystallized minerals results in the evolution of the composition of the residual melt. This important magmatic differentiation process can be confirmed graphically or by numerical modelling, using either major or trace element contents.

9
Microstructures and fabrics of granites

The word 'fabric', originally defined by German metallurgists (*fabrik*), infers size, shape, orientation and spatial distributions of grains in a rock sample. Here, 'microstructure' will be reserved for the microscopic scale of size, shape and mutual arrangement of grains, and 'fabric' for defining shape and lattice orientation of mineral grains. Fabric is therefore synonymous with preferred orientation, i.e. the average orientation of shapes (shape fabric) or lattice (lattice fabric) of a population of crystals in a rock sample.

The word 'texture', frequently used by petrologists, is commonly restricted to size, shape and arrangement of grains, leaving aside anisotropy or preferred orientation. Note that 'texture' is also used in metallurgy to describe lattice preferred orientation of crystals in metals. Hence it is equivalent to the lattice fabric generated in metal sheets after rolling or stamping at the metal factory.

Magmatic rocks always have a fabric, no matter how faint. Hence, the shape and/or lattice preferred orientation of the rock-forming crystals is (almost) never random. In this chapter, it is explained how fabrics develop in a granite before and after magma crystallization. Microstructures will be presented first in this chapter. They provide information concerning the mechanisms of deformation. The 'structural framework' (foliation, lineation; x, y and z axes) will be defined with respect to the particular fabrics being considered. Using a combination of classical structural measurements and microstructures, it will be shown how to reveal the internal structure of a pluton, first demonstrated by early workers such as Cloos (1925) and Balk (1937) Structural maps of plutons are now routinely obtained through the study of the anisotropy of magnetic susceptibility (AMS) described in Chapter 10. Finally, fabric analyses using image processing will be briefly presented.

9.1 Granite microstructures

The simplest scenario for the formation of a magmatic microstructure is when the continuing growth of early crystals, progressively nourished by the crystallizing intergranular

160 *Microstructures and fabrics of granites*

liquid, forms a healed crystal skeleton. However, during the last stages preceding total crystallization, the already formed crystal framework may experience some deformation in the presence of a residual melt, if subjected to a high enough level of stress. If no further deformation took place, the resulting microstructure is often called 'submagmatic' (i.e. very close to 'purely' magmatic). It can be easily deduced by observing thin sections under the microscope. After total magma crystallization, the rock may continue to deform, subjected to temperature and stress conditions that will be reflected in the microstructure by deformation features in the solid state. Therefore it is claimed that microstructures provide information about the temperature, stress and deformation conditions that affected the magma, both during the ultimate stages before complete magma crystallization and after crystallization.

Quartz (Fig. 9.1), the easiest mineral to deform by intracrystalline plasticity, is used as the main marker for recording the deformation of its host-rock. In the sequence shown in Fig. 9.3, microstructures are presented from totally undeformed in the solid state (magmatic: upper-left) to significantly deformed (mylonitic: lower-right). Note that the lower-right microstructures have suffered all the microstructural stages.

At the beginning, when quartz crystallizes from the melt, its dislocation density is minimal and the dislocations (see Box 9.1) are homogeneously distributed throughout the grains. The reader is invited to review the theory concerning dislocations, for example in Nicolas (1989).

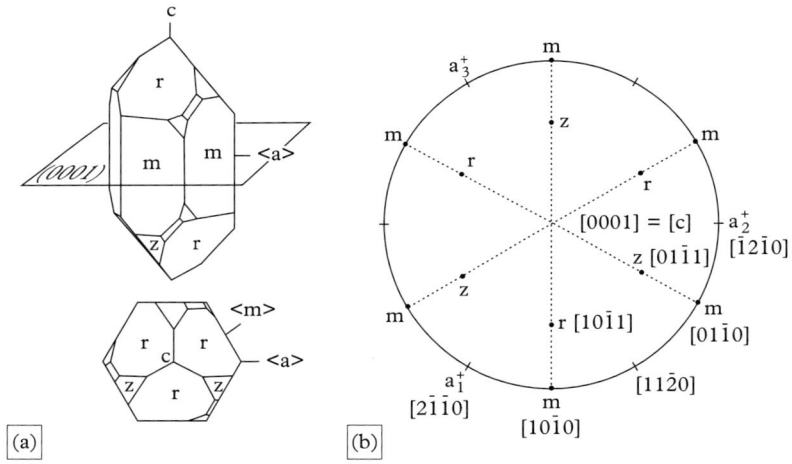

Figure 9.1 *Quartz crystallography. (a) Prism faces, [c] axis, and (0001) or basal plane. (b) Stereoplot: projection onto (0001) = plane perpendicular to [c]. Close to the solidus temperature ('very high'-T), slip is achieved along prismatic planes ({m} or {a}) parallel to [c]. More common slip systems are (0001) <a> (whole range of temperatures), and {m}, {r} or {z} <a> (high temperatures). Axes and planes are noted conforming to the Miller indices (see Baronnet, 1988): [] for an axis; < > for a family of axes of identical nature (here repeated by the threefold symmetry around [c]); () for a plane, and { } for a family of planes.*

Box 9.1 A tale of dislocation

Dislocations are defects in crystals that form lines on TEM (transmission electron microscope) photographs (see Fig. 9.5). Displacement in the lattice of these defects is due to the application of a differential stress and results in a permanent deformation of the crystal. In three dimensions (3D), these linear defects can be seen as the borders of additional atomic planes (planar defects), as shown in Fig. 9.2d. Before the advent of TEM, these defects were just conceptual objects made necessary by the awareness of the amount of energy needed to deform a crystal by slip along entire atomic planes, thus implying the simultaneous rupture of an enormous number of atomic bonds. On the contrary, analogous to a caterpillar (Fig. 9.2a), the elementary displacement of the upper-half crystal of Fig. 9.2b and c implies the step-by-step displacement of a dislocation, each step being associated with the breaking of a single atomic bond. The Burgers vector **b** is the elementary vector, usually corresponding to the smallest interatomic distance along which the crystal is locally displaced. Dislocation lines are edge (Fig. 9.2b) or screw (Fig. 9.2c) dislocations depending on whether these lines are perpendicular or parallel to **b**. A dislocation is usually of a mixed type (edge or screw), forming a loop which grows under the effect of differential stress (Fig. 9.2d) and finally leaves the crystal which is then displaced by b.

A quartz grain newly crystallized from a granitic melt contains 'a few' dislocations (showing a cumulated length of 10^7 cm per cm^3, i.e. a rather low dislocation density of 10^7 cm^{-2}) and these dislocations are homogeneously distributed in the crystal. As soon as a differential stress is applied to the crystal, its dislocation density increases

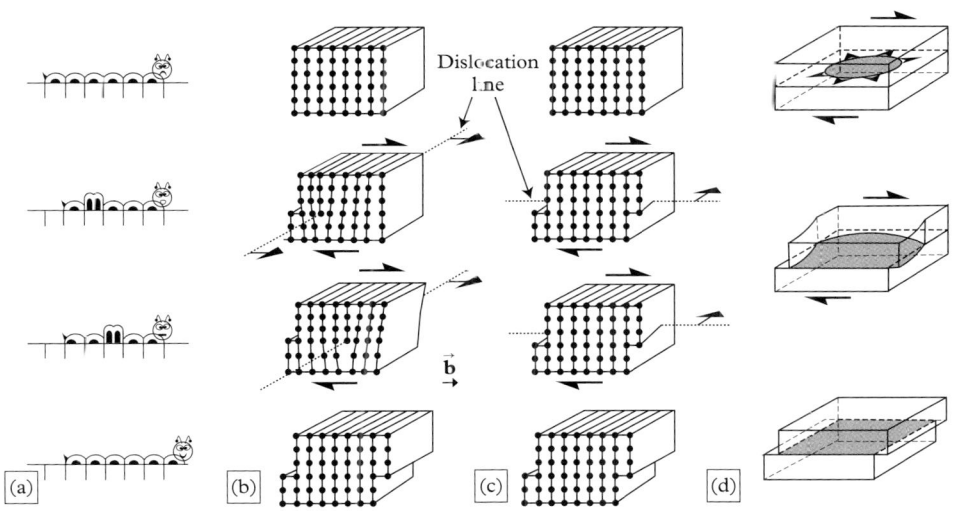

Figure 9.2 *Dislocations. The caterpillar analogy (a). Displacement of an edge (b), of a screw (c), and of a loop (d) ending by the permanent deformation of the crystal by an interatomic distance according to **b**, the Burgers vector. After Nicolas (1989).*

> **Box 9.1** *continued*
>
> (up to 10^9 cm^{-2} for example), the dislocations move and tend to gather into dislocation walls responsible for undulose extinction. Further application of stress leads to formation of subgrain boundaries (the 'substructure') and then of grain boundaries around new grains, that characterize recrystallization. Hence, 'magmatic' quartz crystals remain intact if no stress is applied; otherwise they transform into sets of new grains during the post-magmatic stage of granite cooling.

9.1.1 Magmatic microstructures

Large monocrystalline quartz grains indicate that no plastic deformation has affected them since their crystallization. When viewed under the optical microscope with crossed polars, their extinction is the same over the whole grain and no substructure can be observed due to the low dislocation density. Curved quartz–quartz boundaries are features inherited from crystallization. The absence of deformation is also confirmed by other mineral phases that show magmatic textural relationships. This is magmatic microstructure par excellence (Figs 9.3a and 9.4a).

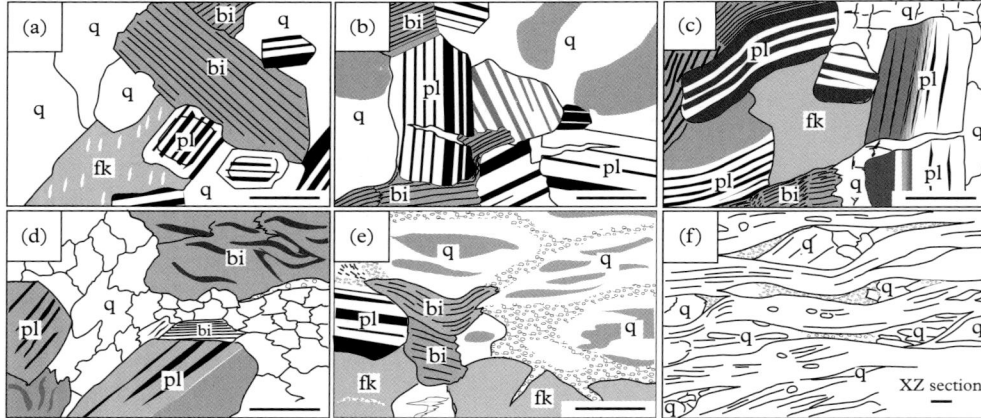

Figure 9.3 *Microstructures observed in the Mont-Louis-Andorre granite (Eastern Pyrenees). Scale: 1 mm. From upper-left to lower-right: (a) Magmatic sensu stricto: large, undeformed quartz grains with curved boundaries. (b) Submagmatic: note the local microfractures in plagioclase confirming its brittle nature at near-solidus temperatures. (c–f) Solid-state deformation. (c) Slight deformation at high-T bent plagioclase and biotite crystals, and resulted in 'chess-board' substructure in quartz. (d) High- to medium-T deformation: recrystallization in quartz. (e) Low-T and/or high stress deformation: undulose extinction (at low strain) in porphyroclasts, and fine-grained recrystallized quartz around porphyroclasts confirming strain localization. (f) Mylonitic deformation: development of C/S structures: fine-grained layers concentrating strain and marking the shear plane (C), oblique to the average elongation (S) of the large grains. After Gleizes (1992).*

Box 9.2 From microstructures to nanostructures

Minerals larger than 2 μm (2–3 times the average wavelength of light, λ) can be identified under the optical microscope. In quartz, the microscope helps to distinguish between basal and prismatic sub-boundaries. Now consider a basal sub-boundary (Fig. 9.4a). Since it is

Figure 9.4 *A few common microstructural examples found in granites. (a) Typically magmatic (in Bouchez et al., 2006). (b) Chess-board pattern in quartz (Qc) indicating low strain at near-solidus temperature (in Auréjac et al., 2004). (c) Quartz-infilled microfracture in plagioclase indicating that 'submagmatic' (near-solidus) deformation took place before cataclasis and recrystallization of the whole matrix (in Sadeghian et al., 2005). (d) Rather high-T deformation marked by quartz recrystallization (in Bouchez et al., 2006). (e) High-temperature (bent feldspar) high-stress (nucleation in quartz) deformation (in Auréjac et al., 2004). (f) Localized deformation at low temperature (in Auréjac et al., 2004).*

continued

> **Box 9.2** *continued*
>
> perpendicular to the [c] axis (i.e. to the highest refractive index), by inserting a gypsum plate into the microscope that modifies the optical path by ±λ, a quartz grain will become yellow in colour (down Newton's scale) under crossed Nicols if +λ of the gypsum plate is parallel to the basal sub-boundary; conversely it will become blue if +λ is inserted perpendicular to the basal sub-boundary.
>
> A sub-boundary results from the planar accumulation of dislocations, parallel to the (0001) plane (plane normal to [c]) in the case of a basal sub-boundary that accompanies the 'very-high'-*T* slip system, i.e. having [c] as the slip direction. Viewed under the TEM, basal sub-boundaries are tilt walls that result from stacking of edge dislocations gliding parallel to [c] along (m) planes (Fig. 9.4b). Prismatic sub-boundaries derive from the stacking of <a> dislocations corresponding to lower-*T* slip systems. Under the TEM, dislocations having [c] as Burgers vector are commonly observed to be present together with dislocations having <a> as a Burgers vector (Fig. 9.4b and c).

This latter case above is rather uncommon, however. Even before full crystallization of the magma, the already formed quartz grains may suffer some plastic deformation. Deformation may be due to local modification of the geometry of the crystal skeleton, for example due to melt migration from one site to the other inside the crystalline framework. More frequently, quartz suffers plastic deformation after complete crystallization of the magma.

At near-solidus temperatures, ~700–800 °C, accommodation of deformation in the quartz lattice is performed through the (very) high temperature slip system, namely prism [c] (prismatic slip plane and [c] slip direction). This mechanism leads to the formation of tilt sub-boundaries (or boundaries between subgrains) that are parallel to the (0001) plane, also called the basal plane. Identification of basal sub-boundaries is rather easily done under the petrographic microscope (Fig. 9.5). To summarize, the presence of basal sub-boundaries in quartz indicates that some solid-state deformation took place at a temperature close to the solidus.

Basal sub-boundaries (i.e. due to prismatic [c] slip) are seldom observed alone in quartz. If deformation continues, either in intensity or at a lower temperature, the common <a> slip systems re-appear with their accompanying prismatic sub-boundaries. At near-solidus temperatures, the commonly observed coexistence of prismatic sub-boundaries with basal sub-boundaries, results in paved quartz surfaces made of square subgrains, a feature also called 'chess-board' microstructure (see Fig. 9.3b). Presence of chess-board patterns in quartz points to the existence of near-solidus temperatures of deformation, but does not indicate whether deformation took place in the presence of a melt or not.

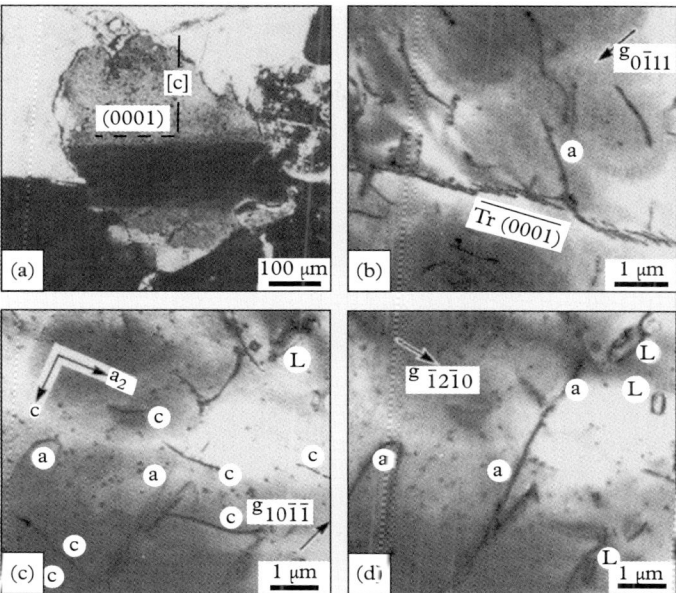

Figure 9.5 *[c]-slip in quartz at near-solidus temperatures. (a) Optical microscope image of a quartz grain in a migmatite from the Vosges massif (France) showing a basal sub-boundary confirming slip parallel to [c]. (b–d) Transmission electron microscope (TEM) images of a quartz grain showing basal sub-boundaries in a granite with a magmatic microstructure. (b) Tilt sub-boundary of basal orientation almost perpendicular to the image, composed of edge dislocations, gliding parallel to [c] in the prismatic (m) = (1010) plane. (c, d) View of the same (c, a_2) crystallographic plane, observed under distinct diffracting vectors (g). The different contrasts between (c) and (d) observed on the free dislocations help to identify these dislocations as edge dislocations: in **a** they glide parallel to [a] (Burgers vector = [a]) and in **c** they glide parallel to [c] (Burgers vector = [c]). The latter dislocations demonstrate the existence of slip at very high temperature. In L, dislocations loops elongate parallel to [c], suggest that the slip rate parallel to [c] is higher than parallel to [a]. After Blumenfeld et al. (1986).*

9.1.2 'Submagmatic' microfractures

At temperatures close to the solidus, plagioclase crystals have a tendency to behave in a brittle fashion. If the crystal framework is subjected to high differential stress while a residual melt is still present in the pores between crystals, a few crystals that are pressed against each other may be dissolved at their contact, particularly at stress edges

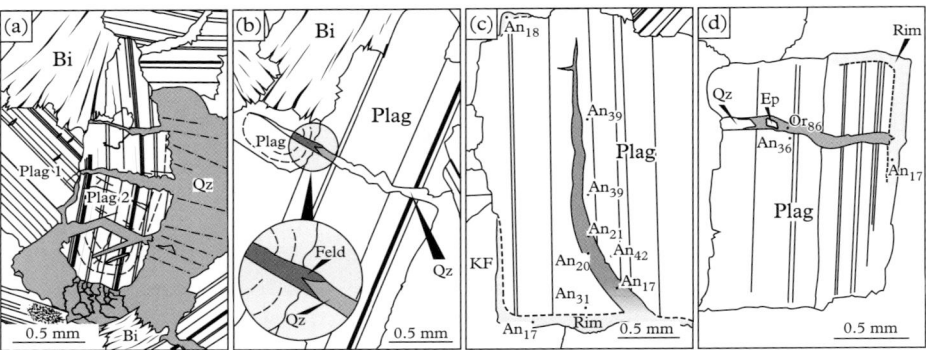

Figure 9.6 *Submagmatic microfractures observed in granites. Qz: quartz; Plag: plagioclase (An: anorthite content; Or: orthoclase content) Feld: feldspar; KF: potash feldspar; Bi: biotite. (a) Plag2 has been indented by Plag1; dashed lines in quartz: prismatic sub-boundaries (tonalite from Foix pluton). (b) Microfracture filled with quartz and feldspar (granodiorite from Bassiès pluton). (c) Microfracture filled by a plagioclase ($An_{17}-An_{21}$) enriched in albite with respect to its host ($An_{39}-An_{42}$); note the overgrowth around the host-plagioclase with no discontinuity with respect to the material filling the microfracture. (d) Microfracture filled with orthoclase, quartz and a grain of epidote, itself sealed by an overgrowth of plagioclase richer in albite than its host (granodiorite from Mont-Louis-Andorre pluton). After Bouchez et al. (1992).*

(Fig. 9.6). Some feldspars may bend while others may break, the microfractures being immediately filled by the melt. The infilling material is principally made of quartz, but feldspar is also present, the whole being close to the eutectic in composition. The presence of such infilled microfractures (incorrectly called 'submagmatic') demonstrates that the crystal framework was deformed in the presence of a residual melt, hence before complete crystallization of the magma. Naturally, such microfractures usually survive during solid-state deformation (Fig. 9.2c). Unless made unrecognizable by subsequent deformation, these microfractures appear as transgranular tension cracks, i.e. parallel to the local σ_1 (or perpendicular to the local σ_3). A statistical treatment of such features makes it possible to infer the orientation of the principal stress components responsible for submagmatic deformation (Fig. 9.7).

9.1.3 Solid-state deformation microstructures

Leaving the domain of near-solidus temperatures, deformation of granites continues to be mainly overcome by the plastic behaviour of quartz that uses exclusively the <a>-axis as the principal slip direction. The corresponding low- to moderate-T slip systems generate prismatic sub-boundaries usually observed under the microscope, and easy to identify using a gypsum plate (colour becomes blue under crossed polars when the $+\lambda$

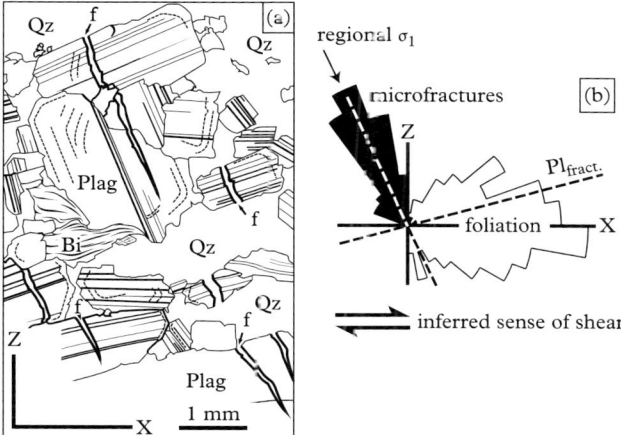

Figure 9.7 *Submagmatic microfractures as potential indicators of sense of shear before full crystallization. (a) Thin section image showing microfractures at the submagmatic state; each microfracture (f) is parallel to the local σ_1 (tension fracture); X: whole-rock magmatic lineation; Z: axis perpendicular to the magmatic foliation plane. (b) Rose-diagram showing the orientation of the microfractures (in black) and the orientation of the plagioclase crystals or trace of foliation plane (in white); note that the fractured plagioclases are more-or-less perpendicular to the microfractures. The mean orientation of the 'local σ_1' therefore allows a dextral sense of magmatic shear to be deduced, based on 120 measurements of cracks affecting 106 plagioclase grains, and 477 measurements of plagioclase orientation. After Bouchez et al. (1992).*

of the inserted gypsum plate is parallel to the sub-boundary; see Box 9.2). Different microstructural 'landscapes' appear, however, according to the temperature reached at the end of deformation, and to the intensity of deformation subjected by the rock in a given temperature domain.

(i) Minor deformation at high-T. In a granite, deformation at high temperature (i.e. higher than 500 °C) usually begins as soon as the magma becomes fully crystallized and settles in a hot, highly plastic environment. Plastic deformation of the quartz grains are commonly observed, tens to hundreds of metres across pluton margins, particularly when the country rocks continue to deform after pluton emplacement. Due to the high mobility of dislocations at such temperatures, basal sub-boundaries in quartz are rapidly transformed into prismatic sub-boundaries with well-defined lattice misorientations between subgrains or between grain boundaries (Fig. 9.4d).

168 *Microstructures and fabrics of granites*

(ii) Large deformation at high-*T*. The large magmatic quartz grains progressively become ovoid to elongate. Subgrains become numerous with increasingly large angular misorientations, then give rise to new recrystallized grains (or neoblasts), invading the original grains (Fig. 9.3). If high temperatures last long enough after the end of deformation, the neoblasts become arranged into polygonal mosaics, with straight to slightly curved boundaries and junctions at 120°. In theory, the magnitude of the stress applied to the rock can be derived as inversely proportional to the recrystallized grain size, as physical metallurgists have claimed. However, in the usual case of slow temperature decrease, subsequent grain growth due to grain-boundary migration may result in underestimation of the imposed level of stress (see 'Structural piezometers' in Nicolas, 1992). The other granite-forming crystals also display plastic deformation features, such as twisted plagioclase and biotite, and micro-shear zones, rich in small micas and quartz grains, that may appear at the borders of larger and less-deformed grains (Fig. 9.4d). Granitic rocks having such features may be interpreted as orthogneissic or gneissified.

(iii) Minor deformation at low-*T*. At temperatures less than ~450 °C (under usual stress conditions) quartz will display the well-known undulose extinction corresponding to ill-defined prismatic sub-boundaries, i.e. spaced and ill-formed dislocation walls resulting from the low mobility of dislocations. Since quartz is more easily affected by deformation when the resolved shear stress is high onto its low-*T* slip system ((0001) <a>), it can be verified that the magnitude of the lattice misorientation of the undulose extinction zones in a specific grain depends on the orientation of the grain lattice with respect to the direction of compression. Small deformations at low-*T* may belong to a continuum with respect to the previous cases (i) or (ii), or be independent from them. Whatever the situation, undulose extinctions are simply superimposed onto the former microstructure with no significant disturbance.

(iv) Large deformation at low-*T*. This type is identified when large lattice misorientations in quartz become ubiquitous, sometimes evolving into kink-bands. Phyllosilicates become creased and feldspars fractured. It is also rather common to observe the effects of late-magmatic fluids, by the formation transgranular fluid inclusion trails in quartz, strips or clouds of alteration in feldspar and chloritization in biotite. Larger deformations, often associated with strain localization and high stress environment, are characterized by recrystallization in quartz giving small, equant to elongate neoblasts, depending on the relative rates of deformation versus recrystallization. Still smaller quartz grains may nucleate in local high-stress areas such as along kink-band boundaries affecting larger quartz grains, or along their margins (Fig. 9.4e). Micro-shear zones and C/S features may also appear (Fig. 9.4f). Macroscopically, such rocks are called 'orthogneiss' (Fig. 9.4e) if the quartz grains remain ovoid in shape with a low density of micro-shear zones, and 'protomylonites' (Fig. 9.4f) if the quartz grains are elongate to ribbon-shaped, a situation often associated with a high density of C/S features.

9.2 Fabrics in granites

Fabric elements will be defined first, then the origin of mineral shape fabrics will be depicted using two-dimensional (2D) concepts. Measurement techniques allowing us to obtain fabrics from 2D sections will be presented at the end of this section. Direct three-dimensional fabric measurement using magnetic techniques will be presented in Chapter 10.

9.2.1 Foliation and lineation

The statistical, mean or preferential disposition of platy crystals, defines the foliation (xy plane). Within the foliation plane, the preferred linear arrangement of elongate crystals defines the lineation (x). This definition is valid for minerals, but also for enclaves which may be flattened and elongated (Fig. 9.8). Foliation and lineation define a shape fabric. The term 'subfabric' is used if the fabric of a specific crystal species is to be discussed. For example, the angular differences between the subfabrics of feldspar and biotite may be used to provide information about the sense of shear in the crystallizing magma. The directional aspect of a fabric, i.e. its orientation, has now been defined.

The intensity, or scalar aspect of a fabric is less easy to define. Observations under the optical microscope help to distinguish between strong fabrics, observable and easy

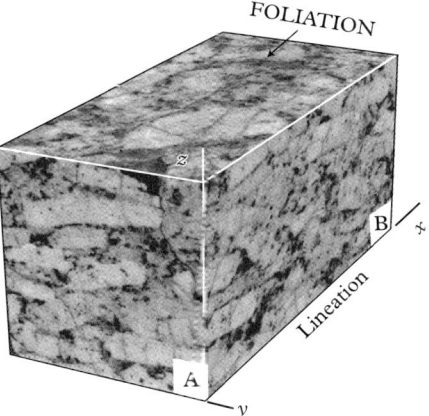

Figure 9.8 *Magmatic foliation and lineation of a porphyritic granite (i.e. containing feldspar megacrysts). The z-axis is perpendicular to the foliation; in the foliation plane (xy), the x-axis represents the magmatic lineation. Face A: yz plane (perpendicular to lineation). Face B: xz plane (perpendicular to lineation and foliation).*

to measure in the field using a compass, and weak ones. In the field, linear fabrics are always difficult to define since outcrop sections parallel to the foliation are required before statistical orientation measurement of crystals can be undertaken. A granite having a magmatic microstructure always has a low fabric intensity, rendering direct fabric measurement in the field almost impossible. Superimposed solid-state deformation (Section 9.1.3) usually increases the fabric strength, hence facilitating the measurements of orientation.

In principle, fabric measurements are more precise when performed in the laboratory, based on orientated specimens collected in the field. However, if such measurements are done with the same 'visual' technique as in the field (Fig. 9.9), a better precision of the orientation measurements cannot be expected. In fine-grained granites, lattice fabric determinations may be improved by using the optical microscope equipped with a 'universal' stage (Fig. 9.10a).

However, complete fabric determination, both in orientation and intensity, requires the use of quantitative techniques, such as digital imagery (Section 9.4) and magnetic techniques. See Chapter 10 for further details.

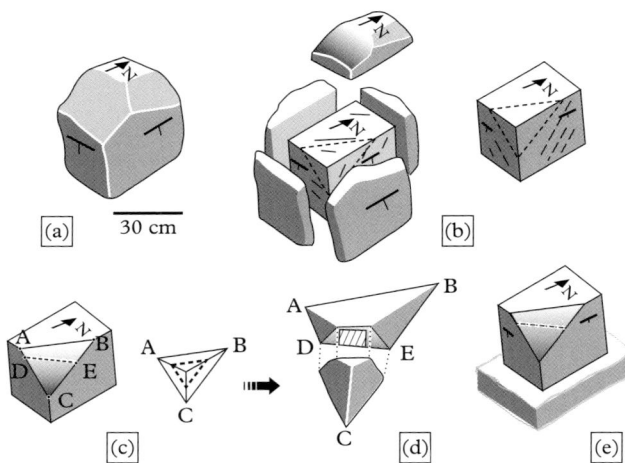

Figure 9.9 *'Traditional' determination of magmatic fabric in a granite. (a) An orientated granite block is collected. (b) Various faces of the block are saw-cut while maintaining the orientation marks; the intersections of the foliation plane with these faces are determined. (c) Sections parallel to foliation are cut in order to determine the lineation (dashed). (d) Thin sections are prepared (parallel to xz) for the optical microscope (U-stage) and/or digital image study (see Section 9.4). (e) Placed on modelling clay, the block is reorientated with respect to the North and the horizontal plane, allowing the original orientation of foliation and lineation to be retrieved. In Bouchez, 1997.*

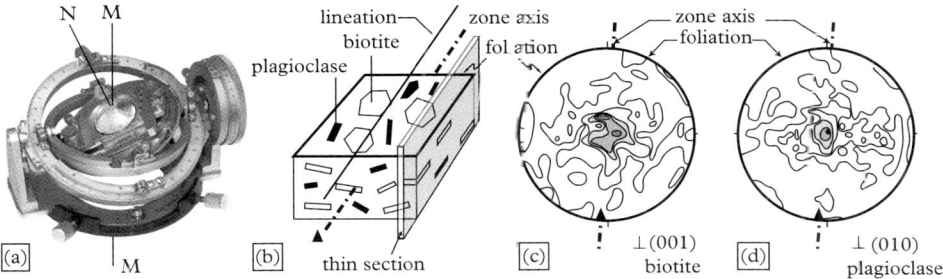

Figure 9.10 *Mineral fabric determination using a universal stage (U-stage). (a) Conventional U-stage (five different rotation axes) placed onto the microscope stage and allowing (partial) exploration in 3D (M: built-in microscope axis; N: axis perpendicular to thin section). (b) XZ thin section, i.e. perpendicular to the foliation and parallel to the lineation. Macroscopically, the biotite (cleavage plane (001)) and plagioclase grains (polysynthetic (010) twin plane) appear to be preferably arranged parallel to the foliation plane (bold line). (c, d) Guérande granite massif (France; in Bouchez et al., 1981): contoured stereo plots (200 measurements rotated into the foliation plane) of the poles (or normal directions) of (001)-biotite cleavage (c) and (010)-plagioclase planes (d). (001)-biotite and (010)-plagioclase are indeed statistically organized parallel to the foliation, with a clear tendency to rotate around a common axis called the 'zone axis' (full triangle).*

9.2.2 Origin of magmatic fabrics

Above all, anisometric crystals, i.e. presenting elongate and/or flattened shapes, are required to define a crystal shape fabric. Biotite and amphibole are typically anisometric, hence well-adapted to develop shape fabrics. However, if the orientation distribution of these crystals is random, no shape fabric can be expected.

In a flowing magma, a shape fabric results from the rotation of the anisometric crystals in the melt. The rotation rate of the crystals must not be constant, otherwise the shape fabric remains random. It is reasonably assumed that simple shear, or laminar flow, is the main deformation mode of a magma. Due to shear, the crystals are therefore subjected to a couple of forces inducing crystal rotation (see Fig. 9.11 in Box 9.3), and the rotation rate is higher when the long axis of the crystal is at a high angle with the shear plane (Fig. 9.12).

In the case of a *single elliptical particle* having an initial orientation α_0 with respect to the shear plane, its resulting orientation α as a function of the shear strain γ (= tangent of the shear angle) is given by:

$$\text{tg}^{-1}(A\,\text{tg}\alpha) = \text{tg}^{-1}(A\,\text{tg}\alpha_0) + B\gamma$$

where $A = [(1-K)/(1+K)]^{1/2}$ and $B = (1-K^2)^{1/2}$, in which $K = (n-1)/(n^2+1)$ is the shape parameter of the particle, where n is the shape ratio of the particle (long axis/short axis) (Fernandez et al., 1983). The variation of α as a function of γ is given in Fig. 9.13.

Box 9.3 Pure shear? Simple shear?

Consider a spherical object and deform it in pure shear (Fig. 9.11a). The object will become flat and elongate. If the deformation is homogeneous the sphere will become an ellipsoid whose long axis represents the extensional direction and the short axis the shortening direction. As illustrated in two dimensions in Fig. 9.11a, the particular feature of pure shear is that these (principal) axes remain parallel to themselves during the progression of deformation. This kind of deformation is therefore called 'coaxial'. Now, let us deform the sphere in simple shear (Fig. 9.11b). The object will also be transformed into a flat and elongate ellipsoid, but its principal axes will rotate with the progression of deformation, hence the term 'non-coaxial' given to this deformation.

In geology, simple shear is often used as a reference deformation mode in which the y intermediate axis does not vary, meaning that all displacements take place in the xz plane, a deformation also called planar. Simple shear is frequently imaged using a stack of playing cards (Fig. 9.11c) in which θ is the shear angle and $\theta = tg(\gamma)$ the shear strain. The long axis of deformed objects represents the direction of finite extension, or lineation (x axis), and the short axis is perpendicular to the flattening plane, or foliation (xy plane, with x perpendicular to y).

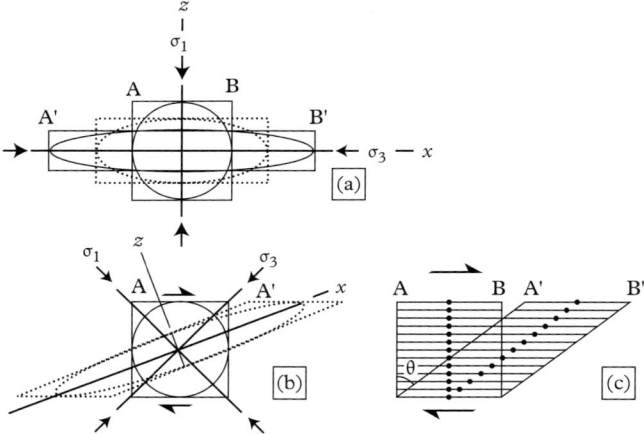

Figure 9.11 *Pure shear (a) and simple shear (b, c) in two dimensions. In this representation, the principal stress components ($\sigma 1$) and ($\sigma 3$) are fixed: we observe that they remain parallel to the deformation axes (x and z) in a coaxial deformation (a), and that they rotate during deformation in a non-coaxial deformation (b). (c) A pack of playing cards illustrates deformation by simple dextral shear (θ : shear angle), each card sliding parallel to the shear plane. The very first extension direction (for a very small θ) is at an angle of 45° to the shear plane. After Nicolas (1989).*

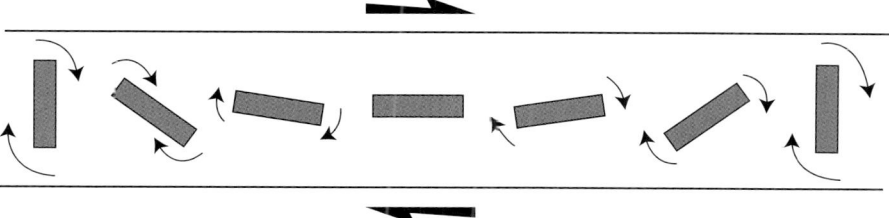

Figure 9.12 *Rotation of an elongate and rigid particle in simple shear (dextral laminar flow). Rotation rate is proportional to the shear couple (length of arrow) applied to the particle. Note that, since a particle is three-dimensional, its rotation axis is not necessarily always perpendicular to the plane of representation. In Nicolas (1989).*

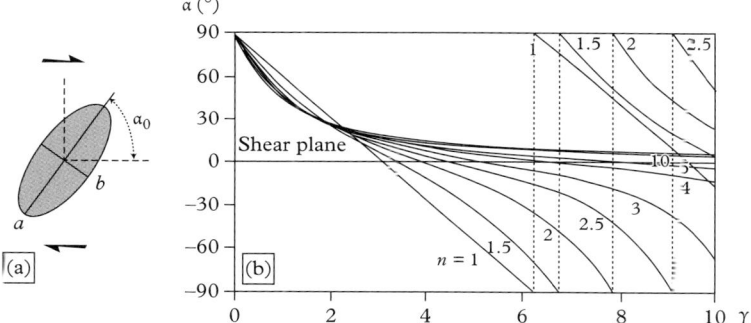

Figure 9.13 *Rotation of a rigid particle subjected to simple shear. (a) Shape ratio ($n = a/b$) and initial orientation α_0 of the particle. (b) Evolution of α (final orientation) as a function of γ (shear strain) for different n values (assuming $\alpha_0 = 90°$). Rotation rate is uniform for a circular particle ($n = 1$), becomes zero in case of a large deformation and for infinite particle length ($n = \infty$, 'passive markers'). For intermediate n values a cyclic rotation is observed with a periodicity T ($\gamma_T = 4\pi/(1 - K^2)^{1/2}$). After Fernandez et al. (1983).*

In the case of a *population of particles*, the fabric will result from the population of α values which, in two dimensions, may be represented by a histogram, or more spectacularly by a frequency rose-diagram of the angular distribution of the particle long axes. Figure 9.14 illustrates the results of analog modelling of magmatic deformation leading to shape fabrics. More quantitatively, the principal axes of the ellipse (or ellipsoid in 3D), representing both the orientation and the intensity of the shape fabric, can be calculated from the α values as follows:

$$F = \frac{1}{m}\begin{bmatrix} \sum_i \cos^2 \alpha_i & \sum_i \cos \alpha_i \sin \alpha_i \\ \sum_i \sin \alpha_i \cos \alpha_i & \sum_i \sin^2 \alpha_i \end{bmatrix}$$

174 *Microstructures and fabrics of granites*

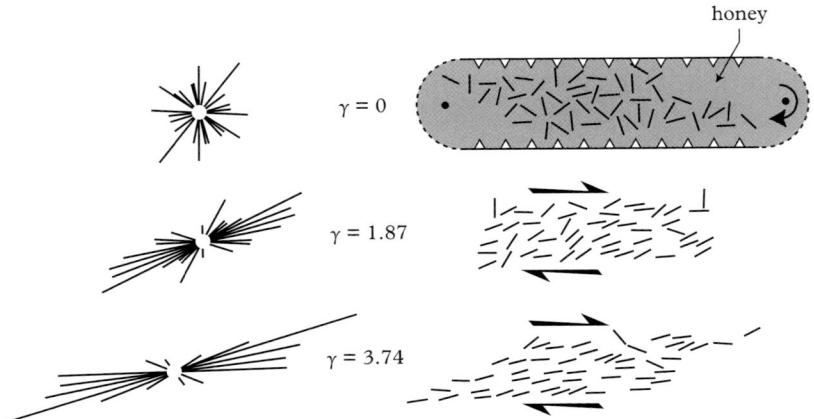

Figure 9.14 *Analog modelling of shape fabric formation in a 'magma' subjected to dextral simple shear, for n = 5 particles' shape ratio. Toothbrush bristles (!) are laid onto a layer of honey (material having an almost Newtonian rheological behaviour) with initially ($\gamma = 0$) a random orientation distribution. An elliptical conveyor-belt deforms the viscous layer in simple shear. Rose diagrams represent the shape fabrics at $\gamma = 0, 1.87$ and 3.74 (respectively, 0°, 62° and 75° shear angles). After Ildefonse and Fernandez (1988).*

where m is the number of particles and α_i is the final orientation of particle i ($i = 1, m$). The orientation of the long axis of the fabric ellipse (i.e. the magmatic lineation or, in two dimensions, the foliation trace) is given by the orientation of the 'largest' eigenvector. The fabric intensity is given by the ratio between the largest and smallest eigenvalues. Analysis of the fabric ellipse is similar to that of the strain ellipse in simple shear, both in orientation (α) and intensity (I_F; Fig. 9.15). A random shape fabric, considered to represent the initial distribution before deformation ($\gamma = 0$), can be imaged by a circle of unit radius ($I_F = 1$). The first increment of deformation leads to an almost circular ellipse ($I_F = 1 + \varepsilon$) whose long axis makes an angle $\alpha_F = 45° - d\alpha$ with the shear plane. Subsequent increments bring the long axis of the ellipse closer to the shear plane, and the fabric maximum (I_{Fmax}) is reached when $\alpha_F = 0°$.

9.2.3 The non-cyclicity of magmatic fabrics

When applied to a population of particles, the Fernandez et al. (1983) formula (Section 9.2.2) leads to cyclic fabrics developing and disappearing with increasing shear strain, at a period depending on the shape ratio of the particles (Figs 9.13 and 9.15). In reality, this behaviour does not seem to apply to granitic magmas. No cyclic variation of the fabric (or subfabric) intensity is observed, for example when approaching the margin of a granite pluton, locations where the shear strain undergone by the

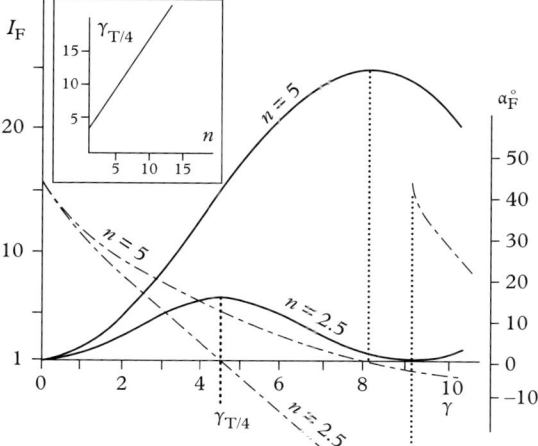

Figure 9.15 Shape fabric cyclicity in a virtual suspension deformed in simple shear; non-interacting particles of identical shape ratios (n) and random initial shape fabric. Solid line: fabric intensity (I_F) for $n = 5$ and $n = 2.5$. Dashed line: fabric orientation α_F (angle between the fabric ellipse long axis and the shear plane) for $n = 5$ and 2.5. Note that α_F varies from $-45°$ to $+45°$ while the individual particles rotate from $-90°$ to $+90°$. γ: shear strain; $\gamma_{T/4}$: γ value for which the shape fabric is at the maximum ($I_F = I_{Fmax}$) and $\alpha_F = 0$. The shape fabric becomes random ($I_F = 1$) at each $\Delta\gamma = 2\gamma_{T/4}$. Box: $\gamma_{T/4}$ values as a function of n. After Fernandez et al. (1983).

magma should vary substantially. As developed below, the mechanical interactions between crystals, and the shape ratio distribution of the crystals explains the lack of fabric cyclicity.

(1) *Interactions between crystals and the fabric acquisition window.* Since the solid fraction (S) of the magma progressively increases during magma cooling, crystal rotation slows down ($S > 0.4$–0.5) then stops ($S > 0.7$–0.8). Within this 50–80% window of crystal content, magmatic shape fabrics appear, grow and become permanently recorded in the rock. For such crystal contents, the tendency of crystals to rest upon each other while rotating is called *tuilage*, or tiling (Fig. 9.16). The effect of tiling is to slow down crystal rotation since the shape ratio of the aggregate (n) is larger than the shape ratio of each isolated crystal. When observed in appropriate xz sections in quarries of porphyritic granites, a statistical study of tiling features allow one to determine the sense of shear undergone by the magma during its fabric acquisition window (Fig. 9.16).

176 *Microstructures and fabrics of granites*

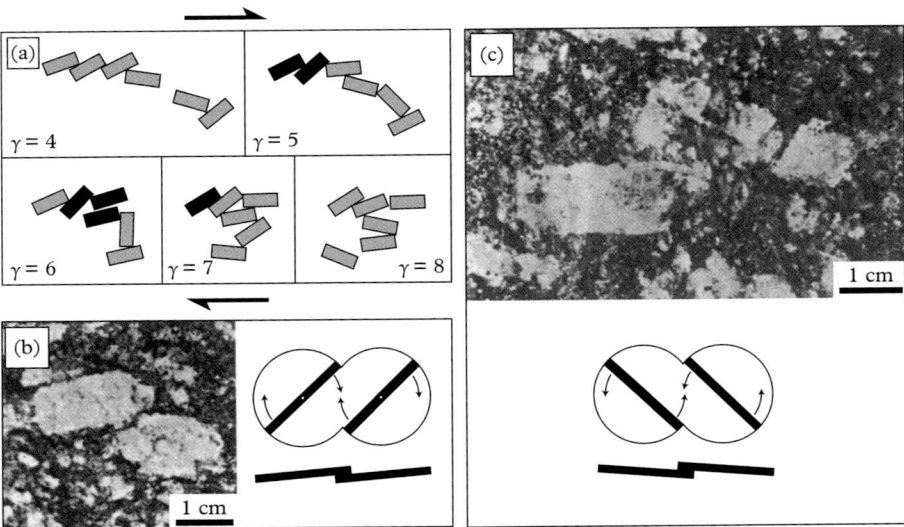

Figure 9.16 *Mechanical interactions between crystals. (a) As observed in analog models (dextral simple shear) for different shear strain intensities: most crystals rotate dextrally. Examples of dextral (b) or sinistral (c) 'tilings' (tuilages) of K-feldspar megacrysts from the Barbey-Seroux granite pluton (Vosges, France). After Blumenfeld and Bouchez (1988).*

(2) *Shape ratio distribution and shape fabric modelling.* Based on a 'natural' distribution of shape ratios for biotite and amphibole, obtained from optical microscope observations (Fig. 9.17a and c), the shape fabric can be calculated using the Fernandez model (Fig. 9.17b and d). Following a transition of a few γ units during which fabric intensity (I_F) and orientation (α) rapidly vary, the intensity is observed to stabilize at a rather low level while the orientation remains around 0°. This implies that the magmatic foliation becomes rapidly parallel to the shear plane. In natural suspensions, it is easy to imagine that the 'stabilizing' effect of the mechanical interactions between crystals further impedes the cyclic variations of I_F and α. This numerical model explains why natural shape fabrics are so stable both in orientation and intensity (see also Chapter 10). As a consequence, it also suggests that the intensity of the shape fabric cannot be used as a marker of deformation intensity undergone during the magmatic stage.

9.2.4 Magmatic lineation and finite extension direction

It can be seen that the magmatic lineation is parallel or close to the finite extensional direction recorded by the magma during its acquisition of a fabric when its crystal load was 50–80% of crystals. Note that submagmatic microfractures (Fig. 9.7) support an increment of extension (perpendicularly to the microfractures) in the presence of no

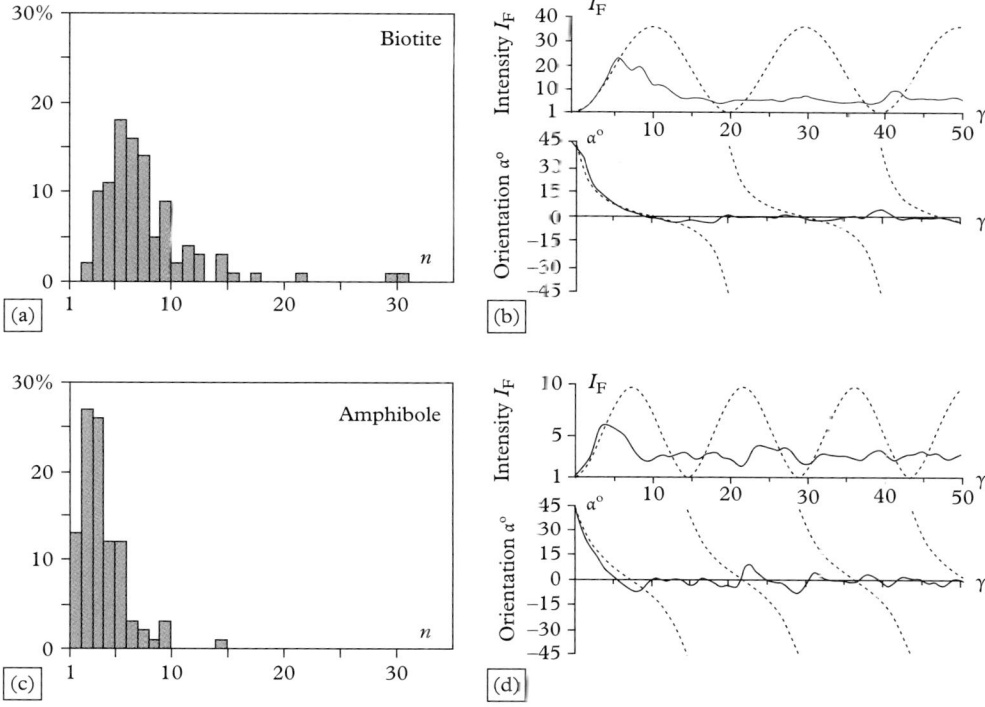

Figure 9.17 *Natural distributions of biotite (a) and amphibole (c) shape ratios determined from thin sections, and corresponding fabrics (b, d) according to the model of Fernandez et al. (1983). Each frequency histograms is established from 100 grains. Solid curves: fabric intensity (I_F) and orientation (α_F) calculated using the 'natural' shape ratio distributions. Dashed curves: I_F and α_F calculated using the average shape ratio of biotite (n = 6.14) or amphibole (n = 3.1). After Ildefonse et al. (1997).*

more than a few percentage of melt. These microfractures are more-or-less perpendicular to the magmatic lineation (Fig. 9.7). It can be concluded that the finite extensional direction recorded between 20% and 50% of melt (that leads to a magmatic lineation) is close to the 'infinitesimal' extensional direction recorded by the microfractures at the end of crystallization. Such a (near-)parallelism suggests that, except in cases of multiple injections or in proximity to a feeding zone, the finite extensional direction does not vary substantially with time. This may be true if the time spent by the magma between the end of fabric records ($S > 0.8$) and its complete crystallization has a short duration. This is the case for leucogranite magmas which have compositions close to the minimum melt hence tend to crystallize rapidly, particularly at low melt fractions. The example of the Sidobre granite (Chapter 10) shows that the parallelism between the extensional direction and the magmatic lineation can be observed at the scale of the whole massif.

9.3 Fabric of the Brâme–St Sylvestre–St Goussaud complex

Since Cloos (1925), many fabric studies have been performed in plutons, mostly based on field measurement of the preferred planar orientation of feldspar and biotite. Their analysis in terms of fabric patterns has been reviewed by Paterson et al. (1998). The case of the Brâme–St Goussaud (Fig. 9.18) complex (western Massif Central, France; Guéret 1 : 80,000 map) is original since the orientation measurements of its foliations and lineations have been entirely realized with the 'traditional' technique described in Fig. 9.9. From each of the 256 locations, orientated blocks of granite weighting several kilograms were collected. The conception of one thin section per sampling site helped to establish that the microstructures are magmatic throughout, except along the north–south

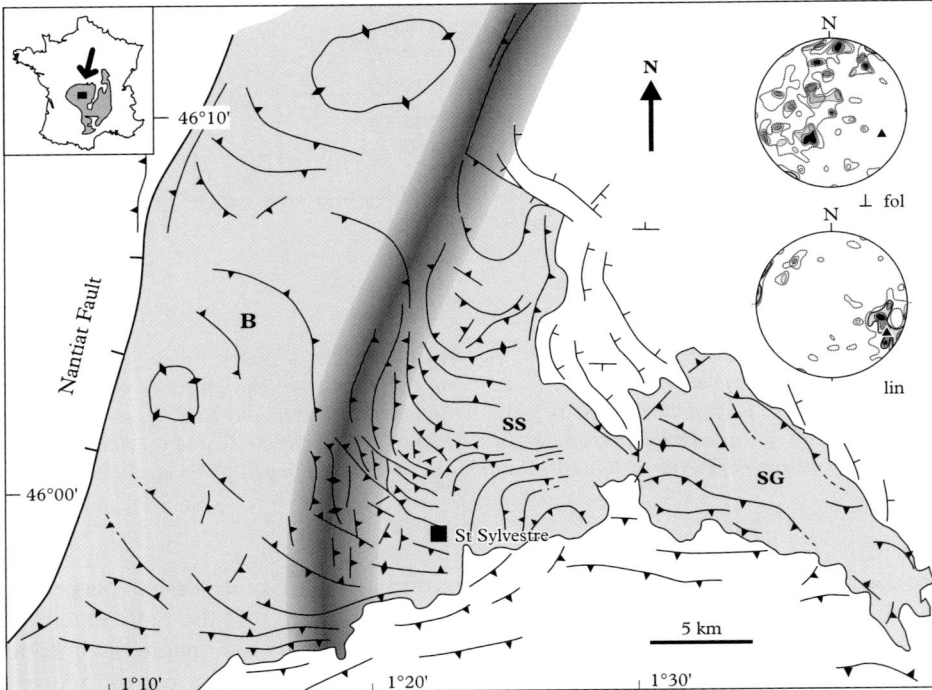

Figure 9.18 *Magmatic foliation trajectories in the Brâme–St Goussaud (French Massif Central) leucogranite complex, mapped from 256 regularly spaced strike/dip measurements performed in the laboratory (Fig. 9.8). Upper-right corner: stereo plots (lower hemisphere, Lambert projection) of the magmatic foliation poles (~ fol: 256 meas.) and lineations (lin: 141 meas.). In the foliation pole diagram, the full triangle represents the pole of the best-fit plane; its orientation (trend/plunge: 115°, 29°) is close to the best axis of the lineations (full triangle of the lineation plot: 117°, 18°). This demonstrates that, statistically, the foliations rotate around the mean lineation. Grey: (dextral normal) magmatic shear zone revealed by this study and which hosts significant uranium concentrations (Section 14.2.2). After Mollier and Bouchez (1982) and Mollier (1984).*

trending, several kilometres wide shear zone that separates the Brâme massif from the St Sylvestre massif (grey in Fig. 9.18) where a high-temperature solid-state deformation was superimposed onto the magmatic microstructure.

This complex of about 800 km² was emplaced at ~320 Ma (Late Hercynian) within Devonian (early Hercynian) mica schists and gneiss of the Limousin nappes. In the complex, the foliations have shallow dips, particularly to the west of the NNE-trending magmatic shear zone. The latter has steep foliations and high-T microstructures, and its sense of shear is underlined by map-scale sigmoids of the foliation trajectories on the flanks of the shear zone (Fig. 9.18). In the St Sylvestre massif, the more-or-less concentric shapes of the foliation trajectories have been identified as being related to late granite intrusions within the incompletely crystallized complex. As described in Fig. 9.9, the magmatic lineations of the complex were determined by statistical counting of the elongate feldspars from samples collected in the field and forwarded to the laboratory to make sections parallel to the foliation plane. Almost all the lineations have NW–SE trends, and their plunge is close to the horizontal. Therefore, statistically, the foliations have a 'zonal' organization around the average lineation, i.e. in the stereo plot of Fig. 9.18, the foliation poles occupy a great circle girdle whose pole is close to the lineation best axis. Such a zonal distribution of foliations around a common lineation direction is extremely frequent in rocks having undergone large plastic or magmatic deformation.

This example also illustrates that microstructures and shape fabrics cannot be separated. In the present case, the fabrics associated with magmatic microstructures, also called 'magmatic fabric', have recorded the emplacement stage of the granitic complex. Such a dynamic emplacement of a several kilometres-thick laccolith into the horizontally foliated pile of nappes corresponds to a late-Hercynian 'flat' shearing event toward the NW, attributed to post-orogenic thinning by Faure (1995). By the end of its emplacement, the laccolith was itself intruded by more differentiated granitic magma which developed concentric, as well as cross-cutting, magmatic foliations on cooling. The complex was finally subjected to high-temperature dextral shearing whose deformation became localized along the wide NNE-trending shear zone of Fig. 9.18, with a strong normal-faulted displacement which also affected the underlying metamorphic pile. The partly mylonitic, ductile normal fault of Nantiat ultimately developed a gravity-driven detachment at the source of the exhumation of the complex from its western host (see Section 14.2.2).

9.4 Fabrics and digital imagery

The rather recent and easy use of digital images, i.e. images made of arrays of pixels, led to the development of new methods of shape fabric quantification. They have rapidly superseded other types of representation, such as rose diagrams (Fig. 9.14) which cumulate the orientation directions of particle long axes into angular classes, a rather rough method of representation since, for a given elongation, the crystals do have different sizes, shapes and spatial distributions which are not taken into account. The images are acquired from saw-cut and eventually polished sections carefully chosen with

180 *Microstructures and fabrics of granites*

respect to the structural framework. If the geologist is looking for a lineation, a section parallel to the (already defined) foliation plane (*xy* plane) will be used. The section is digitized, either with a camera or with an office scanner. Using software such as Photoshop™, the objects to be measured are isolated using classical thresholds and multispectral techniques (Launeau et al., 1994), and eventually decorated using the staining technique of Laduron (1966) that helps to separate K-feldspar (yellow) from plagioclase (red). If one is not satisfied by this automatic grain separation or identification, because of altered or contiguous grains, hand-drawn corrections must be made, such as contour modification or removal of incorrectly identified grains. Once separated, the phase of interest will be numerically tagged, for instance by giving an identifiable colour to the phase.

9.4.1 The intercept method

Following principles first developed by Saltykov (1958), Underwood (1970) and Coster and Chermant (1989), the method is based on counting, inside a counting-window, the number of phase transitions or intercepts (no-X → X and X → no-X) met by the straight lines of a counting grid, X being the phase of interest (Launeau, 1990). The counting grid made of one-pixel spaced straight lines is plated onto the digital image and rotates from $\alpha = 0°$ to $180°$ by (for example) $2°$ increments (Fig. 9.19a). The mean intercept number by unit length as a function of α, or $N(\alpha)$, is obtained by dividing the number of phase transitions by twice the total length (in pixels) of the 'exploration' lines. Since the maximum number of intercepts is perpendicular to the mean object elongation (Fig. 9.19b), the graphical representation of $N(\alpha)$ is called an *inverse* intercept rose,

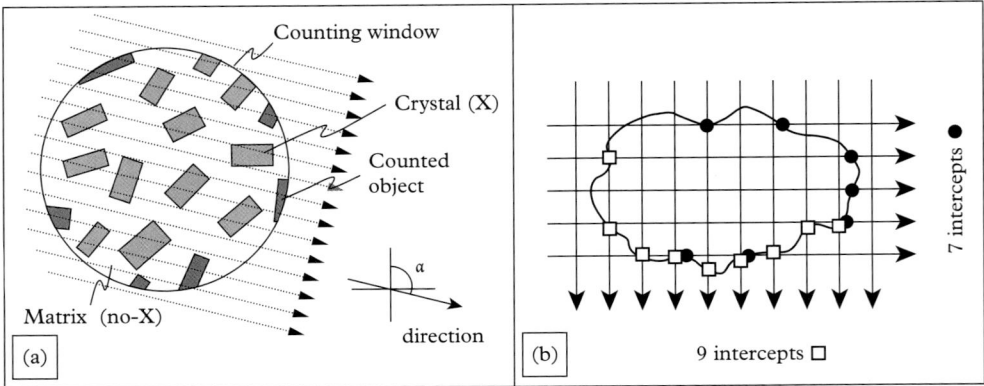

Figure 9.19 *Principle of the intercept technique. (a) The numbers of transitions $N_{noX\text{-}X}(in)$ and $N_{X\text{-}noX}(out)$ are counted in the α direction inside a circular window and along a set of parallel and equidistant straight lines. The number of intercepts $N = 1/2(N_{noX\text{-}X} + N_{X\text{-}noX})$ takes into account the section effect due to the window (grey objects). (b) Example of counting $N_{X\text{-}noX}$ applied to a crystal, for $\alpha = 90°$ and $180°$. After Launeau and Bouchez (1992).*

or more simply intercept rose. The rose has the shape of a circle in the case of random orientation distribution of objects (crystals or aggregates), and is more-or-less elliptical or potato-shaped in other cases. The long axis of the (inverse) intercept rose gives the mean elongation of the population of crystals, and the ratio between the long and short axes (which are not necessarily exactly perpendicular to each other) is a measure of the mean shape ratio of the crystals.

This explanation hides difficulties related to pixels grid of such as the 'staircase' contour lines of the crystal boundaries for α values different from 90° (*modulo* 90°). Such difficulties were solved by Launeau and Robin (1996) who proposed an online freeware at <http://www.sciences.univ-nantes.fr/lpgnantes/SPO>. In addition, these authors have refined the analysis of the intercept rose using Fourier transforms that help to underline the principal harmonic modes of the fabric, hence facilitating the interpretation. They also compared the results obtained from their technique with the auto-correlation method developed by Panozzo-Heilbronner (1992), which uses vectorial images for the objects instead of a matrix representation.

The intercept technique is widely used, as confirmed by abundant scientific literature on the subject. For example, a *pseudo* three-dimensional subfabric of K-feldspar in a site of the Sidobre granite massif (Massif Central, France) has been determined in a quarry where favourably oriented rock-faces were easily accessible (Fig. 9.20). A remarkable similarity is observed between the fabric of K-feldspar and the magnetic fabric that reveals the biotite subfabric (see Chapter 10).

This example emphasizes the main problem of images being 2D. Such a problem can be overcome if, as in the example of Fig. 9.20, the rock-sections are close to the principal section planes of the fabric ellipsoid, and if the number of crystals is sufficient for 'good' statistics to be effective. The long axes obtained from each 2D section will be close (both in orientation and modulus) to the principal axes of the fabric ellipsoid. Otherwise, the analysis must be refined in order to obtain an ellipsoid.

Determination of true ellipsoids from sets of three more-or-less perpendicular 2D elliptic sections (Robin, 2002) helps to overcome this problem. The procedure has recently been illustrated by Launeau et al. (2010), who obtained well-defined shape fabrics from digital camera images of the Pocinhos granite pluton (Paraiba State, NE-Brazil).

9.4.2 2D wavelet analysis technique

The fabric element remaining to be taken into account and quantified is the distribution mode of crystals (or aggregates) constituting a population, at least in 2D sections. The wavelet analysis technique uses a 2D filter resembling a 'Mexican hat', which can be adjusted according to four parameters (Fig. 9.21). Its basic concepts are described by Ouillon et al. (1996). In addition to giving a quantitative description of crystal sizes, shapes and orientations, this filter allows the simultaneous study of spatial distributions of crystals at scales ranging from a few pixels up to the entire image. The convolution between the digital image and the filtering function is mathematically described by Gaillot et al. (1999). These authors show how the wavelet coefficients are normalized

182 *Microstructures and fabrics of granites*

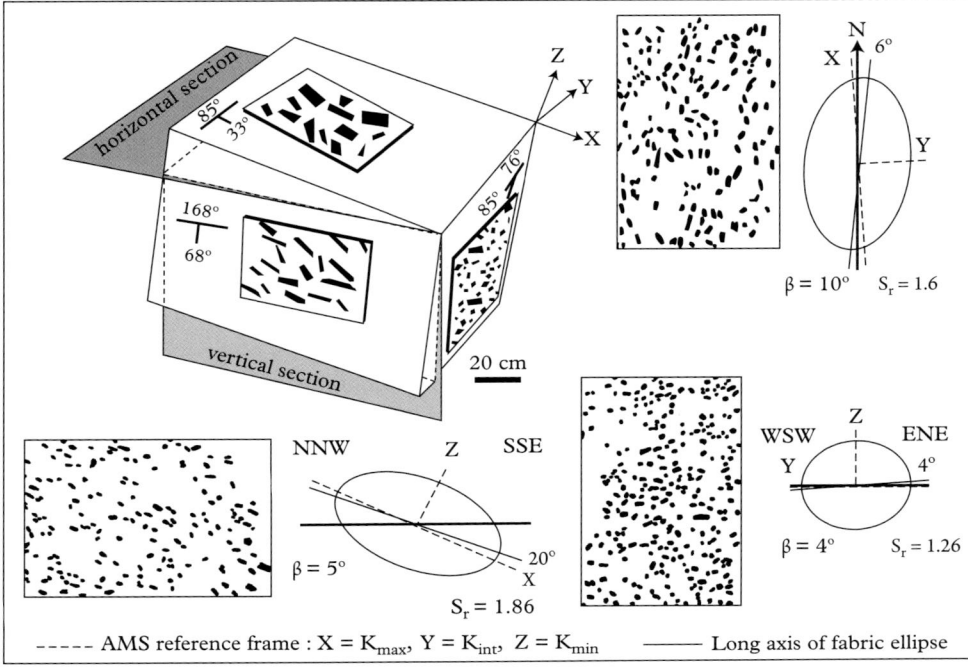

Figure 9.20 *Shape fabrics of K-feldspar megacrysts in the La Trivalle quarry (Sidobre granite massif; see Fig. 10.22). From three almost orthogonal rock-faces in the quarry, with orientations close to the principal section planes of the magnetic (AMS) ellipsoid, the K-spar megacrysts have been hand-copied onto transparent sheets (~0.5 m²), then downsized and digitized. Analysed by the intercept method, the images yielded smoothed elliptical intercept roses of the three principal sections planes (XZ, XY, YZ; X = lineation; Z = normal to foliation). On each section, the K-spar average elongation is revealed to be close to the long 'magnetic' axis (K_{max}), as determined by AMS (biotite subfabric). S_r: shape ratio of the fabric ellipses. After Darrozes et al. (1994).*

and filtered during exploration of the digital image while resolution (a), anisotropy (σ) and orientation (θ) are changed step by step according to nested loops. This operation is easily performed with a personal computer using the program devised by Darrozes et al. (1997).

A quality of the wavelet technique is to reveal the different scales of organization. From the section of a rock sample coloured with Laduron's technique then digitized, it becomes easy to understand how the size, distribution and orientation of K-feldspar crystals are organized (Fig. 9.22). At low resolution (a = 10 mm), the wavelet analysis reveals elongate clusters of K-feldspar crystals orientated at N34°E. At high resolution (a = 3 mm), the average elongation of the megacrysts that represent the magmatic lineation is close to N–S, as in Fig. 9.18. At even higher resolution (a = 0.5 mm), the tension gashes filled with K-feldspar appear to be statistically orientated at N84°E, more-or-less

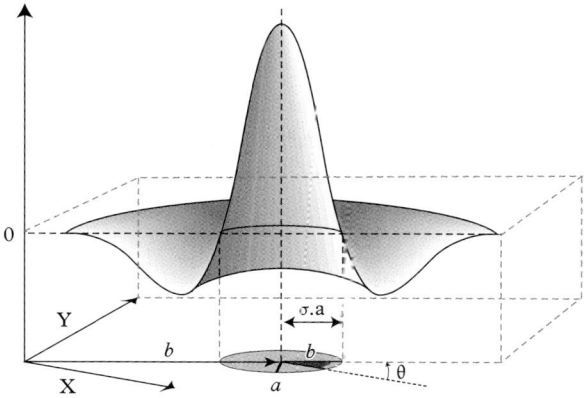

Figure 9.21 Section of the anisotropic wavelet filter (Mexican hat) used for the study of fabrics and distributions. The four adjustable parameters are size (a) on which depends the resolution of the analysis and allows one, by contraction or dilation, to explore several scales of organization; step of displacement in the image (b); shape ratio (σ) and orientation (θ), which help in determining the anisotropy of objects, clusters and tension gashes (see Fig. 9.20). After Gaillot et al. (1999).

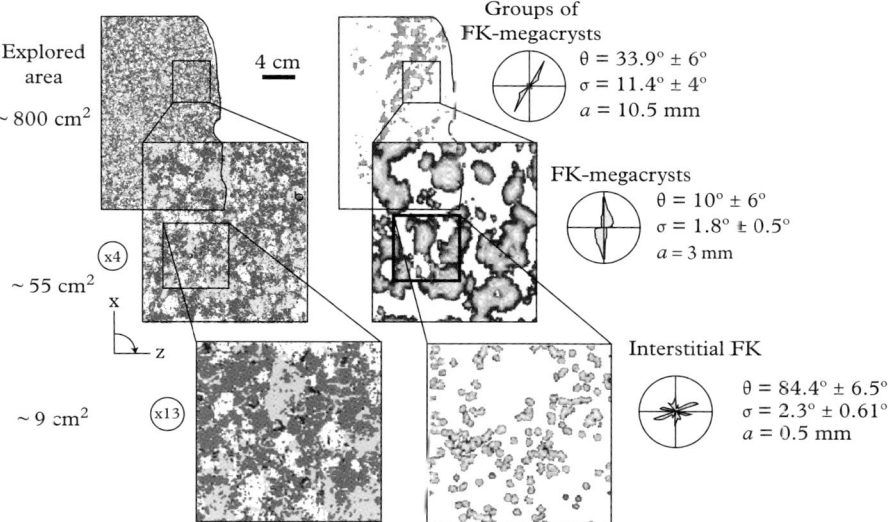

Figure 9.22 K-feldspar fabric of a chemically coloured granite section (technique described in Bailey and Stevens, 1960, and Laduron, 1966) from the Sidobre massif (Massif Central, France). K-feldspar is here in light grey (instead of yellow), quartz remains in white, plagioclase is dark grey (instead of red) and biotite remains in black. Three windows, explored at finer resolutions, help to isolate the megacryst layering, or cluster organization, elongate parallel to N34°, the shape fabric of the megacrysts at N10°, and the preferred orientation of the interstitial K-feldspar at N84°. After Gaillot et al. (1999).

perpendicular to the magmatic lineation, i.e. perpendicular to the stretching direction of the magma during the late stages before complete crystallization.

9.5 Conclusions

In granite studies, shape fabrics (or preferred orientations), which cannot be separated from microstructures at the scale of the optical microscope, are rich in information concerning stress and deformation conditions endured by the magma during its crystallization and after its emplacement. Microstructures in quartz are the most precious markers (chess-board, basal and prismatic subgrains, recrystallization, fluid inclusion trails, etc.). Together with flexured, fractured or altered feldspars, and kinked or chloritized micas, they carry information that completes the history of magma deformation.

Shape fabrics result from rotation, due to shear flow, of the population of anisometric crystals present in the magma. As confirmed by numerical modelling, the preferred orientation of the crystal long axes tend to align with the extension direction due to shear flow, and, for a given population of crystals, the intensity of the fabric tends to be constant. Crystal shape fabrics in granites have a great significance in terms of magma flow (deformation) by the end of its emplacement in the crust. They provide information about the regional geology and especially about geodynamics, as discussed in Chapter 10.

Several sophisticated techniques, particularly those using digitized imagery, help to define the shape fabrics. They facilitate quantitative descriptions and comparisons between fabrics, hence refining the geological models that can be deduced from them.

10
Magnetic fabrics in granites

The previous chapter has shown that crystal shape fabrics are rich in information concerning magma deformation, and hence are worthy of being studied. However, shape fabrics in granites are faint and their measurement tedious, inaccurate and difficult to achieve using traditional methods (Fig. 9.9). This is why the anisotropy of magnetic susceptibility (AMS) technique has emerged as a powerful tool in fabric analysis of granitic rocks. Long known for its potential in structural geology (Graham, 1954), this technique begins with the collection of orientated drill-cores in the field, followed by a quantitative description of the fabric of 'magnetic' minerals contained in the granite. In association with magnetic mineralogy studies and measurements of other magnetic properties, AMS was originally developed by research teams in palaeomagnetism. AMS has therefore, in the 21st century, become a valuable method to explore the structure and geodynamics of granite emplacement into the crust.

10.1 Magnetic properties of materials

10.1.1 Basic concepts

Physics distinguishes the induced magnetic field vector, of magnitude B (in tesla, T), from the inducing field vector, of magnitude H (in A/m). The relation between B and H is: $B = \mu_0 H$, where μ_0 is the magnetic permittivity of free space $(= 4\pi \times 10^{-7})$. In a material, $B = \mu_0(H + M)$, where M (also written as J, in A/m) is the induced magnetization, corresponding to the magnetic moment divided by the volume of the material. M can be written as $M = KH$, where K, which is the magnetic susceptibility of the material (Fig. 10.1), has no units (in the International System, written as SI units).

With the exception of diamagnetism, magnetic properties result from atoms or groups of atoms that carry permanent magnetic moments. They belong to the first transition series of Mendeleev's periodic classification, principally Fe, the most abundant magnetic element in rocks, but also Mn, V, Cr, Co and Ni. The induced magnetization at low-field strength (a few 10^{-4} T under usual laboratory conditions) must be distinguished from

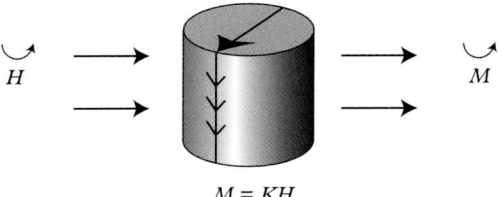

Figure 10.1 *Magnetic susceptibility K. When placed inside the coil of a susceptometer, the sample acquires a magnetization (M) parallel to the magnetic field (H) that is applied in the device, and from which its susceptibility (parallel to the coil axis) is derived. K is a scalar if the material is isotropic. It is a second rank tensor otherwise, and hence necessitates several measurements for different orientations with respect to the magnetic field (H).*

high-field (≥ 1 T) magnetization frequently used for magnetic mineralogy, and from the natural remanent magnetization (NRM, or fossil magnetization) that characterizes ferromagnetic minerals, particularly for its potential in palaeomagnetic studies. In this domain, note the excellent text book in English and available online, entitled *Paleomagnetism* by R. F. Butler (1992).

Some facts about the magnetic properties of minerals will be considered before AMS itself. A few field examples will provide an idea about the amount of information that can be obtained from the study of magnetic fabrics. Concepts concerning the anisotropy of remanent magnetization, which help to isolate the fabric of the ferromagnetic minerals, will also be discussed in this chapter.

10.1.2 Types of magnetic behaviour

In addition to the always present diamagnetism, a material may also be paramagnetic, antiferromagnetic or ferromagnetic (*sensu lato*, i.e. including ferromagnetism and ferrimagnetism, Fig. 10.2a). Diamagnetism is due to the rotation of electrons, whose (faint) magnetic field tends to counteract the effect of the induced magnetic field. Therefore, the diamagnetic susceptibility is negative and very weak, on the order of -10^{-5} SI, independent of the magnitude of the inducing field and also independent of temperature. In rocks, every mineral carries a diamagnetic susceptibility, which is considered as isotropic with a value equal to the susceptibility of quartz: $K_{dia} = -14 \times 10^{-6}$ SI, or -14 μSI.

In paramagnetic behaviour, since there is no interaction between the magnetic 'moments' (or elementary magnetic vectors carrying the magnetization) these moments are randomly arranged in the absence of a magnetic field. When subjected to a magnetic field, the magnetic moments statistically tend to become parallel to the induced field,

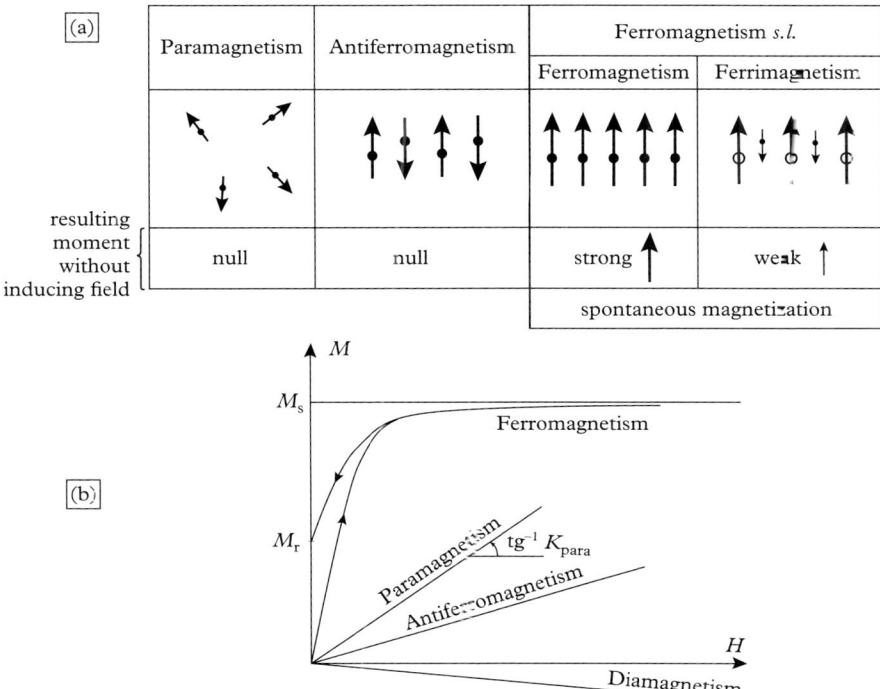

Figure 10.2 *Types of magnetic behaviour in a material. (a) Representation of the permanent magnetic moment carried by atoms (or groups of atoms) in the absence of an inducing field. (b) Magnetization (M) as a function of the applied field (H). For ferromagnetism, the curve rising from zero is the first magnetization (for example, magnetite in a basalt cooling in the Earth's magnetic field); otherwise magnetization at zero field is the remanent magnetization value (M_r).*

yielding a weak and positive susceptibility. Paramagnetism is independent of the magnitude of the inducing field, hence is represented by a straight line on the $M = f(H)$ plot (Fig. 10.2b). Paramagnetism is sensitive to thermal motion of atoms (Curie law), resulting in a paramagnetic susceptibility inversely proportional to temperature. Paramagnetic minerals obey the Curie–Weiss law: $K_{para} = (K - K_{dia}) = Cd/(T - \theta)$, where K_{dia} is the (always present) diamagnetic susceptibility, C the Curie constant, d the density, T the temperature (in kelvin) and θ (in kelvin) the paramagnetic Curie temperature (Fig. 10.3a). The θ value is a few kelvins for minerals poor in iron, such as muscovite, and about 10 K for the iron-bearing silicates such as biotite (Fig. 10.3b). The Curie constant is equal to 744×10^{-3} K and 983×10^{-3} K for Fe^{2+} and Fe^{3+}, respectively (see Fig. 10.5). Silicates containing iron in their lattice are the most common paramagnetic minerals, i.e. mainly biotite, chlorite, iron-bearing muscovite and amphibole in granites, but also pyroxene, garnet, tourmaline and cordierite.

188 Magnetic fabrics in granites

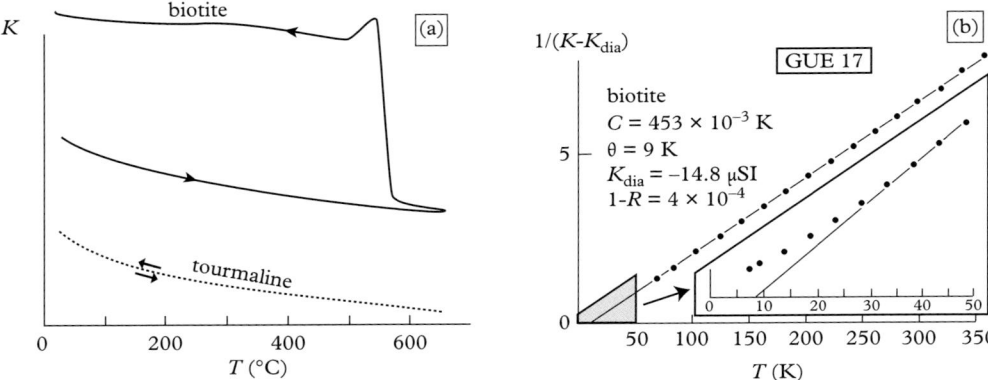

Figure 10.3 *Paramagnetic behaviour as a function of temperature. (a) Susceptibility decreases in a hyperbolic fashion with temperature, as exemplified by tourmaline from the Manaslu granite (Himalayas) and by biotite from a pegmatite. Note that in the case of biotite, fine-grained magnetite (supported by the susceptibility peak at the Curie temperature, 560–580 °C during cooling) grew during heating above 600 °C (see also Fig. 10.4a). (b) At low temperatures, the magnetic susceptibility (measured down to 6 K) does not follow the Curie–Weiss law (straight line) below 40 K (see enlarged zone). Maximizing the correlation coefficient (R) for the fitting line representing the data above 50 K helps to calculate the values of C, K_{dia} and θ (biotite from the Guéret granite massif, Massif Central, France. After Jover et al. (1989).*

Box 10.1 The ferromagnetic behaviour of magnetite

Magnetite (Fe_3O_4) is the main ferromagnetic mineral present in granites, with susceptibilities between 1 and 3 SI units. In the absence of an applied field, interactions between magnetic moments are responsible for a spontaneous magnetization, called remanent magnetization (Fig. 10.4b: M_r). Around the Curie temperature, $T_C \sim 580$ °C for pure magnetite, thermal motion is responsible for a drastic decrease in susceptibility which then disappears along with the remanence (Fig. 10.4b). Another specific property of ferromagnetism is given by magnetization reaching saturation, hence cancelling the susceptibility (Fig. 10.2b) at high applied field strength (0.2–0.4 T for magnetite). Variation of the applied field between $-H$ and $+H$ results in magnetization that follows a hysteresis loop, a central property of the ferromagnetic behaviour (Fig. 10.2b). Hysteresis loops have various shapes whose characteristics (M_r/M_s and H_{cr}/H_c ratios) are used for magnetic mineralogy and magnetic grain-size determinations (Fig. 10.4c). In granites, primary magnetite is coarse-grained, made of several magnetic domains (multidomain magnetite) contrary to fine-grained magnetite (a few microns in size) that is present in sedimentary rocks, particularly magnetite that is secreted by bacteria.

continued

Box 10.1 *continued*

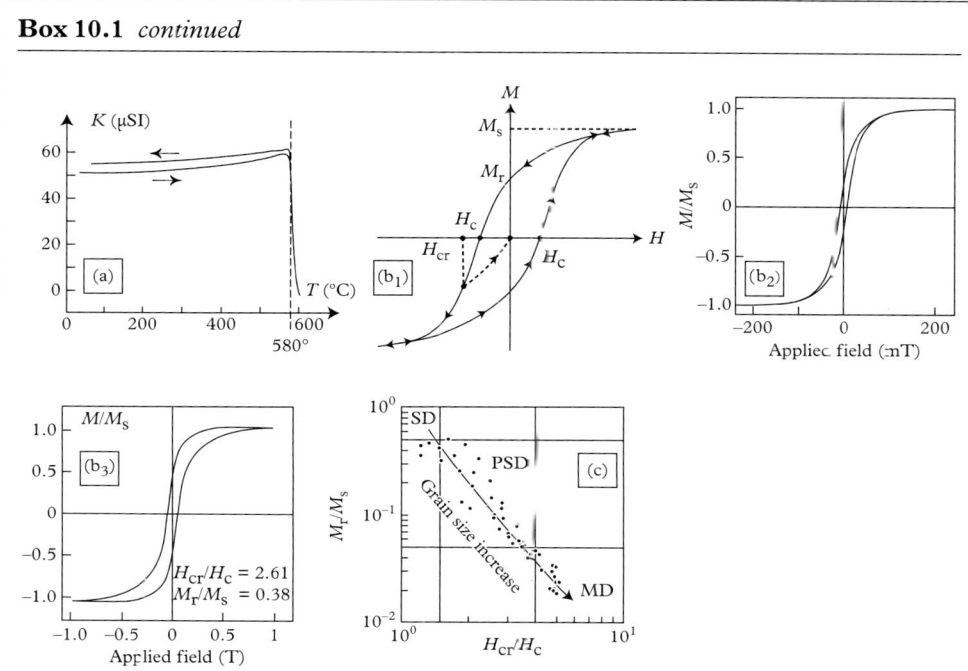

Figure 10.4 *Ferromagnetic behaviour. (a) Susceptibility versus temperature plot: during cooling, non-perfect reversibility is due to the formation of new ferromagnetic species resulting from oxidation by heating in air (magnetite from biotite breakdown: see Fig. 10.3a). (b) Hysteresis loop: (b_1) definition of M_s, M_r, H_c and H_{cr}, the coercivity of remanence, or magnitude of the magnetic field to be applied to return to zero magnetization (dashed lines); (b_2) typical loop for a multidomain magnetite: very weak coercivity; (b_3) 'wasp-waisted' hysteresis loop, characterized by variable slope during magnetization and variable width of the loop, attributed to a mixture of magnetite and haematite (Tana granite, Corsica); both H_{cr}/H_c and M_r/M_s ratios are no longer typical of magnetite alone due to the presence of haematite (Trindade et al., 2001). (c) Day plot (Day et al., 1977) showing the sorting of magnetite into monodomain or single domain (SD), pseudo-monodomain (PSD) and multidomain (MD) grains. SD grains (one magnetic domain per grain) are very small (0.1–10 μm according to grain shape ratio) with $M_r \approx M_s$ and $H_{cr} \approx H_c$ in these grains. MD grains are made of several magnetic domains, and are much larger and more common in granitic rocks (Butler, 1992).*

Ferromagnetism (see Box 10.1) is characterized by a high susceptibility at room temperature and at a low magnetic field (<1 T), the magnetic moments being parallel to the applied field. Ferrimagnetism appears when the two sublattices of a mineral acquire unequal magnetization. This leads to behaviour slightly different from ferromagnetism, particularly for the variation of magnetization with temperature.

190 *Magnetic fabrics in granites*

Antiferromagnetism corresponds to a weak ferromagnetism up to the Néel temperature (T_N) above which the material becomes paramagnetic (low and positive susceptibility). In granites, the common antiferromagnetic minerals are ilmenite (FeTiO$_3$; $T_N \sim 56$ K), haematite (α-Fe$_2$O$_3$; $T_N \sim 950$ K) and goethite (α-FeOOH; $T_N \sim 395$ K), all having very low susceptibilities at room temperature and at low field strength.

10.2 Magnetic susceptibility of granitic rocks

Magnetic properties of a granite result from the summation of the magnetic properties of its constituent minerals. This cumulative property is valid for the magnetic susceptibility $K_{total} = K_{dia} + K_{para} + K_{antiferro} + K_{ferro}$. Magnetic susceptibilities are measured at room temperature and at low induced field ($H < 1$ mT). Since the K_{dia} and $K_{antiferro}$ contributions are very weak (see above), the main carriers for magnetic susceptibility and anisotropy in granites are the iron-bearing silicates (paramagnetic) and magnetite (ferromagnetic). Paramagnetic minerals having much lower 'intrinsic' susceptibilities than magnetite (Fig. 10.5), the 'bulk' susceptibility of a granite greatly differs according to the presence or absence of magnetite. This significance was recognized by Ishihara (1977), who classified granitic rocks into 'ilmenite series' (dominantly paramagnetic) and 'magnetite series' (dominantly ferromagnetic). Such a bimodal sorting of the susceptibility reflects the existence of an oxygen fugacity threshold above which magnetite easily precipitates in the melt (Fig. 10.6a).

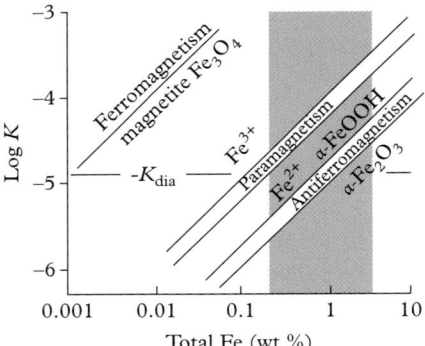

Figure 10.5 *Magnetic susceptibility of rocks as a function of their iron content (wt %) for different mineralogies and for a density of 2.65. Typical granite iron contents belong to the shaded zone. The K_{dia} straight line represents the absolute value of the diamagnetic contribution. After Rochette (1987).*

10.2.1 Paramagnetic granite and ferromagnetic granite

A granite is said to be paramagnetic when its magnetic susceptibility is carried almost exclusively by the iron-bearing silicates. In summary, when magnetite is absent. In this case, the total susceptibility of granites is close to the value of K_{para}, itself directly related to the content in total iron in granite according to the Curie–Weiss law (at a given temperature): $K_{para}(\mu SI) = K_{dia} + d(25.2t + 33.4t')$, where d is the rock density and t and t', respectively, are the contents (wt %) in Fe^{2+} and Fe^{3+} (Rochette et al., 1992).

Hence, without magnetite, leucogranites (i.e granites poor in biotite) do not exceed 100 μSI, and tonalites, rich in biotite and hornblende, rarely exceed 400 μSI in

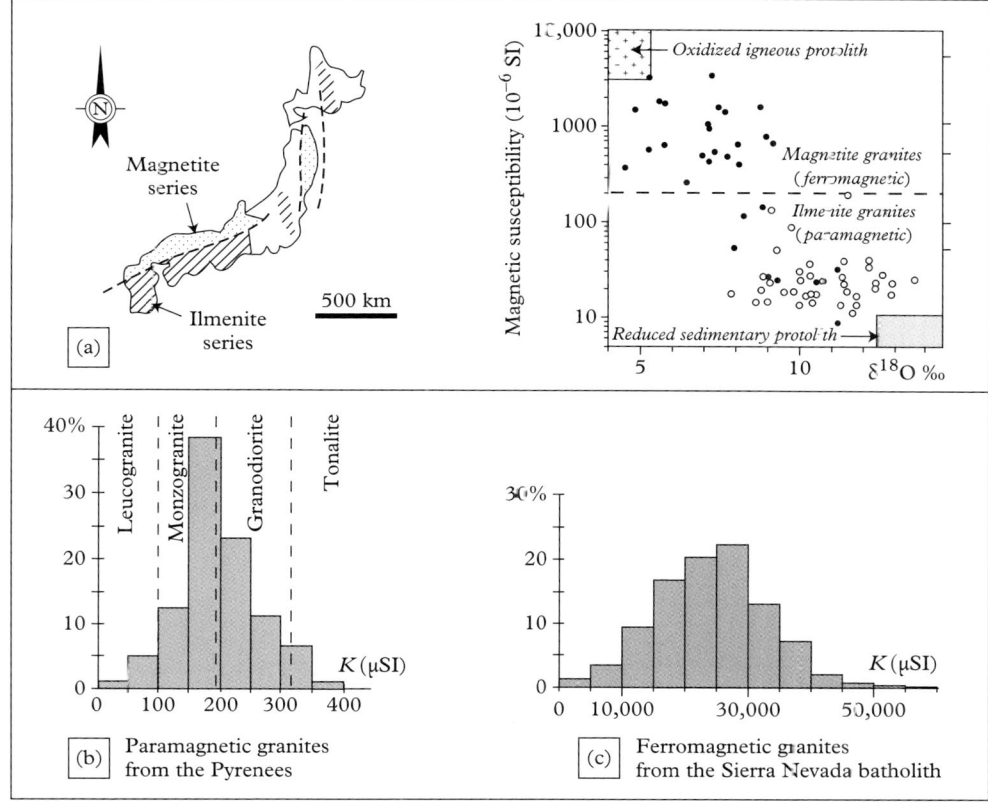

Figure 10.6 *Magnetic susceptibility of granites. (a) Susceptibility zoning of granites from Japan, a map, which supported Ishihara's (1977) classification into 'ilmenite series' and 'magnetite series', and relationship between susceptibility and nature of the protolith based on $\delta^{18}O‰$. (b) The magnetite-free (paramagnetic) Mont-Louis-Andorra granite pluton (Pyrenees; 254 stations of susceptibility measurements) and corresponding petrographic types. (c) Magnetite-bearing granites (ferromagnetic) of the Sierra Nevada (California; 359 measurement stations). Note the two orders of magnitude difference in susceptibility for similar iron contents.*

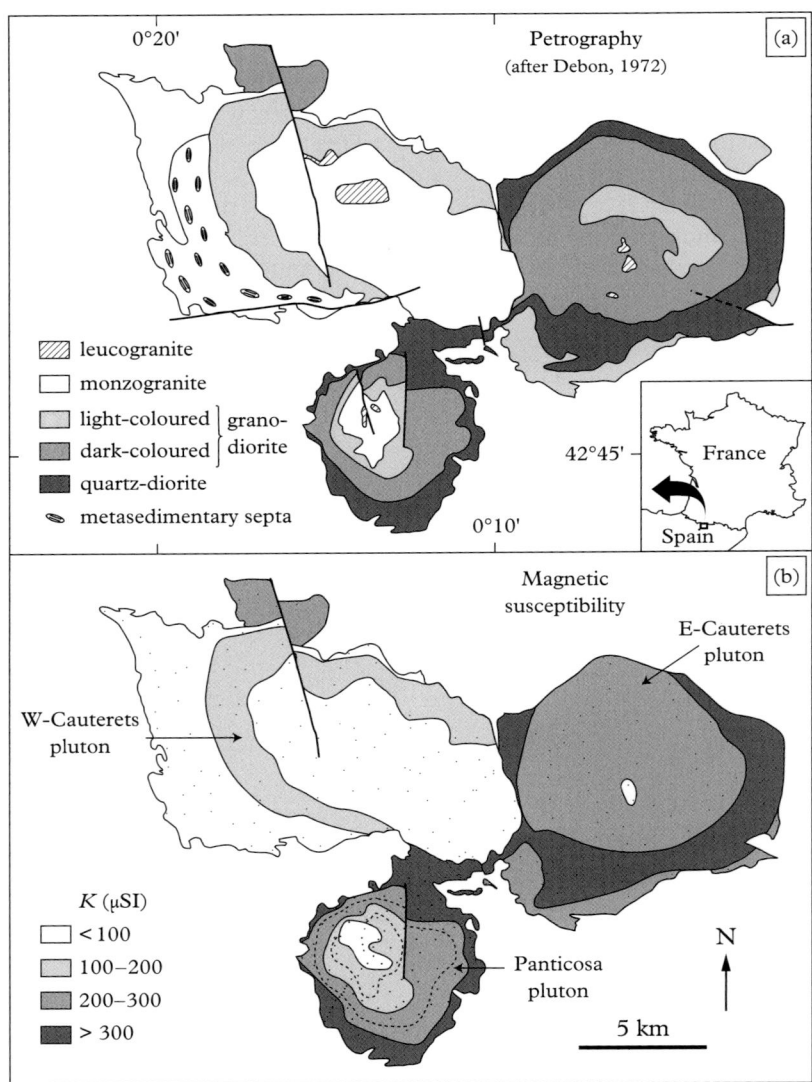

Figure 10.7 *The paramagnetic granite complex of Cauterets–Panticosa (French and Spanish Pyrenees). Comparison between (a) Debon's map (1972), representing the various petrographic types, and (b) the map of Gleizes et al. (1998) representing the iso-susceptibility contour lines obtained from 310 measurement stations, here represented by dots. In the Panticosa pluton, the dotted contours represent 150 and 250 μSI.*

susceptibility (Fig. 10.6b). Therefore, in the example from the Cauterets–Panticosa massif (Fig. 10.7), magnetic susceptibility measurements are used to delineate the petrographic types of paramagnetic granite plutons on the basis of their iron-bearing silicates. Owing to the intrinsic susceptibility of magnetite, more than two orders of magnitude higher for the same iron content than for iron-bearing silicates (Fig. 10.5), a few magnetite grains are enough to confer a high susceptibility on a granite. When the 'ferro' contribution dominates, the granite is said to be 'ferromagnetic'. Magnetic susceptibilities of such granites, on the order of 10^{-2} SI, depend on their content in magnetite (Fig. 10.8). In addition to susceptibility contrasts between 'para' and 'ferro' granites, contrasts in natural remanence (NRM) due to the presence of magnetite (see Box 10.1) are favourable factors to perform airborne surveys (see Fig. 11.4).

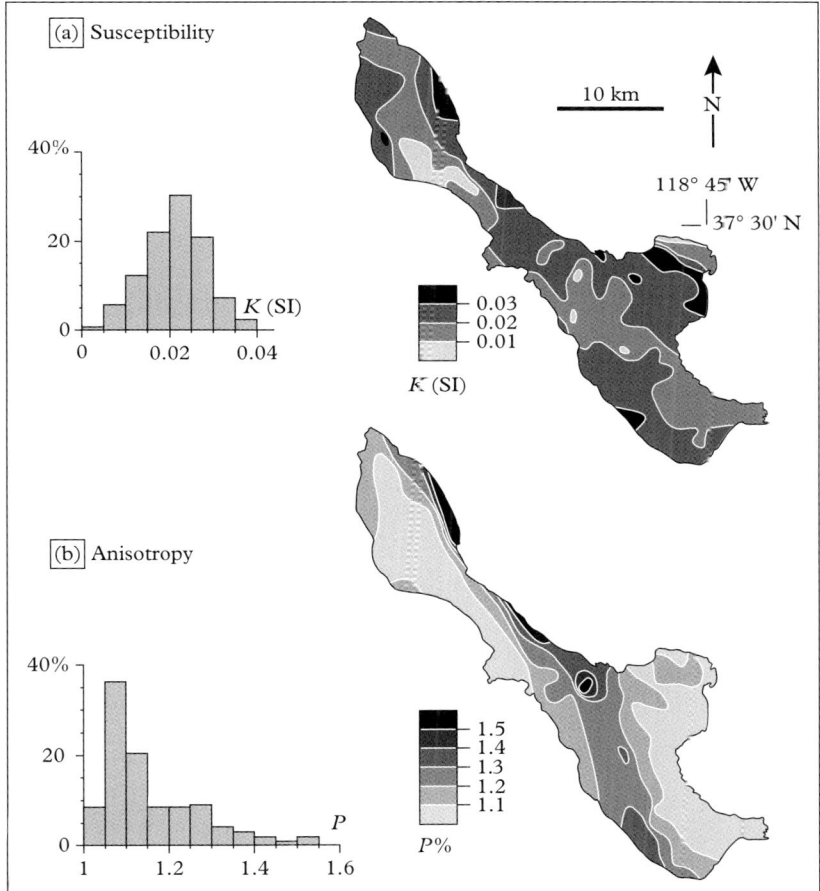

Figure 10.8 *The ferromagnetic pluton at Mono Creek (California). (a) Magnetic susceptibility (K). (b) Anisotropy degree (P). Maps and frequency histograms obtained from 183 regularly spaced sampling sites. From Saint-Blanquat and Tikoff (1997).*

10.2.2 The AMS ellipsoid

If the magnetic susceptibility of a rock does not depend on its orientation with respect to the inducing field, K is a number (scalar). Otherwise, in the situation of an anisotropic material, magnetic susceptibility is at a first approximation a second rank tensor, K_{ij}, relating H (the inducing field) to M (the magnetization). The geometrical representation of such a tensor is an ellipsoid, called 'the AMS ellipsoid'. Mathematically, this object is similar to the stress or strain ellipsoids, made up of three principal axes $K_1 \geq K_2 \geq K_3$ (or K_{max}, K_{int} and K_{min}), which are the eigenvectors of the (3×3) matrix representing the tensor, each having a direction and a modulus. The K_1 axis represents the magnetic lineation, and the K_3 axis the perpendicular to the magnetic foliation plane. The moduli of K_1, K_2 and K_3 give, respectively, the maximum, intermediate and minimum susceptibilities of the tested specimen, and their average value, $K = 1/3(K_1 + K_2 + K_3)$, is the total susceptibility, also called bulk susceptibility.

Ratios between moduli provide a number of magnetic 'parameters':

- the anisotropy degree: $P = K_1/K_3$, or anisotropy percentage $P\% = 100(P - 1)$;
- the linear anisotropy: $L = K_1/K_2$ (or $L\% = 100(L - 1)$);
- the planar anisotropy: $F = K_2/K_3$ (or $F\% = 100(F - 1)$);
- the Flinn parameter: $P_{Flinn} = (L - 1)/(F - 1)$, a rarely used shape parameter due to its variation from 0 to 1 for planar ellipsoids, and from 1 to infinity for linear ones;
- the parameter of Jelinek (1981), which is preferred (Fig.10.9): $T = [\ln(F) - \ln(L)]/[\ln(F) + \ln(L)]$, which varies from $T = +1$ (pure planar, or oblate) to $T = -1$ (pure linear, or prolate), through $T = 0$ (triaxial ellipsoid).

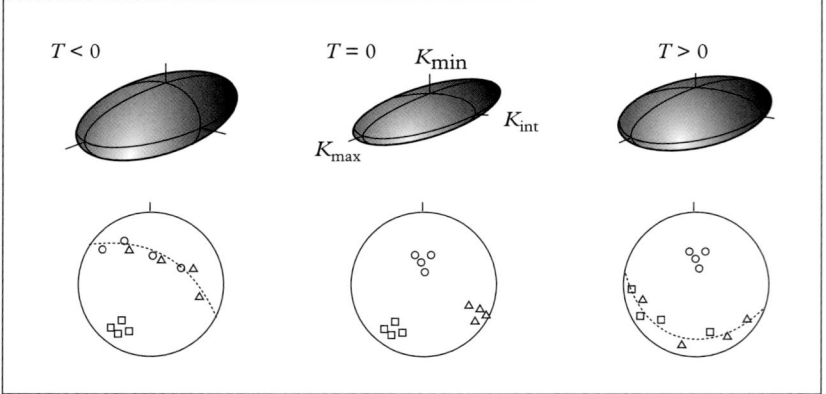

Figure 10.9 *From elongate (T < 0) to oblate (T > 0) ellipsoids, via the triaxial ellipsoid (T = 0), and corresponding stereo plots of their axes (K_1: squares; K_3: circles) T: Jelinek's shape parameter varying from −1 to +1.*

10.3 Magnetic fabrics in granites

10.3.1 Fabrics in paramagnetic granites

The magnetic susceptibility axes of phyllosilicates (including biotite) are parallel to their crystallographic axes. The minimum axis of the individual grain (k_3) is perpendicular to the cleavage plane (001) of the phyllosilicate and, in the (001) cleavage plane, the intermediate and maximum axes are almost identical ($k_2 \approx k_1$). The (intrinsic) magneto-crystalline anisotropy degree of a crystal of biotite is $k_1/k_3 \approx 1.3$. Therefore, in a granite containing only biotite as the iron-bearing phase (the simplest case of a paramagnetic granite), K_3 is perpendicular to the magnetic foliation plane marked by biotite crystals (or biotite subfabric). The magnitude of K_3 is the vectorial sum of all the individual k_3 (Fig. 10.10a). The anisotropy degree, or fabric strength, would have the maximum value of $1.3 = P = K_1/K_3$ (or 30% anisotropy percentage) if all biotite cleavage flakes were perfectly parallel to each other.

In summary, biotite is an excellent marker of the foliation as it may be observed in the field by geologists. Owing to its magneto-crystalline anisotropy, which mimics its shape anisotropy, the foliation derived from AMS measurements in biotite granites is also a perfect marker of their magmatic foliation. The long (K_1) axis of the magnetic ellipsoid, also called magnetic lineation, represents the 'zone axis', or preferred axis around which the biotite cleavages rotate (see Section 9.2).

Therefore, magnetic fabrics of biotite granites give a perfect image of the crystalline fabrics of biotite, at least in orientation. In intensity, the anisotropy percentage varies between 0% for a random orientation distribution, to about 30% for a perfect planar orientation of all biotite crystals. In practice, the crystalline fabrics observed in granites correspond to anisotropies of a few per cent, rarely larger than 10%, as illustrated by the frequency histogram of the anisotropy degrees in granites from the Pyrenees (Fig. 10.11a).

Note that, in case of very low anisotropies, $P_{para} = (K_1 - K_{dia})/(K_3 - K_{dia})$ is used instead of $P = K_1/K_3$, K_{dia} (≈ -14 μSI) being the ubiquitous diamagnetic contribution considered as isotropic, the way to avoid artificially high anisotropy degrees when K_3 tends to zero.

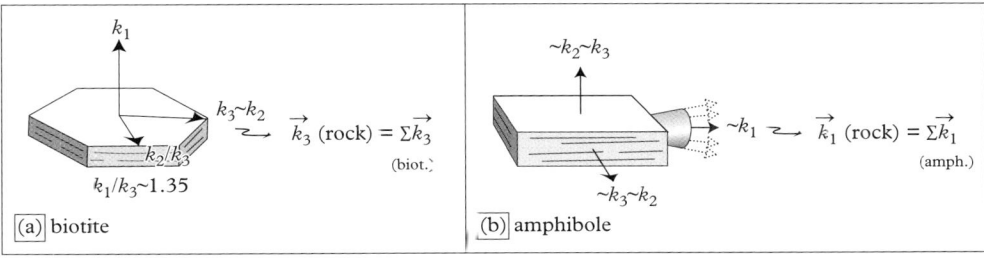

Figure 10.10 *Magnetocrystalline origin of paramagnetic granite fabrics. (a) Case of a biotite-granite. (b) Case of an amphibole-granite.*

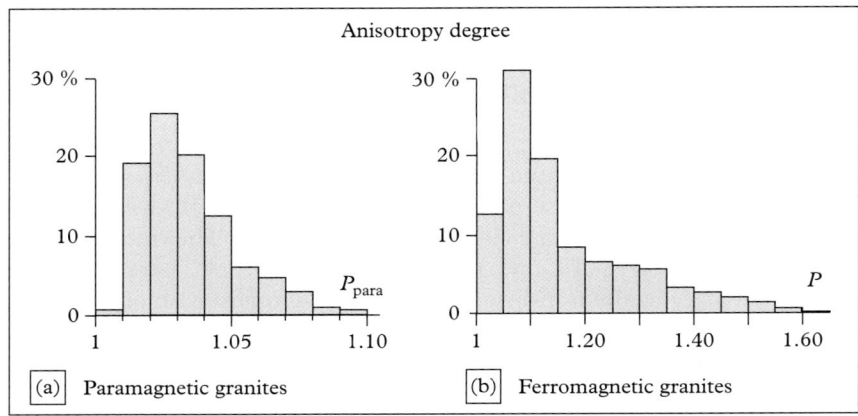

Figure 10.11 *Degrees of anisotropy in granites. (a) Paramagnetic granites (>1000 different sites in the Pyrenees; from Gleizes, 1992). (b) Ferromagnetic granites (400 different sites from the Sierra Nevada batholith; in Bouchez (2000), from Saint-Blanquat (pers. com.).*

10.3.2 Fabrics in ferromagnetic granites

In ferromagnetic granites, primary, i.e. magmatic, magnetite is close to the pure iron end-member (very poor in titanium) and dominates the iron-bearing silicates in both susceptibility and anisotropy. The magnetic anisotropy of magnetite is dominated by its shape, the magnetization being maximum parallel to the long axis of the grain, along which the demagnetizing field is minimal (Stacey and Banerjee, 1974). Therefore, the anisotropy carried by magnetite reflects the shape of this mineral rather than its magneto-crystalline anisotropy, which is negligible. Indeed, from a syenite rich in coarse-grained magnetite (>100 μm), the shape anisotropy of magnetite, obtained from digital image analyses (see Section 9.4), was shown to be identical to the rock magnetic fabric ellipsoid, both in orientation and intensity (Fig. 10.12).

The intensity of the anisotropy in ferromagnetic granites varies from isotropic ($P = 1$) for random orientation distribution of magnetite, to values that can be 'very' high since nothing prevents magnetite being both elongate and well orientated. In practice, the degree of anisotropy may exceed 1.5, as illustrated in granites from the Sierra Nevada batholith, California (Figs. 10.8b and 10.11b).

Since magnetic interaction is a law acting as $1/d^3$, magnetite grains behave as small magnets that interact as soon as their mutual distance (d) is smaller than the grain size. In this case, interactive anisotropy may exaggerate or reduce the anisotropy due to the shape fabric of the magnetite grains or clusters of grains (Fig. 10.12). This effect may be important, particularly in basalts (Hargraves et al., 1991), where densely distributed magnetite grains of small size increase their chance of forming clusters, or groups of grains, subjected to magnetic interaction. In granitic rocks, the rather low density in the

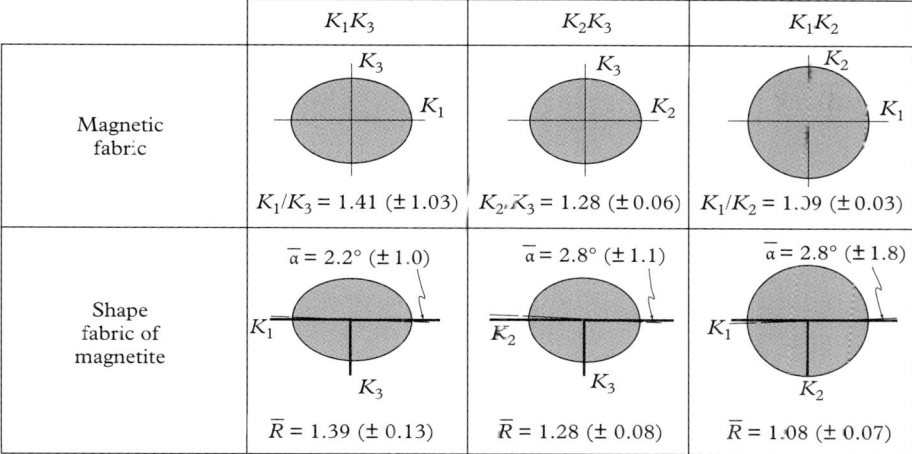

Figure 10.12 *The AMS ellipsoid of a syenite from Madagascar is similar to the shape fabric ellipses for magnetite. Top line: the three principal magnetic fabric ellipses: K_1K_3, K_2K_3 and K_1K_2. Bottom line: shape fabrics of magnetite obtained through digital image analyses of 200 grains (~0.75 mm mean size) in thin sections cut along the three magnetic reference planes. Average ratios (R) and angular difference (α) between axes are mentioned. In Grégoire et al. (1998).*

distribution of magnetite grains, as well as their large size, make them less likely to have magnetic interactions (Fig. 10.13).

Note that ferromagnetic grains of very small size (<<10 μm), made of a single magnetic domain (SD grains), due to the around-grain distribution of the electric charge, have their maximum susceptibility parallel to their short axis. Thus, 'inverse' magnetic fabrics are made possible (Stephenson, 1994). In granites, such very fine magnetite grains are seldom encountered, making their eventual contribution to the rock anisotropy negligible.

10.3.3 Fabrics with mixed magnetic mineralogy

Such fabrics are encountered when several magnetic species are present such that one species does not supercede the others. The resulting magnetic fabric is the addition of at least two subfabrics. Separating the respective fabrics may become an important concern.

In a granite, a mixed para/ferro fabric type appears when tiny quantities of magnetite are added to the iron-bearing silicate(s), making the rock sample a ferromagnetic type. By separately analysing the shape fabrics of biotite and magnetite using the intercept technique (Section 9.4), Archanjo et al. (1995) have shown that the fabric ellipses are coaxial, i.e. having almost identical orientations of their principal axes. The fabric ellipse axial ratios, however, differ greatly due to the different habits of these species: e.g. biotite shows elongate sections but magnetite is stubby and frequently forms anisometric

198 *Magnetic fabrics in granites*

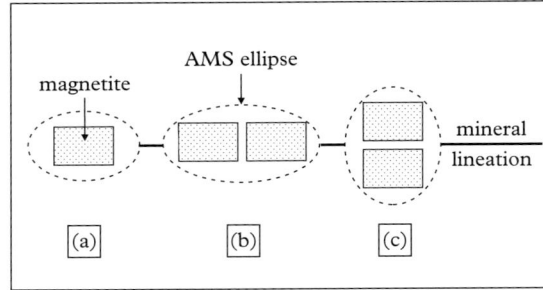

Figure 10.13 *Magnetic interactions between magnetite grains. (a) Isolated magnetite grain and corresponding AMS ellipse. (b, c) Interacting magnetite; the aligned configuration (b) parallel to grain elongation (mineral lineation) increases the magnetic anisotropy (as well as the susceptibility). On the contrary, the adjacent configuration (c) lowers and may even reverse the resulting anisotropy. After Bouchez (2000).*

sections. This coaxial feature is attributed to magnetite grains imitating the shape fabrics of other crystalline phases, through preferred collage, or crystallization of magnetite along grain boundaries, or exsolution of magnetite along biotite or amphibole cleavage planes.

A lack of correlation between para- and ferromagnetic fabrics would require magnetic susceptibility measurements under high applied fields in order to saturate the ferromagnetic fraction. The resulting cancellation of K_{ferro} would isolate the susceptibility of the paramagnetic fraction. This technique would need large (high-field) magnets and is difficult to implement for large specimens, hence is not normally used. Conversely, the anisotropy measurement of the remanent magnetization of the ferromagnetic fraction, AAR (see Section 10.3.5), helps to isolate the ferromagnetic fabric. Note that the AAR ellipsoid can be approximated from measurements of the magnetization vectors (see Louis et al., 2004). However, the AAR ellipsoid, which is not a true ellipsoid, is not equivalent to the AMS ellipsoid, particularly in the magnitude of its principal axes. This makes the calculation of the paramagnetic AMS ellipsoid difficult from both bulk-rock AMS and AAR measurements.

A biotite + amphibole mixture is the most frequent case in granites, resulting in mixed para/para fabrics. Biotite granites have been discussed earlier. Hornblende in granites forms elongate prisms. As such, it is a good marker for magmatic lineation. Due to its monoclinic symmetry, the long axis of its magneto-crystalline ellipsoid makes an angle of 10°–15° with its elongate axis. However, when averaged over a whole population, the long axis of the susceptibility ellipsoid (K_1) is expected to represent the average orientation of amphibole, i.e. the lineation (Fig. 10.10b). Subfabrics of biotite and amphibole are considered as coaxial, making the bulk magnetic fabric well-defined both in foliation (plane ⊥ K_3) and lineation (K_1), but not in intensity, which depends on the intrinsic anisotropy and respective amounts of each mineral species.

Tourmaline or cordierite, with their 'inverse' magneto-crystalline paramagnetic anisotropy (k_3 parallel to the prism axis), may interfere with the traditional anisotropy markers (biotite, amphibole, iron-bearing muscovite, . . .) if present in substantial amounts. This is the case for tourmaline and muscovite leucogranites from the Himalayas that show typically inverse magnetic fabrics due to the dominance of tourmaline (Fig. 10.14). In favourable situations, such as the Carnmenellis granite in Cornwall, the signal from tourmaline can be removed by heating the specimens up to the breakdown of biotite (~620 °C) which become partly replaced by mimetic magnetite (Fig. 10.7).

10.3.4 AMS in practice

Rock specimens are collected in the field with a portable drill (Fig. 10.15a). The collected cores are one inch (~ 25 mm) in diameter, a standard for most susceptometer sample-holders. They are orientated *in situ* with respect to the geographical framework using the instrument shown in Fig. 10.15b, by noting the azimuth and plunge of the core-axis measured with a compass, taking care of the eventual magnetic interaction between the compass and the granites' natural magnetization. Precision of such orientation measurements may be better than ±2°. In the case of

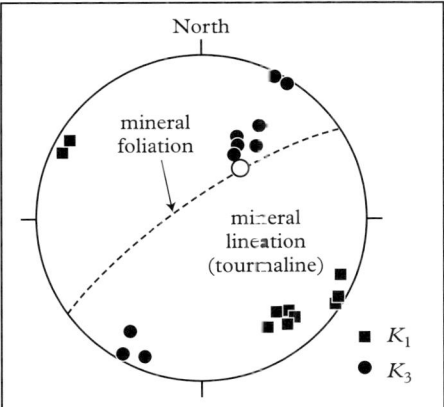

Figure 10.14 *Inverse magnetic fabric of the tourmaline-bearing leucogranite from the Himalayas. The short axis of the AMS ellipsoid (K_3) plots close to the tourmaline lineation as measured in the field, and the long axis (K_1) close to normal to the foliation. In Rochette et al. (1992).*

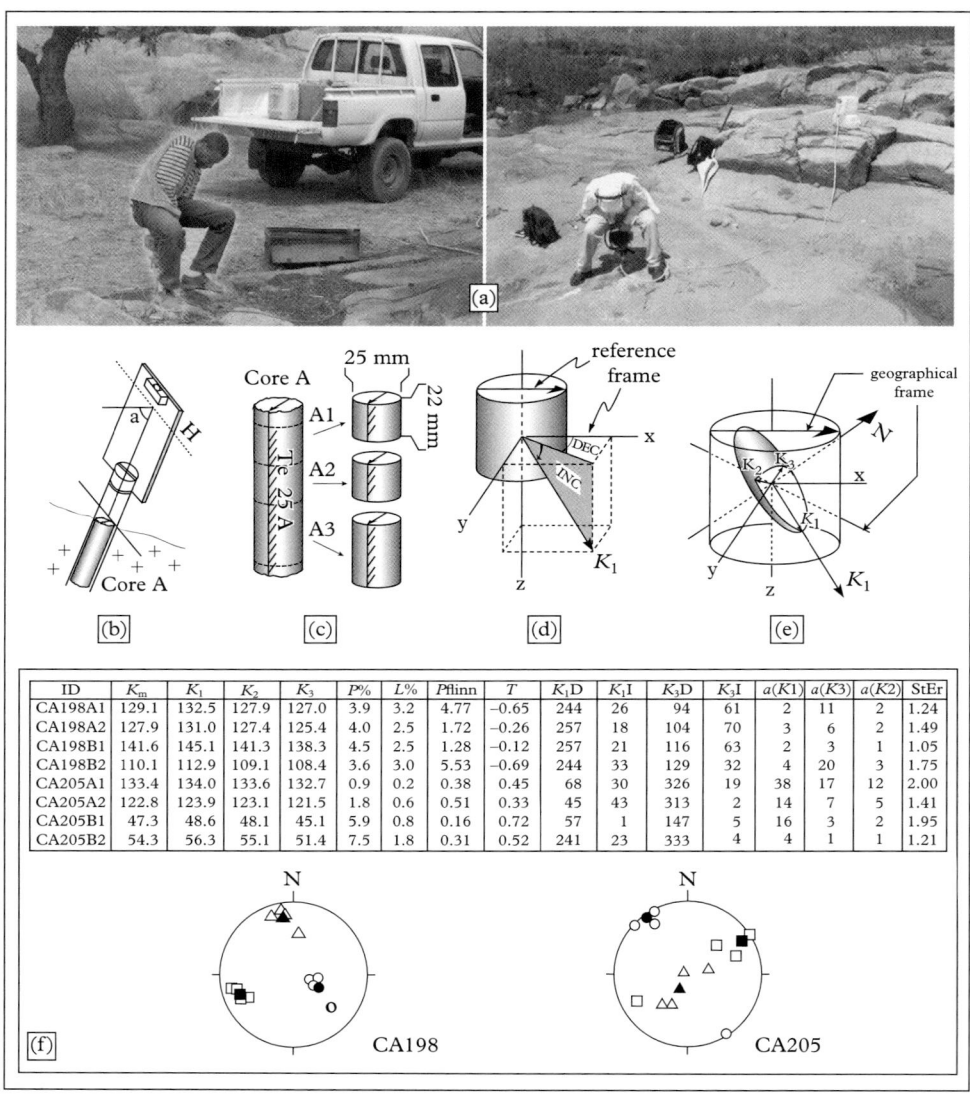

Figure 10.15 *Practical magnetic fabrics. (a) Sampling using a portable drill. (b) In-situ orientation of the rock core: H: horizontal line whose azimuth is at 90° to the vertical plane containing the core axis; a: core-axis plunging angle. (b–e) Orientating procedure and conventions. (c) The vertical plane containing the top arrow (azimuth) and the core axis is indicated. Then, the drill core is sectioned into 2 or 3 specimens. (d) Definitions of 'Dec' and 'Inc' in the reference framework of the specimen. (e) AMS ellipsoid in the geographical framework. (f) AMS data and corresponding stereo-plots for two sampling stations, 2 cores per station and 2 specimens per core (horizontal plane orientated due north (N); lower hemisphere, Schmidt net); D: azimuth; I: plunge; a: confidence angle; square: K_1; triangle: K_2; circle: K_3; open symbols: individual specimens; solid symbols: tensor averages.*

highly magnetic rocks, a solar needle used for palaeomagnetic studies, or alignment with a GPS, may be useful for determining the azimuth. In a given outcrop, two to three rock-cores are collected, each separated by a few metres. In a specific pluton, sampling locations are regularly spaced apart by one or two kilometres, depending on the project.

In the laboratory, the cores are sectioned into ~22 mm-long specimens, giving the best approximation of a sphere (for one-inch cylindrical sections). The off-cuts are used to make orientated thin sections. Four to six specimens per location are prepared. The AMS of each specimen is measured using a susceptometer. Most AMS studies use the KLY (Kappabridge) instrument manufactured by Agico (Czech Republic) which applies a low (0.1 mT) alternating (920 Hz) field to a coil in which a specimen is inserted. Susceptibility measurements are made by compensating the induced field using a bridge (zeroing technique), for successive positions of the specimen aided by an ingenious sample-holder that allows rotation of the specimen. The susceptibility tensor is calculated online with respect to the geographical coordinates using the *in-situ* orientation data of the core from which the specimen was cut (Fig. 10.15e). AMS data are presented in a file, and stereo-plots of the K_i axes, including the tensor averages, are automatically produced using (for example) the Agico software (Fig. 10.15f).

10.3.5 AAR fabrics

Anisotropy of the anhysteretic remanent magnetization (AAR), made popular by Jackson (1991), helps to isolate the fabric of the remanent fraction, hence the average shape ellipsoid of magnetite. This technique has been applied by Mamtani et al. (2011) to demonstrate that, in a magnetite-bearing granite pluton, the quality of the magnetic fabric is due to a better arrangement of the magnetite grains assisted by their rigid body rotations in the matrix, rather than to their hypothetic changes of their individual shapes. This makes the magnitude of the AAR fabric a significant marker of strain intensity.

The anhysteretic magnetization is obtained by (magnetically) 'shaking' the magnetic domains of a specimen subjected to a decreasing alternating field, from $\pm H_c$ to zero, while at the same time a DC-field (H_1) imparts a magnetization to the specimen (Fig. 10.16). This procedure needs the use of instruments that are common in palaeomagnetism (AC-demagnetizer and DC-magnetizer), aimed at magnetizing the specimen in the direction of the applied DC-field under the chosen coercivity 'window' ($\pm H_c$). The corresponding remanence vector is measured in a shielded room (under zero magnetic field). Replication of such measurements for different directions of the imparted field allows the development of the AAR 'ellipsoid' for all orientations in space. As already indicated in Section 10.3.3, it is a pseudo-ellipsoid, contrary to the AMS ellipsoid, which is built from scalars. In conclusion, both ellipsoids can be compared with respect to the orientations, but not to the magnitudes of their axes.

Total AAR is when $\pm H_{c1} = \pm 100$ mT, the usual maximum demagnetizing field. It is used to study the anisotropy of all ferromagnetic phases, i.e. when most coercive minerals

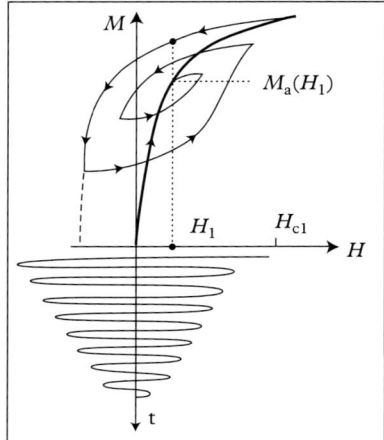

Figure 10.16 *Principle of the anhysteretic remanent magnetization and its anisotropy. The anhysteretic curve (= with no hysteresis and no inflexion point), always situated above the first magnetization curve, is characteristic of a specimen. Each point ($M_a(H_1)$) on this curve is obtained through imparting a direct field (H_1) to the specimen and, at the same time, an AC-demagnetization under an applied field decreasing from $\pm H_{c1}$ to zero. The anisotropy of anhysteretic remanence (AAR) consists of the measurement of $M_a(H_1)$ according to several directions in order to built up the AAR 'ellipsoid'. In practice, the total AAR is obtained with $\pm H_{c1} = \pm 100\ mT$, by (almost) saturating all the coercive phases. The partial AAR concerns only a chosen 'window' of coercivity, i.e. a family of magnetic grains. For example, AAR (0–50 mT) is obtained through imparting a magnetization in the centre of the 0–50 mT window during $\pm H_{c1} = \pm 50\ mT$ demagnetization, and the AAR (50–90 mT) is obtained by applying a magnetization in the middle of the 50–90 mT window during $\pm H_{c1} = \pm 90\ mT$ demagnetization, then by demagnetizing under $\pm H_{c1} = \pm 50\ mT$. After Trémolet de Lacheisserie (1999) and Trindade et al. (1999).*

are saturated. Partial AAR is used specifically to study the anisotropy of a chosen family of magnetic grains belonging to the $\pm|H_{c2} - H_{c1}|$ window of coercivity (Fig. 10.16).

- AAR and mixed para/para minerals. In a biotite + tourmaline (paramagnetic) granite, the magnetic signal of tourmaline may be reduced by heating the specimen at atmospheric pressure up to about 650 °C. This procedure develops a population of magnetite grains, as suggested by the $K(T)$ diagram in Fig. 10.3a. Comparison of the AMS and AAR fabrics before and after heating supports the concept that tourmaline is overwhelmed by magnetite which grew 'mimetically' after the biotite grains (Fig. 10.17). This mimetic fabric of magnetite is attributed to its epitaxial growth onto the cleavage planes of biotite.

- Partial AAR and mixed ferro/ferro minerals. Measurements of partial AAR fabrics may reveal eventual mixtures of ferromagnetic grains. In the Tourao granite (Brazil) coarse-grained magnetite forms the primary, magmatic fabric, that trends N–S to NNE–SSW, as suggested by the AMS (Fig. 10.18a). The partial AAR fabric

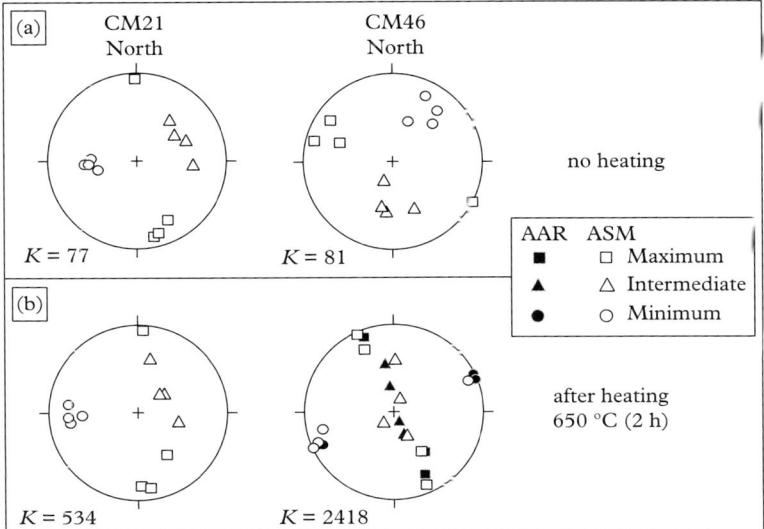

Figure 10.17 *Magnetic fabrics in the biotite ± tourmaline granite at Carnmenellis (Cornwall). K is the magnetic susceptibility in μSI. (a) AMS fabrics before heating two representative specimens: no remanence (= no magnetite). (b) Through in-air heating, magnetite develops at the expense of biotite, giving a progressive susceptibility increase (after 2 h, 650 °C). The AMS fabric diagram of specimen CM21 does not change: it is concluded that this specimen does not contain (much) tourmaline and that the fabric of magnetite (now dominant) mimics the fabric of biotite. By contrast, the AMS fabric of CM46 changes due to heating: it is concluded that tourmaline disturbed the fabric of biotite before heating; the (almost) similar AMS and AAR fabrics after heating suggest that magnetite is responsible for the AMS fabric; accepting that magnetite mimics biotite, this fabric represents the subfabric of biotite. After Mintsa et al. (2002).*

(AAR 50–90 mT) is revealed to be close to E–W in trend (Fig. 10.18c), which is precisely the orientation of the microfractures of hydrothermal origin, in which very fine and coercive magnetite grains have precipitated.

10.4 The conspicuous structural homogeneity of granites: examples and implications

The high level of coherence of magnetic fabrics, particularly in granitic rocks, stimulated AMS studies worldwide. Coherence is particularly exemplified by the magnetic lineation maps that can be found in this book. The first example shows that lineations are the best markers of stretching undergone by granitic magma during its emplacement. The other

204 Magnetic fabrics in granites

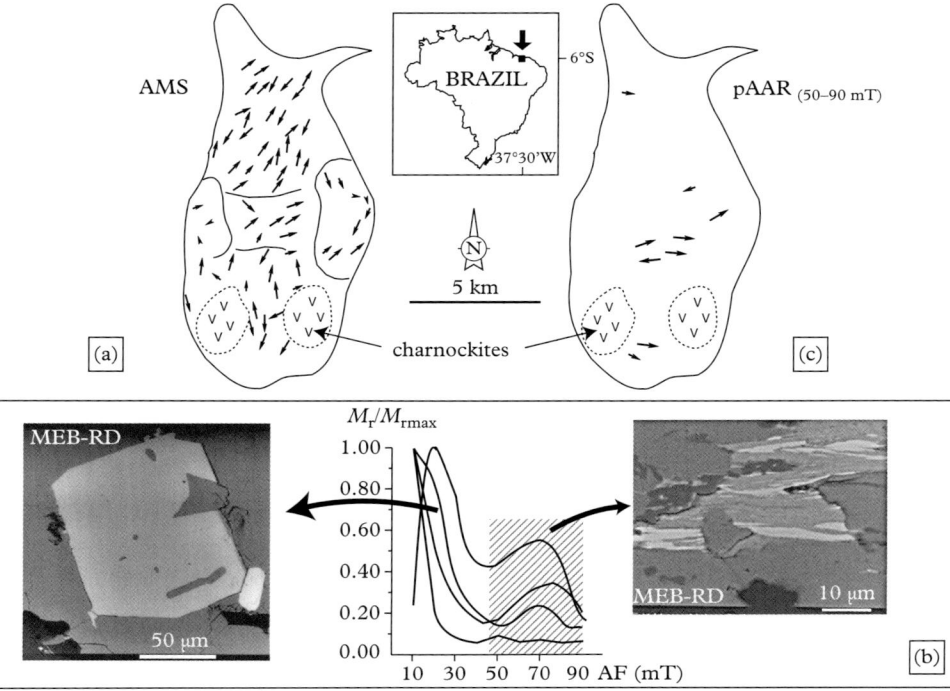

Figure 10.18 *Comparing AMS and AAR fabrics in the Tourao granite pluton (Brazil). This pluton is a Braziliano (592 Ma) porphyritic biotite + magnetite monzogranite from the north-east, intrusive into a Palaeo- to Neoproterozoic basement (migmatites and metasediments). Two small stocks of charnockites (540 Ma) intrude the Tourao pluton to the south. (a) AMS lineation map. (b) Coercivity spectra of magnetite (remanent magnetization measured as a function of progressive AC-demagnetization, and normalized to the maximum remanent magnetization) showing some specimens displaying two families of magnetite: one, with large grains and low coercivity, is primary; the other, secondary and highy coercive, forms inclusions along biotite cleavage planes. (c) Lineation map of the pAAR (50–90 mT) vectors corresponding to the secondary magnetite. After Trindade et al. (1999).*

examples will demonstrate the benefits of a regional approach using magnetic fabrics, particularly for basement geology.

10.4.1 Sidobre granite (Massif Central, France)

To the north of Castres city (43°36′ N/02°14′ E), northwest of the Montagne Noire, France (Fig. 6.14), the Sidobre granite pluton was emplaced 305 Myrs ago into Cambrian schists. The pluton has a weak petrographic zonation and is slightly richer in biotite at its centre (granodiorite) than at its borders (monzogranite). Its microstructures are almost entirely magmatic except at its periphery where an incipient solid-state

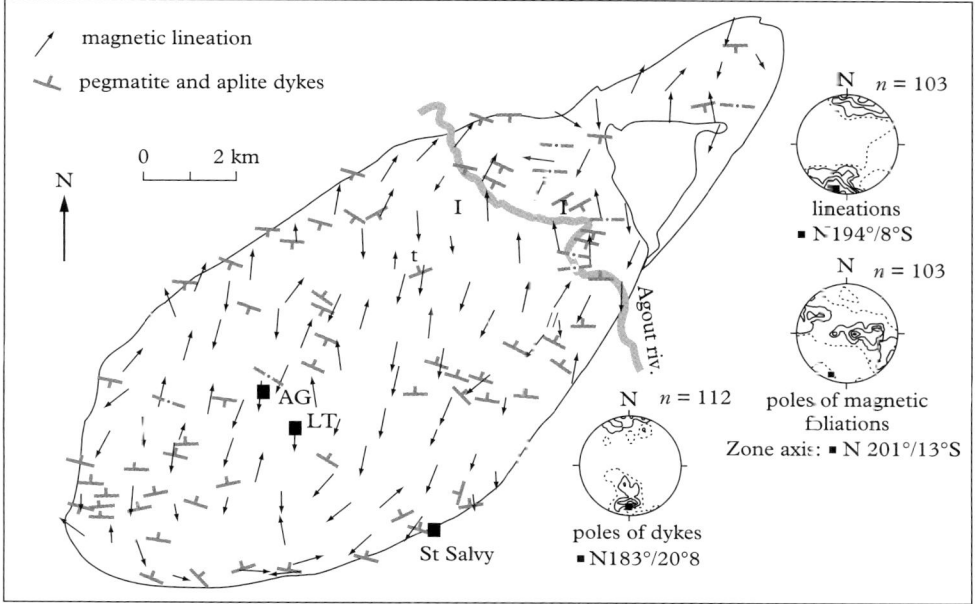

Figure 10.19 *The Sidobre granite pluton (Massif Central; see Fig. 6.14 for location). Map showing magnetic lineations and aplite-pegmatite dykes. Stereo-plots (lower hemisphere): lineations, foliation poles and dykes. After Darrozes et al. (1994); dykes are from Borrel (1978).*

deformation is observed, marked by chess-board substructures in quartz and flexured biotite. Absence of magnetite in the pluton results in the magnetic susceptibility not exceeding 200 µSI. AMS measurements, collected from 103 different locations, give a very homogeneous pattern of the internal structure of the pluton, particularly exemplified by the lineation map in Fig. 10.19. Shallow plunging to the NNE or SSW, the lineations are oblique to the elongation of the pluton, except along the margins where they tend to be parallel.

The main lesson from this example comes from the comparison between the lineation map of Darrozes et al. (1994) and Borrel's map (1978) reporting the azimuths of the aplite and pegmatite dykes. These dykes, a few centimetres in width, correspond to late magmatic fractures that collected the ultimate magmatic melt (aplite), more-or-less associated with the fluid fraction (pegmatite) released just before final crystallization of the pluton. The direction perpendicular to these dykes therefore indicates the orientation of the extension (or the extensive stress direction) that prevailed just before complete crystallization. The approximate parallelism observed between the poles of these aplo-pegmatitic dykes (average orientation: N183°/20°S) and the lineations (average orientation: N193°/8°S) demonstrate, as already discussed in Section 9.2.4, that the magmatic lineations represent the direction of the extensional deformation to which the magma was subjected just before complete crystallization. This study also shows that the foliations,

206 *Magnetic fabrics in granites*

organized more-or-less concentrically, 'rotate' around the average lineation direction, as confirmed by the best fit axis of the foliation poles at N201°/13°S close to the average lineation direction (Fig. 10.19).

The emplacement of the Sidobre pluton, with its magmatic extension oblique to pluton elongation, itself parallel to the regional structures, is interpreted as resulting from a regional transtensional dextral shear along a direction close to the pluton's long-axis. Such a conclusion is compatible with the late-orogenic extensional tectonic models for the French Massif Central developed by van den Driessche and Brun (1989) and Echtler and Malavieille (1990).

10.4.2 Tesnou granitic complex (Hoggar, Algeria)

Lying 250 km to the NW of Tamanrasset, and 25 km to the west of the 4°50'E lineament separating Central Hoggar from the Pharusian chain, this complex consists of a series of coalescent monzogranite to syenogranite intrusions over an area of about 500 km^2 along a dextral late-Pan-African shear zone (~550 Ma; Fig. 10.20a). Magnetic susceptibilities, not exceeding 200 µSI, are characteristic of paramagnetic granites whose magnetic fabrics are dictated by biotite, except in the north-east (T-in-Akkor lobe: Fig. 10.20b) where susceptibilities reach 10^3 µSI. Microstructures are magmatic, except along the western margin of the complex and the N–S trending dextral shear zone where the granite was slightly foliated by the end of the emplacement event (Fig. 10.20b). As in the previous example, the following can be observed: (1) similar orientation (NW–SE) of almost all lineations, oblique with respect to the elongation of the pluton (Fig. 10.20c); (2) late-magmatic dykes (aplites) perpendicular to the lineations (here NE–SW: Fig. 10.20e); and (3) magmatic foliations rotating around the average lineation direction (Fig. 10.20d).

In this example, the high angle between the average lineation direction and pluton elongation is close to the incremental extensional direction. This suggests that the magmatic fabric has recorded only a fraction of the total deformation undergone by the magmatic intrusion. Considering a homogeneous deformation, Djouadi et al. (1997) calculated that each magma reservoir underwent a shear strain $\gamma = 1.2 \pm 0.2$ that led to a relative displacement of approximately 15 km between opposite walls (Fig. 10.20e). In other words, the magma reservoirs were deforming during shearing, defining a several kilometres-wide shear zone. These epizonal granites rapidly crystallized. Shearing continued after crystallization leading to strain localization into narrow shear zones along the contacts with the country rocks.

10.4.3 Mono Creek granite pluton (California)

Part of the eastern margin of the Sierra Nevada batholith (California), this pluton is a classic example of regional scale structural homogeneity (Fig. 10.21). The pluton consists of a homogeneous, magnetite-rich, porphyritic monzogranite (Fig. 10.8) occupying approximately 600 km^2, emplaced in the Cretaceous period, 86 Myrs ago, during

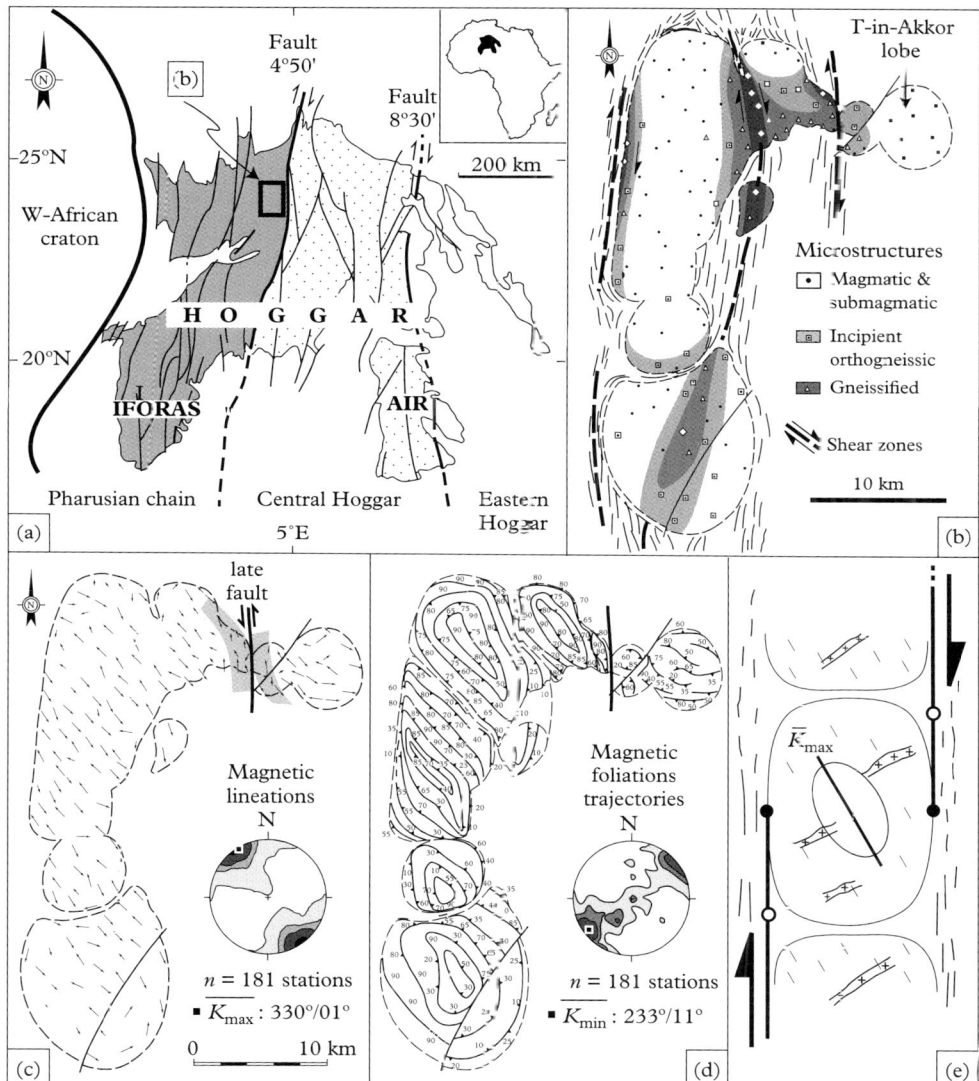

Figure 10.20 *The Tesnou granite complex (Hoggar, Algeria). (a) Location in the Pan-African basement of the Tuareg Shield (southern Sahara). (b) Internal microstructures, and structures in the country rocks (foliations, shear zones). (c) Magnetic lineation map and corresponding stereo plot (181 sampling stations). (d) Magnetic foliation trajectories and foliation poles stereo plot. (e) Interpretation: mean lineation parallel to extension, itself perpendicular to the aplite dykes and oblique with respect to the dextral shear affecting the granitic magma reservoir (arrows and dots: relative displacement of the margins). After Djouadi et al. (1997).*

208 *Magnetic fabrics in granites*

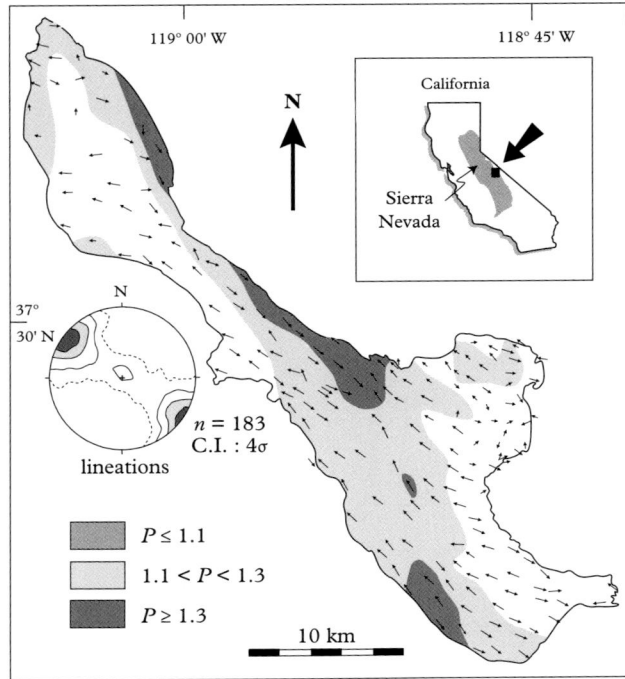

Figure 10.21 *Cretaceous Mono Creek pluton (California). Map and stereo plot (Schmidt, lower hemisphere) of the magnetic lineations (N = 183 sampling sites) and anisotropy degrees (in grey). In Saint-Blanquat and Tikoff (1997).*

regional transpression. The lineations delineate large-scale dextral sigmoids trending from NW–SE in the pluton core to WNW–ESE at the NW and SE extremities, defining a large shear zone called the Rosy Finch (RFSZ) traversing the pluton.

The increasing intensity of magmatic deformation toward the core of the RFSZ is well understood from the anisotropy degree that increases from less than 10% at the pluton extremities to more than 30% along some sectors of the shear zone centre. The microstructures, as observed under the microscope, confirm that deformation ceased during the magmatic stage at the pluton extremities and continued in the solid-state at high temperature within the core of the RFSZ.

The lineations, in Fig. 10.21, exactly reflect the crystalline fabric of the granite through the shape fabric of magnetite. The lineation trajectories thus represent the finite extension paths recorded by the magma at the time of its emplacement. They prove that the emplacement of the Mono Creek pluton took place during regional shearing, and suggest that the RFSZ acted as a magma feeder conduit.

10.4.4 Cauterets–Panticosa granite complex (Pyrenees)

The Cauterets–Panticosa complex, a little more than 250 km² in area located in the heart of the Axial Zone of the west-central Pyrenees, is composed of three juxtaposed massifs that were emplaced during Hercynian times (at ~310 Ma) within Devonian pelites, carbonates and quartzites (Fig. 10.22). The absence of magnetite make the rock iron

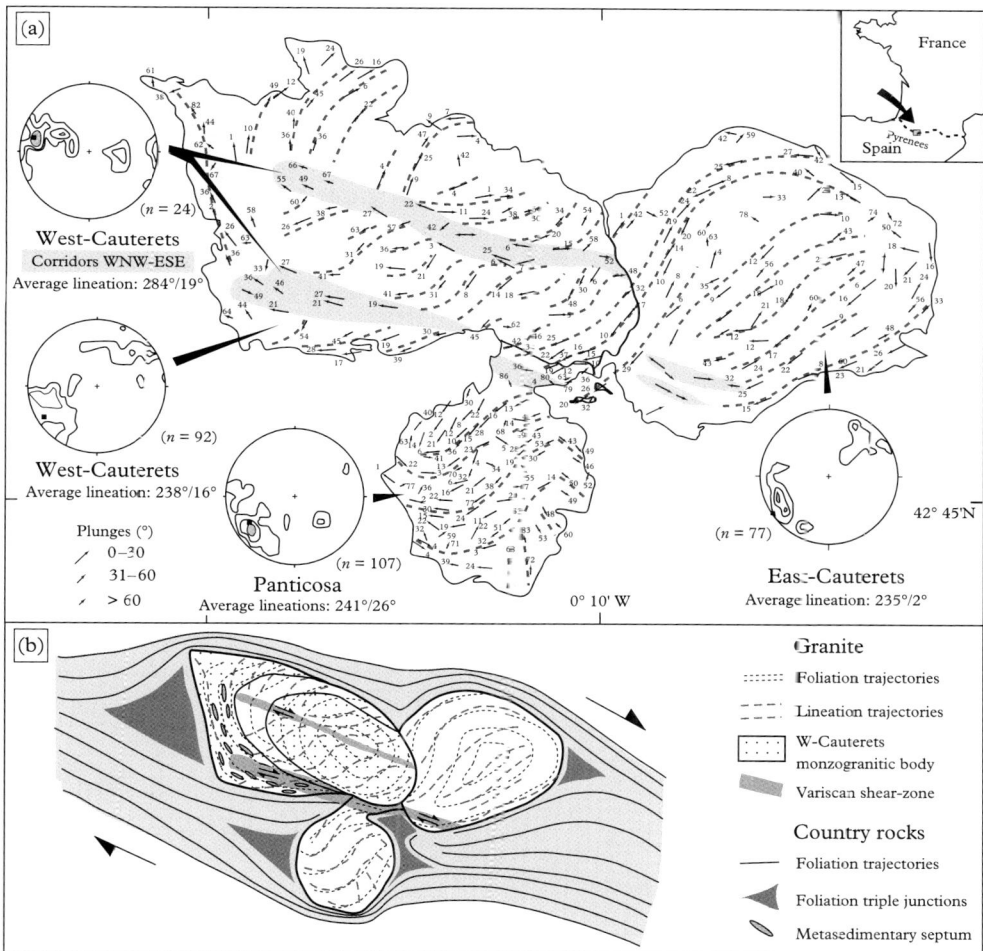

Figure 10.22 *The Hercynian Cauterets–Panticosa granite complex. (a) Lineation map and corresponding stereo-plots (Schmidt lower hemisphere) from different sectors. In grey: magmatic shear zones (magmatic microstructures), ultimate records of the regional dextral shear that prevailed during emplacement of the complex. (b) Proposed regional model. From Gleizes et al. (1998).*

content proportional to susceptibility (paramagnetic behaviour) and help to favourably compare the petrographic types in the complex with the magnitude of magnetic susceptibility (Fig. 10.7). The microstructures are magmatic to submagmatic, and the magnetic fabrics were studied through more than 300 locations. On the map (Fig. 10.22), magnetic lineations display NE–SW trends, oblique with respect to the WNW–ESE trending regional structures and the complex elongation. The remarkable sigmoidal pattern of the lineation trajectories confirms that regional dextral shearing assisted the emplacement in the crust of this complex.

10.4.5 Bassiès pluton (Pyrenees)

This pluton located to the south of Toulouse, close to the Spanish border, shows the evolution of a fabric when the sampling area is progressively reduced (Fig. 10.23). Like many other granite plutons, the Bassiès pluton displays an extraordinary structural homogeneity, marked by NE–SW magnetic lineations recorded in the magmatic state (as depicted by the microstructures), except along the north-eastern and south-western margins where limited subsolidus deformation was observed. Detailed sampling has been performed at one locality (Fig. 10.23) according to nested grids, the small grid (6 m-sided; 49 samples) being a subset of the larger grid (60 m-sided, 49 samples).

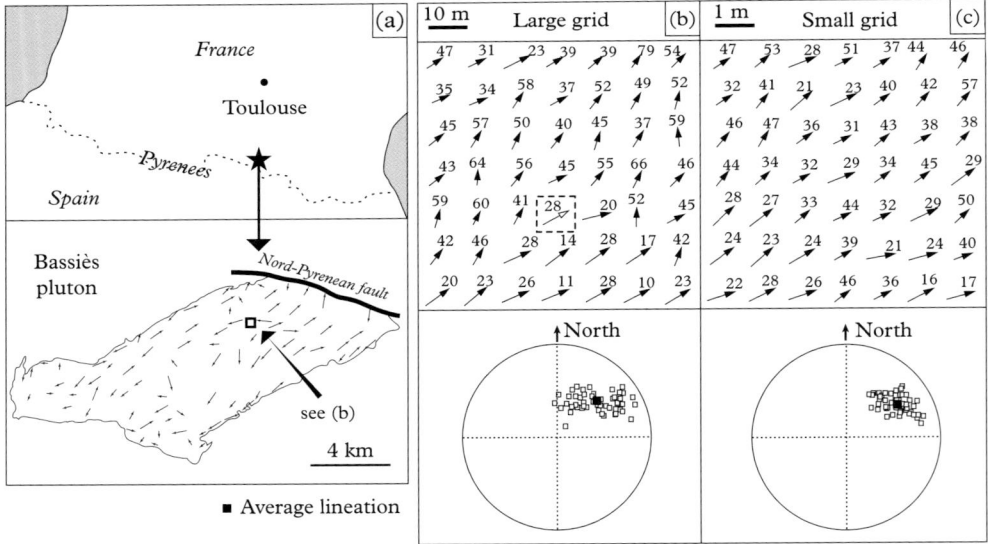

Figure 10.23 *Variability of the magnetic lineations in the Bassiès pluton (location in (a)): maps and corresponding stereo plots (Schmidt, lower hemisphere). At the decametric scale (b), the average lineation is (azimuth/plunge): 47°/E 41° (σ = 20°). At the metric scale (c), i.e. in a smaller grid localized in map (b), the average lineation is 56°/E 35° (σ = 14°). These orientations are close to those reported on the structural map of the massif. After Olivier et al. (1997).*

The resulting lineation trends (K_1 axes) are similar in both grids with naturally a lower variability observed in the small grid. Such a fabric homogeneity, never suspected before, suggests that magma volumes were subjected to homogeneous deformation during their emplacement into the crust.

This 'Russian nesting dolls' experiment, giving virtually the same fabric whatever the sampling size, demonstrates that the rheology of the granitic magmas is close to Newtonian. In other words, granitic magmas and possibly magmas in general, whose deformation does not vary much from one place to the other, are hardly subjected to strain localization.

10.5 Conclusions

AMS reveals unexpected internal structures and is recommended for the study of granitic rocks in their regional environment. Granite is never isotropic. The old distinction between syn- and post-tectonic granites, made mostly on the basis of pluton shape as mapped in the field, and on the presence of observable or not observable internal structures, becomes obsolete. Magnetic fabrics detectable in most plutons make them more-or-less syntectonic.

Relationships between regional deformation and emplacement should now been examined with respect to the internal fabric patterns, microstructures and their distribution within the pluton. The relationships concerning all these elements, and the regional structures in the country rocks, provide new concepts on which regional tectonics and magma pressure can be questioned in terms of forceful *versus* passive emplacement models.

Finally, magnetic studies definitely complement the structural, mineralogical and petrological approach used to document emplacement of plutons and their post-emplacement history. It can be concluded that the internal fabrics of granite plutons are excellent recorders of the kinematics of the regions into which plutons are emplaced, at least for the short duration of their emplacement process. As also concluded by Benn et al. (2001) magmatic fabrics have a strong potential for understanding plate-tectonic reconstruction of orogenic belts.

11
Zoning in granite plutons

In previous chapters, there were examples of plutons displaying internal compositional differences, with a conspicuous spatial distribution known as 'zoning'. This situation is important and justifies here a detailed review of the main types of zoning in granite plutons and an explanation of the origin of zoning.

11.1 Examples of zoned plutons

11.1.1 Concentric zoning

Several plutons display concentric zoning regarded as normal or inverse depending on the location of the less differentiated rocks, either at the margins or in the centre. The Panticosa pluton (Pyrenees) is a typical example of a normally zoned pluton (see Fig. 10.7a). Contacts between the different petrographic types appear gradational over a hundred of metres. As a whole, it suggests *in situ* magmatic differentiation in a chamber, which then crystallized as a pluton. On the other hand, the Toro pluton (Nigeria) displays an inverse zoning with a dioritic core (Fig. 4.5a). Actually, it is a bimodal intrusion where different styles of contacts can be observed corresponding to variable degrees of mixing of two magmas. Systematic measurements of the magnetic susceptibility confirm precisely these petrographic observations (Figs 10.7b and 11.1).

11.1.2 Vertical zoning

In favourable observation conditions, a few plutons also display compositional variations with depth. The Searchlight pluton in Nevada, tilted at nearly 90° because of a normal fault, provides a complete cross-section from its floor to its roof (Fig. 11.2a). This pluton is exceptional due to its size, as it appears to be up to 10 km in thickness (height). It contains a mafic kilometric stock and smaller enclaves ranging from gabbro to monzodiorite in composition, with isotopic signatures differing from the main granite (hence suggesting that these rocks were coeval, but not cogenetic). The remainder of the pluton contains rock-types, whose compositions range from a biotite- and hornblende-bearing monzonite (at the floor) to a biotite-leucogranite (in the upper part of the middle

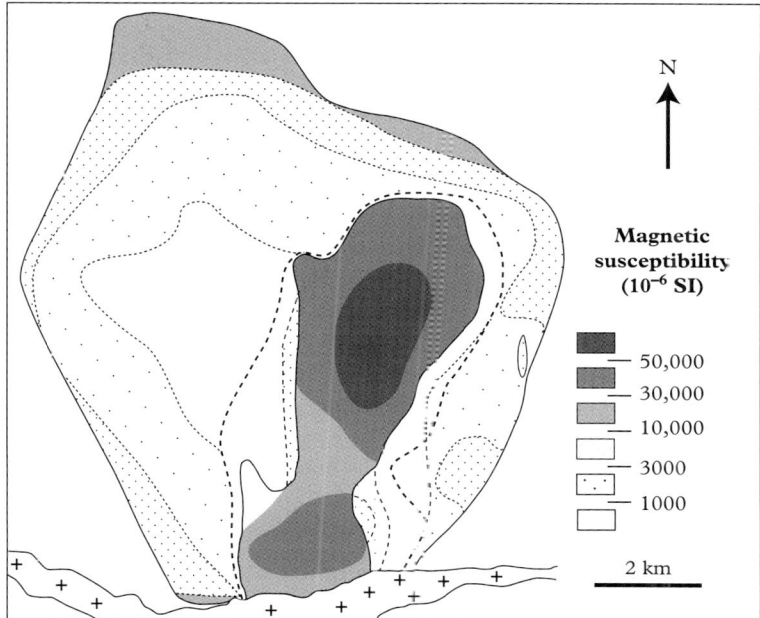

Figure 11.1 *Susceptibility map of the Toro pluton (Nigeria), showing a normal zoning in the granodioritic to granitic part and an inverse zoning of diorite and its surrounding hybrid zone; compare with the lithology map (Fig. 4.5). After Déléris et al. (1996).*

unit). Petrography and geochemistry suggest a magmatic differentiation process by fractional crystallization (Fig. 11.2a and b). The pluton is vertically divided into three units (Fig. 11.2c). The lower unit displays a conspicuous magmatic foliation that is lacking in the other units. It is cut by several undeformed aplitic dykes, of centimetre thickness, resulting from syn- to late-magmatic compaction, which would have expelled an interstitial granitic melt. The upper unit is a fine-grained quartz-bearing monzonite containing less ferromagnesian minerals than the lower unit. It crystallized quickly at the roof of the intrusion, as a chilled margin. The granitic middle unit is the most differentiated, because it is the most silica-rich. It contains lesser amounts of compatible trace elements (such as Sr, which can substitute for Ca in plagioclase). This granitic unit itself displays small compositional variations upwards, suggesting that it was involved in a differentiation process, yielding the most evolved rocks at the top. The lower and middle units correspond to a normal vertical zoning. Contacts between the units are sharp, but gradational on a metric scale, and emphasized by colour (hence modal composition) and textural changes. Aplitic dykes are rooted in the middle unit and extend upwards to the roof zone of the pluton. They represent a late-magmatic residual melt.

Other plutons display compositional variations at depth, which do not result from an *in situ* magmatic differentiation, and will be discussed later.

214 Zoning in granite plutons

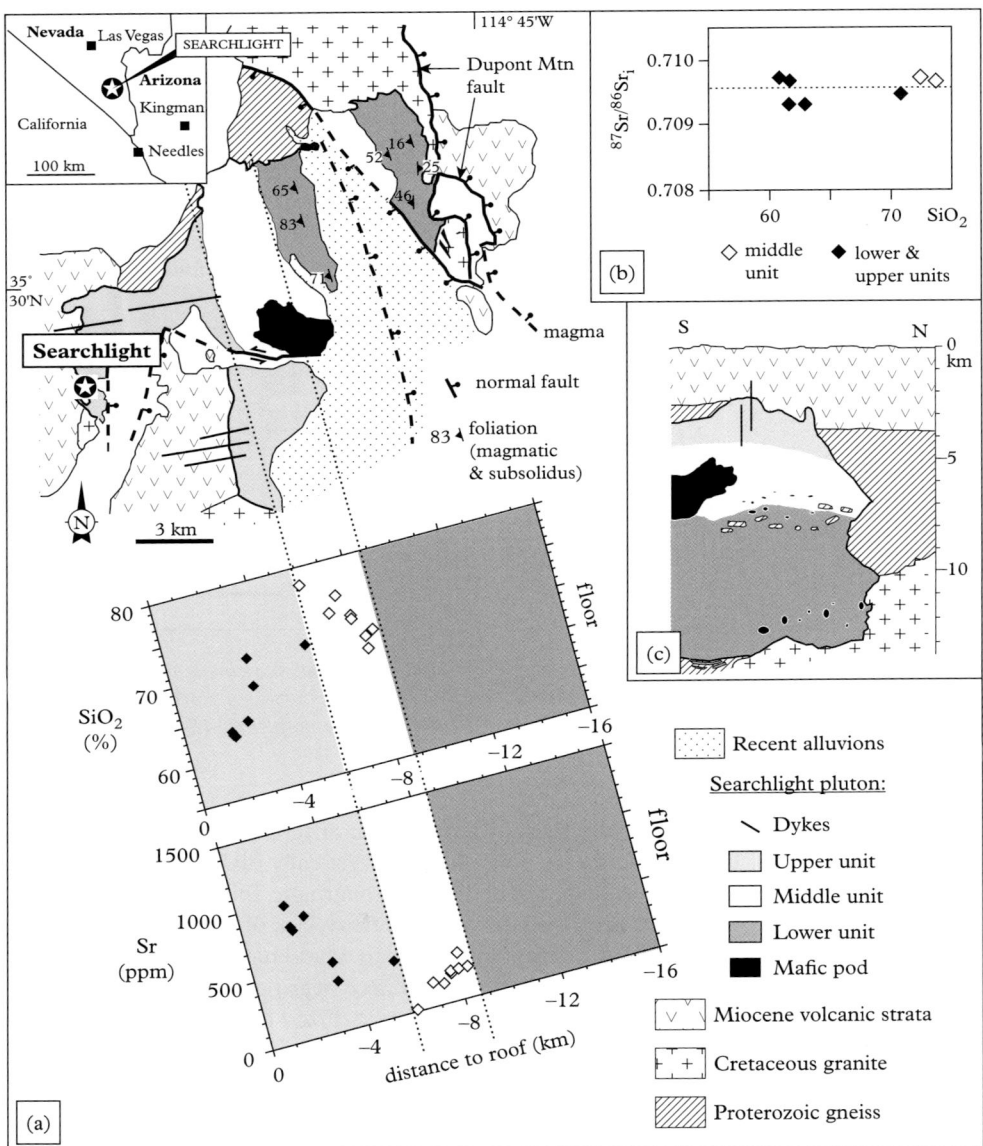

Figure 11.2 *The Searchlight pluton (Nevada). (a) Geological map and variations of SiO_2 (wt %) and of the compatible element Sr (ppm) in pluton cross-section. (b) Diagram Sr(i) vs. SiO_2: the lack of significant variation suggests a unique parental magma for the three plutonic units. (c) Reconstruction of a vertical section before tilting. After Bachl et al. (2001) and Miller and Miller (2002).*

Examples of zoned plutons 215

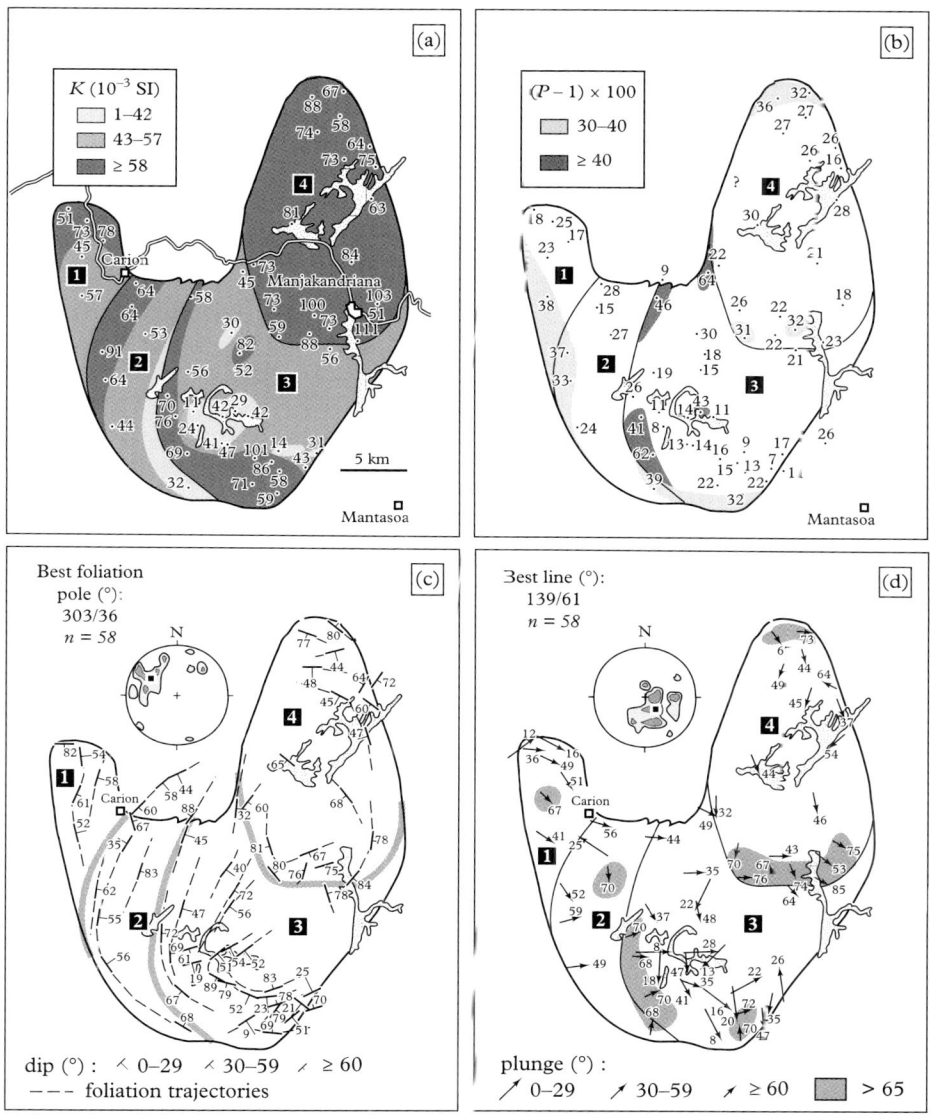

Figure 11.3 *The Carion pluton (Madagascar) and its four sub-units. (a) Magnetic susceptibility (K). (b) Degree of magnetic anisotropy (%). (c) Magnetic foliations. (d) Magnetic lineations. Note the nesting and cross-cutting structures and the higher anisotropy at the contact between sub-units 2 and 3. After Razanatseheno et al. (2009).*

216 Zoning in granite plutons

11.1.3 Complex zoning

A few plutons are characterized by a recurrent normal zoning, sometimes not easy to observe under the microscope. Thus, the Carion pluton (Madagascar), initially described as an intrusion only displaying coarse normal zoning, is actually made of four sub-units corresponding to four successive magmatic pulses as revealed by a detailed AMS study (Fig. 11.3). Three of these sub-units are more or less differentiated, as indicated by their individual normal zoning. The sub-units were 'nested', i.e. intruded one into the other, in such a way that each sub-unit partially cuts the structures and petrographic zoning of the sub-unit emplaced immediately before.

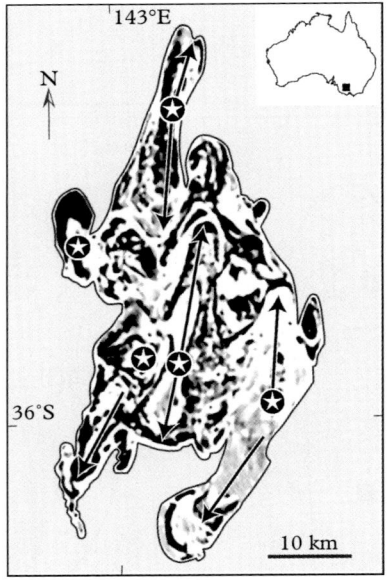

Figure 11.4 *First derivative map of the bulk magnetic intensity measured by aeromagnetic survey: this is a map of the variation of magnetic mineral contents in a Devonian I-type granitic pluton from Victoria state (Australia). Black areas are the richest in highly susceptible minerals. Several magmatic feeders (stars) can be identified, as well as the corresponding directions of magmatic replenishment (arrows). After Clemens et al. (2009).*

Once again, magnetic tools prove their efficiency, even for disclosing cryptic structures. For instance, an aeromagnetic survey may yield a map of the variations in magnetic mineral contents inside a pluton, hence imaging the structure of the pluton. The example in Fig. 11.4 is worth noting, especially as the surveyed pluton outcrops are poorly exposed.

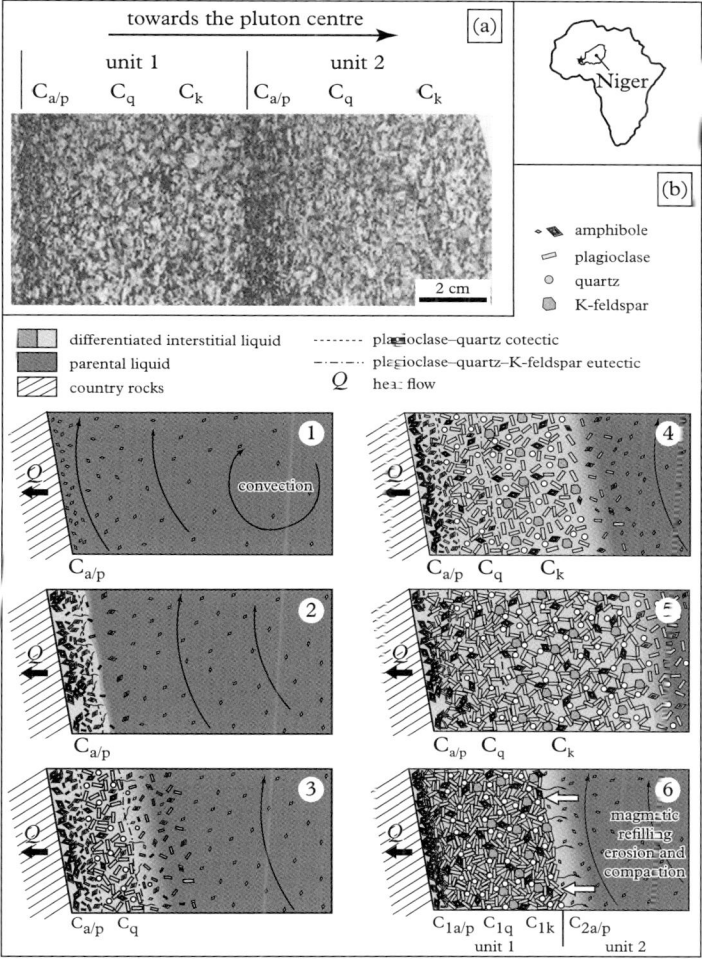

Figure 11.5 *Layering in the Dolbel pluton (SW Niger). (a) Border layering: only two units are displayed, each one made of three layers ($C_{a/p}$, C_q and C_k). (b) Interpretation: early crystals (a and p) were transported by magmatic convection and deposited along the pluton wall, where crystal growth continued; quartz and then K-feldspar crystallized later in this unit. The arrival of a new hotter magma batch eroded unit 1 and was followed by deposition of unit 2. After Pupier et al. (2008).*

11.1.4 Rhythmic zoning or layering

A conspicuous recurrent compositional layering can be observed in some mafic intrusions, such as the Skaergaard complex (Greenland) or the Bushveld complex (see Fig. 12.2). Granitic plutons are often regarded as devoid of any such inner structures, because of the higher viscosities of felsic magmas that would not trigger the development of convection and therefore would hinder an efficient segregation of crystals and residual melt (whether by gravity settling or by dynamic sorting). Actually, this is a preconceived idea! Magmatic layering can be recognized in some granites. The case of the Dolbel pluton (Niger, Fig. 11.5) was studied in detail by Pupier et al. (2008). It is a 2.1 Ga old intrusion of granodioritic to trondhjemitic composition, which belongs to the Eburnian granitoids that will be discussed in Chapter 13 (see Section 13.4.1 and Fig. 13.20). This intrusion has an elliptical shape (7 × 4 km² in size). Over a hundred metres, the pluton border displays magmatic layering, i.e. repeated variations of modal composition at the decimetre scale, whereas the pluton core has a homogeneous composition. Each rhythmic unit is characterized by three successive layer types in the following order (from periphery to core):

- $C_{a/p}$ (80% amphibole, 20% plagioclase);
- C_q (20% amphibole, 60% plagioclase, 20% quartz);
- C_k (10% amphibole, 60% plagioclase, 15% quartz, 15% K-feldspar).

Each layer grew inward and its top appears truncated by the base of the next unit (Fig. 11.5).

Box 11.1 What is a magma chamber?

A magma chamber is a sort of huge container filled with magma. The very existence of magma chambers has been debated, especially in the ductile lower crust. Formation of a magma chamber in the upper crust is easier to imagine. Indeed, there is presently a 'living' magma chamber under the Long Valley caldera in California (Fig. 11.7a, b), as confirmed by recent geophysical imaging and measurements using GPS and radar interferometry (Tizzani et al., 2009). Actually, it is one of the most instrumented places in the world, because of the risk of eruption of such a large magma volume at a shallow depth in the crust (Lowenstern et al., 2006).

Marsh (1989, 2000) developed a theoretical model for the cooling of a magma chamber replenished only once (Fig. 11.6). Magma cools and crystallizes from the roof and from the floor. Conductive cooling is mostly efficient at the roof. The partly upper crystallized zone is unstable, as crystals may detach and sink in the chamber. Early crystals accumulate on the floor because of gravity settling (this is a fractional crystallization process). This process is efficient in low-viscosity liquids, which are also able to undergo convection in the chamber as long as the crystal load is not too high. Compaction of crystals accumulated on the floor results in the expulsion of an interstitial liquid toward the centre of the chamber. Upward migration of this melt is favoured by its low density.

11.2 Origin of zoning

11.2.1 In situ magmatic differentiation

The hypothesis that a granitic pluton represents a magma chamber (see Box 11.1) which evolved by *in situ* fractional crystallization has been often debated. The previously reported example of the Searchlight pluton (Fig. 11.2) closely resembles the case of a magma chamber, active for a time long enough to allow magmatic differentiation by fractional crystallization following the Marsh (2000) theoretical model (Fig. 11.6). The upper unit can be regarded as the crystallization front propagating downward from the top. During ongoing crystallization and before the relevant crystallinity threshold, which is illustrated in Fig. 4.2 (threshold #2, Box 4.1), this upper zone may lose some crystals either by gravity settling or by transport in convection currents resulting from temperature or density differences in the magma. The lower unit represents another crystallization front, propagating upward from the floor, and is also the place for accumulation of crystals removed from the roof. In this mush, crystals acquire a preferential orientation corresponding to the observed foliation. The residual liquid,

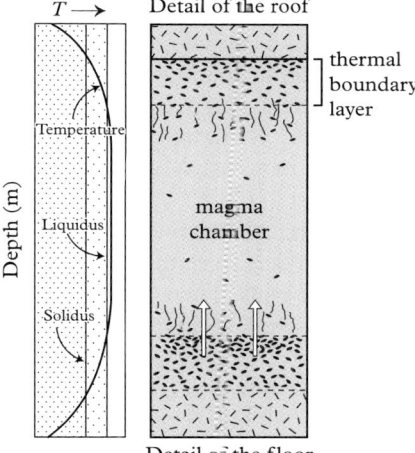

Figure 11.6 *Theoretical model of a crystallizing magma chamber and vertical distribution of temperatures. Crystals accumulate on the floor of the chamber and their compaction results in the expulsion of residual melt upward (white arrows); convection currents (not represented) can quicken or delay the sinking of crystals from the roof. After Marsh (2000).*

still present interstitially, is periodically expelled due to the weight of the accumulated crystals. This residual melt can flow into fractures formed in the crystal framework of the not yet fully solidified magma. The middle unit, enriched in granitic residual melt, is therefore complementary to the cumulative lower unit. This model is in agreement with the petrographic and geochemical data. It was quite likely efficient, because the intrusion is large and thick enough (it is characterized by a large initial vertical dimension), to hinder fast cooling. However, this model is questionable, because it requires that the whole magma chamber volume could have been replenished by only one magmatic pulse.

As a result of *in situ* magmatic differentiation, Hildreth (1979) suggests that a large volume of residual melt, enriched in silica, water and incompatible elements, may grow at the roof (and not at the core) of a large granitic or granodioritic magma chamber. Hildreth's model is derived from his work on the Bishop Tuff, rhyolitic ignimbrites erupted 760,000 years ago during the formation of the Long Valley caldera in California (Fig. 11.7a). These rocks correspond to the expulsion of 600–1000 km^3 of magma by only one cataclysmic eruption, at the same time causing the collapse of the roof of the magma chamber and producing the caldera formation. The thickness of the Bishop Tuff reaches 1500 m inside the caldera (Fig. 11.7b). Vertical zoning can be observed in these ignimbrites. The first erupted magmas are presently at the base of the ash flow; they are the richest in silica (77%) and the poorest in crystals (5%), and were originally expelled from the top of the magma chamber. Throughout the eruption, silica contents decreased (down to 73%) and, at the same time, crystal contents increased (up to 25%) because successively deeper levels of the magma chamber were tapped.

The occurrence of such an efficient magmatic differentiation in a magma chamber questions the lifetime of the chamber. Systematic dating of zircons from the Bishop Tuff and from the pre-caldera rhyolites of Glass Mountain provides the answer as will be explained later.

The pre-caldera rhyolites of Glass Mountain (100 km^3) were emitted by two eruptions: GM1 (900 ka) and GM2 (790 ka) from the same chamber. They are highly silicic (78%) and contain less than 6% crystals. Their zircons yield a maximum age of 1070 ka. In the Bishop tuffs, the zircons yield an age of 850 ka in average (910 ka at maximum). Zircons from the lower tuffs (TB1), i.e. from the chamber roof, and zircons from the upper tuffs (TB2), which represent a deeper level in the chamber, display large differences in uranium (a very incompatible element) contents. However, there is no significant difference between their respective ages. It is concluded that a layer of highly differentiated rhyolitic melt formed after the pre-caldera eruptions. This liquid remained at least for 90,000 years under the roof of the magma chamber before the occurrence of the caldera-forming eruption. The chamber itself persisted for more than 200,000 years without achieving full crystallization. An efficient segregation of the residual rhyolitic melt could occur during these time durations, in as much as the viscosity of this melt is decreased by its higher water content. For a typical water content of 6% in weight, Scaillet et al. (1998) showed that the viscosity of a rhyolitic liquid is 10^4 Pa s. In the case of an initially granodioritic magma chamber undergoing 50% crystallization with grains 2.5 mm in diameter, Bachmann and Bergantz (2004) calculated that the compaction-driven

Origin of zoning 221

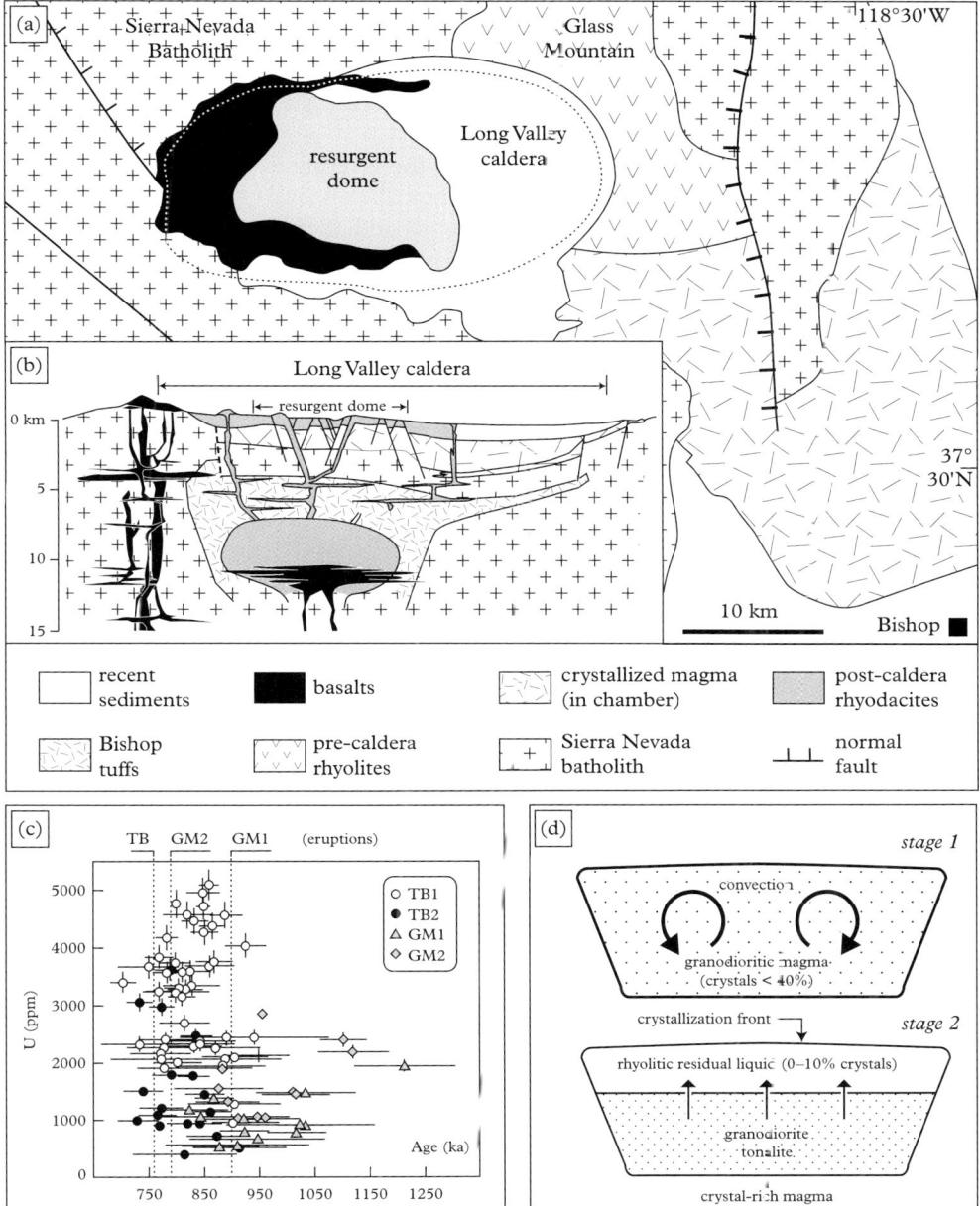

Figure 11.7 *Evolution of the magma chamber under Long Valley caldera (California). (a) Geologic map after Bailey et al. (1976). (b) Cross-section after Fischer et al. (2003); same symbols as in (a). (c) Ages and uranium contents of zircons from Glass Mountain rhyolites (GM) and from Bishop Tuff (TB). After Simon and Reid (2005). (d) Interpretation. After Bachmann and Bergantz (2004).*

segregation of 500 km³ rhyolitic liquid and its migration to the chamber roof would last for about 100,000 years, a result remarkably consistent with the geochronological data derived from zircon dating. These studies provide convincing pieces of evidence for the hypothesis that a few zoned plutons, such as the Searchlight pluton, represent *in situ* fractionating magma chambers.

In the case of the Dolbel pluton (Fig. 11.5), the layering is vertical and not tilted, excluding any gravity settling process. Nevertheless, the mineral crystallization order inside each unit is as expected for this type of magma and, therefore, suggests a fractional crystallization process in the thermal boundary layer at the contact with the outer wall or with the previously solidified unit. Indeed, the first crystallized layer ($C_{a/p}$) contains the liquidus phase. The last-crystallized layer (C_k) of a unit will be eroded by a new magma

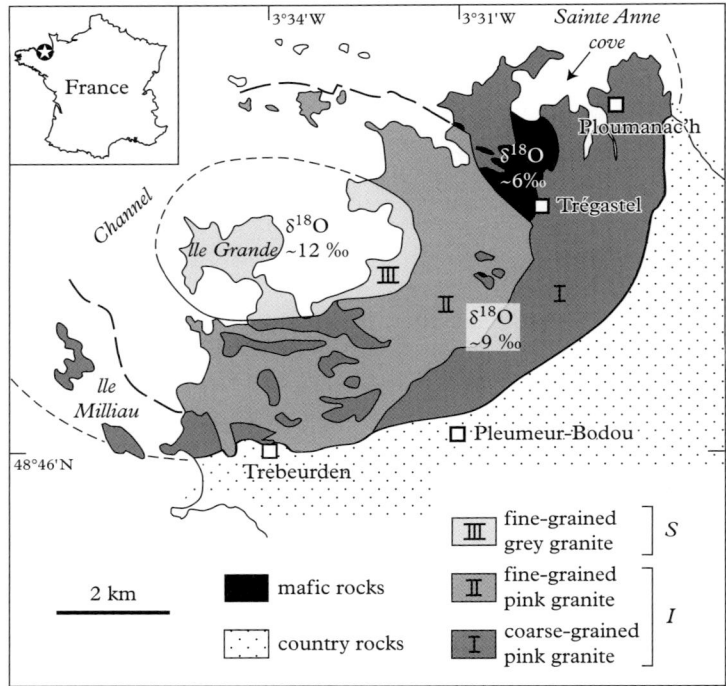

Figure 11.8 *Ploumanac'h late-Hercynian pluton (Brittany) is divided in three zones. The outer zone is made of a coarse-grained pink potassic calc-alkaline (I-type) granite, locally associated with mantle-derived gabbro showing a few mixing-mingling features at the contact. The middle zone is made of a fine-grained pink granite of same geochemical affinity. The inner zone is made of a later-emplaced S-type grey leucogranite. Oxygen-isotope data dismissed any genetic link between the magmas. After Barrière (1977) and Albarède et al. (1980).*

injection and crystallization of another unit will then begin. Amphiboles do not have exactly the same composition in the different units, suggesting the injection of a slightly different magma into the chamber. In this case, *in situ* fractional crystallization is recognized, but has spatially limited effects. In addition, the cryptic evolution of the mineral compositions cannot be explained by fractional crystallization of a unique magma batch. A detailed study of layering in granitoids and its complexity of manifestation and origin is provided by Barbey (2009).

Actually, it is not sure that a whole pluton was built by the emplacement of perfectly coeval magmas during one event. Several intrusions are now regarded as the juxtaposition of successively emplaced small magma batches (Glazner et al., 2004).

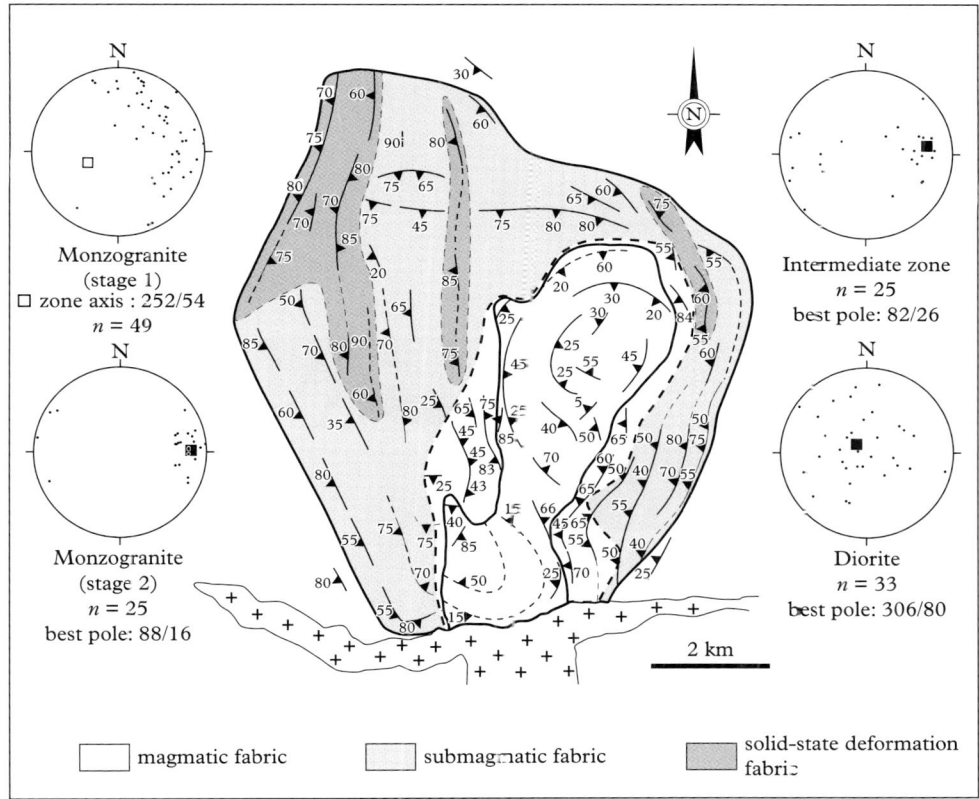

Figure 11.9 *Magnetic foliations and microstructures in the Toro pluton. The concentric magmatic foliations of the granitic mass are cut by N–S shear zones responsible for solid-state deformation. One of these shear zones also affected the hybrid zone, which was still at the magmatic stage, because crystallization of the hybrid magma was delayed due to heat transfer from the hot dioritic magma. After Déléris et al. (1996).*

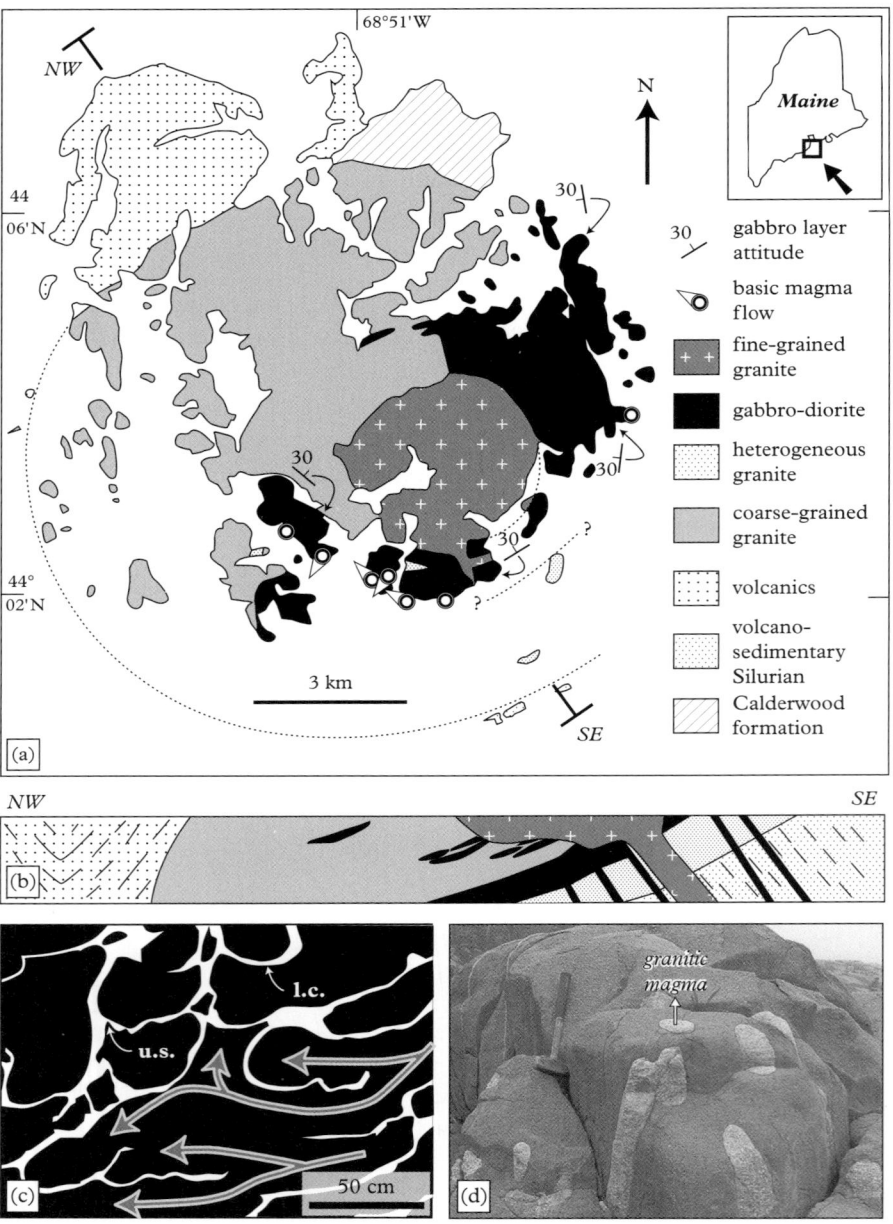

Figure 11.10 *Relationships between felsic and mafic magmas in the intrusion at Vinalhaven (Maine, USA) after Wiebe et al. (2001). (a) Geological map. (b) Schematic cross-section showing that the intrusion has been tilted by about 40° to the north-west. (c) Gabbro-diorite boudins and pillows separated by granite; arrows indicate the lateral motion of the mafic magma, that was ponding at the base of the chamber and crystallizing at the contact with the granitic magma; several polarity criteria can be recognized: load casts (l.c.) and upward spines (u.s.). (d) Tubes containing granitic magma expelled from underneath the crystallizing mafic sill (by courtesy of R. Wiebe).*

11.2.2 Successive magma intrusions

In some cases, pluton zoning results from different magmatic pulses spaced in time. This is the general case of inversely zoned plutons, which actually represent bimodal intrusions, i.e. intrusions fed by non-cogenetic magmas that will eventually mingle or mix (see Chapter 4). These magmas ascended through the same conduit. In the example of Ploumanac'h (Brittany), there are three different magmas as confirmed by isotopic data (Fig. 11.8). The time lag between the emplacement and the crystallization of the different magmatic pulses can be estimated through the detailed study of structures and microstructures, as was shown by the Carion pluton (Fig. 11.3) or the Toro pluton (Fig. 11.9).

An ascending mafic magma can be stopped at different levels when entering a granitic magma chamber. Sometimes, it will pond near the chamber floor. This is beautifully exposed in several granitic plutons from Maine (north-east USA), where good outcrop exposures along the coast enable the reconstruction of respective motions of the mafic and felsic magmas in the chamber, and further behaviour of the granitic magma with respect to the nearly crystallized mafic magma (e.g. Wiebe, 1993; Wiebe et al., 2001; Fig. 11.10).

The successive magmatic pulses may correspond also to cogenetic magmas. This is the case of the nested sub-units that built the Carion pluton. It is even recognized in plutons displaying a normal concentric zoning, which would have been regarded a priori

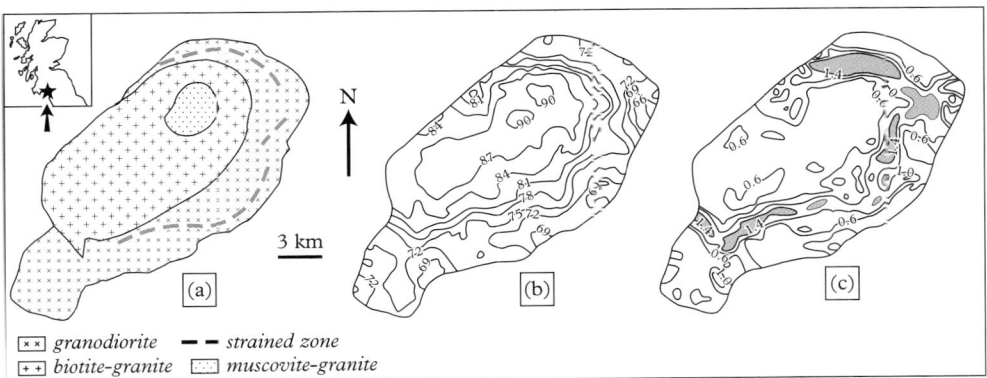

Figure 11.11 *Criffell pluton (Scotland). (a) Lithological map: the boundary between granodiorite and granite corresponds to the first appearance of K-feldspar megacrysts; dashed contour: limit of the deformed zone. (b) Digital contours obtained from 180 measurements of DI (differentiation index of Thornton and Tuttle, 1960, i.e. the sum of normative residual minerals, feldspars and quartz); DI becomes higher when the granite becomes more differentiated. (c) Digital contours of the first derivative corresponding to the variation of DI: a more pronounced gradient can be recognized at the boundary between granodiorite and granite, and suggests emplacement of two successive magmatic pulses; in contrast, the transition between granite and muscovite-bearing granite is gradational. After Stephens and Halliday (1979), Courrioux (1987) and Stephens (1992).*

as resulting from an *in situ* differentiation process. When comparing the structures of the different zones and the nature of their contacts, another interpretation comes to mind. It is suggested that the differentiation process occurred at depth and not at the present outcrop level. Figure 11.11 displays the example of the Criffell pluton, which was built by two successive magmatic pulses: the first one was granodioritic in composition and the second one was granitic. This latter pulse underwent some *in situ* magmatic differentiation producing a muscovite-bearing granite. The outer granodiorite was deformed at the contact by the intrusion of the inner magmatic core.

The recent proliferation of multiple zircon dating in many granitic intrusions has provided other evidence that the building of granite plutons may have required a few millions of years in some cases. This is longer than the lifetime of a single magma chamber in most cases. These data also explain the renewed success of an old idea regarding the incremental assembly of plutons, first suggested by Harry and Richey (1963).

11.3 Conclusions

Zoned plutons are commonly observed and may result from *in situ* differentiation in a closed-system magma chamber. Many plutons display a concentric, or vertical, normal zoning, where the most differentiated rock types constitute the core, or the roof, of the pluton. Such plutons could have been formed by an *in situ* differentiation process. However, structural and/or geochemical studies sometimes lead to the conclusion that zoned plutons may alternatively result from successively emplaced magma batches, either cogenetic or not. Plutons displaying an inverse concentric zoning are typically bimodal intrusions, where closely spaced magma pulses derived from different sources may have undergone mingling or mixing.

12
Granites and plate tectonics

This chapter aims at characterizing granites formed in various geodynamic settings, either intraplate or at plate margins, using representative examples. We will see that it is not possible to establish a simple, or objective, relationship between geodynamic setting and geochemical data. On the one hand, the nature of the magma source and the thermal conditions of melting play a major role in determining the characteristics of magmas, and, on the other hand, identical protoliths (for instance a metapelitic source) are observed in different settings. As a consequence, peraluminous granites may be produced in anorogenic settings as well as in continental collisional settings. Nevertheless, specific (P, T) conditions of each geodynamic setting can explain the formation and composition of granitic magmas in all cases.

12.1 Granites and oceanic accretion

Ophiolites are fragments of oceanic crust tectonically transported and preserved in the continental margins. They sometimes contain very small volumes of granitic rocks, the so-called 'oceanic plagiogranites' after Coleman and Peterman (1975). Actually, these rocks are K_2O-poor trondhjemites or leucotonalites. They are observed as thin veins, dykes or sills (a few centimetres to one hundred metres thick) injected into the uppermost gabbros of the oceanic crust. Two mechanisms can explain their formation: either by fractional crystallization of gabbroic magma or by partial melting of an already crystallized oceanic crust eventually metamorphosed into amphibolites during obduction and collision. An experimental study suggested by the example in Fig. 12.1 favours the mechanism of fractional crystallization.

12.2 Granites and hot spots or continental rifting

Felsic magmas due to intraplate hot spots above mantle plumes are often observed in continental domains, whereas they are extremely rare in oceanic domains. Mantle plumes from the asthenosphere or deep mantle under oceanic plates have been regarded

228 Granites and plate tectonics

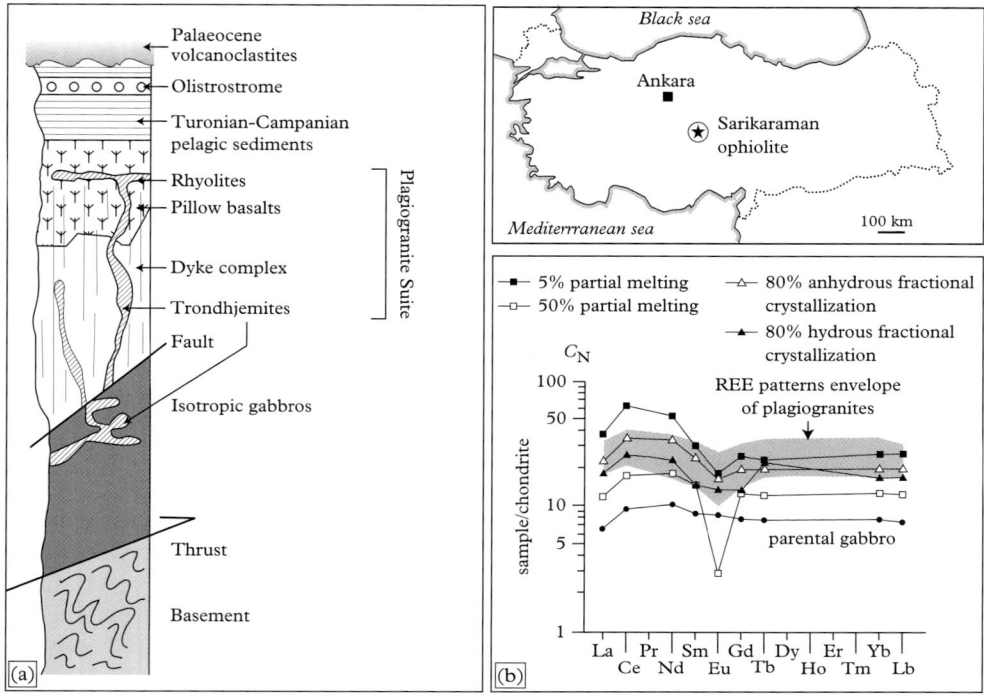

Figure 12.1 *Plagiogranites from the Cretaceous Tethyan ophiolite at Sarikaraman (Turkey). (a) Location of the ophiolite and interpretative lithostratigraphy; plagiogranite (or trondhjemite) dykes are related to a few volcanic effusives with a composition of keratophyre (sodic rhyolite altered by seawater). (b) Rare earth element (REE) distribution patterns typical of plagiogranites are poorly fractionated and display a small Eu negative anomaly (grey shaded area). Petrogenetic model: four hypotheses of plagiogranite formation were tested using the REE patterns of ophiolite gabbros. The best hypothesis (i.e. the model that yields the best fit of the resulting REE patterns with plagiogranite patterns) corresponds to 80% fractional crystallization of gabbroic magma and extraction of an anhydrous mineral assemblage composed of olivine (10%)–clinopyroxene (40%)–plagioclase (50%). After Floyd et al. (1998).*

as responsible for linear chains of volcanic islands, e.g. Hawaii, or for oceanic plateaux, dominantly associated with basaltic magmatism. However, differentiated magmas have been locally described on some oceanic plateaux. For instance, syenitic ring-complexes are known in the Rallier-du-Baty peninsula in the main island of the Kerguelen archipelago. Rhyolitic effusives have been recorded in Iceland (an oceanic plateau presently straddling the mid-Atlantic ridge). They are regarded as products of partial melting of the thick basaltic crust forming the plateau.

Abundant granitic magmatism formed above a mantle plume is typical of the continental realm. Two samples will be detailed hereafter. A-type granites are classically observed in these settings, but often associated with S-type granites.

12.2.1 The Bushveld granite

The Bushveld complex (2054 Ma) is a huge bimodal intrusion, 350 × 200 km² in size, outcropping over 66,000 km² (Fig. 12.2a). It consists of a layered ultramafic to mafic complex (pyroxenite–gabbro–anorthosite), up to 7 km in thickness, the Rustenburg suite,

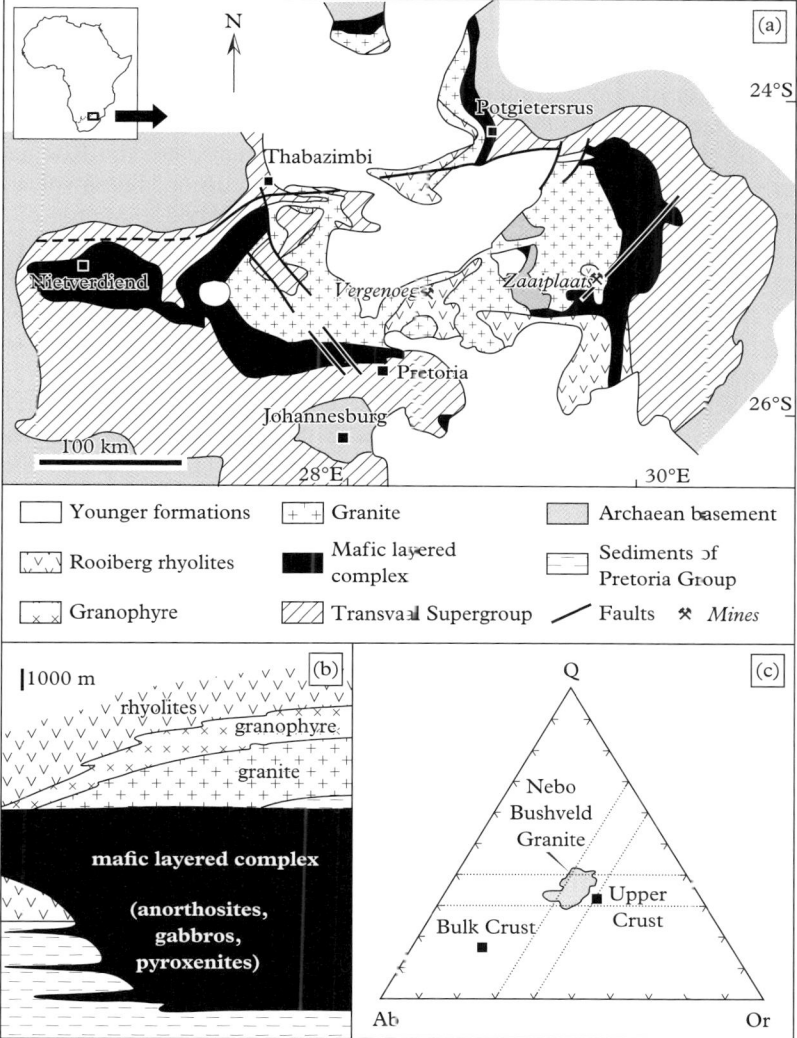

Figure 12.2 *The Bushveld complex (South Africa). (a) Geological map. (b) Vertical lithostratigraphic section. (c) Composition of the Nebo granite with respect to average bulk and upper continental crust. After Kleeman and Twist (1989), Eales and Cawthorn (1996) and Schweitzer et al. (1997)*

overlain by a 2 km thick (or more) zone of anorogenic granite, the Lebowa granite (Fig. 12.2b). The Bushveld complex was emplaced at a high level into the Paleoproterozoic Transvaal Supergroup, a well-preserved sedimentary shelf sequence of 2.65–2.06 Ga age. The sediments underlying the Bushveld igneous complex suffered contact metamorphism down to 5 km below the contact and locally reached the conditions of anatexis. Besides its unusual size, the Bushveld complex is also well known for its substantial ore reserves: it is mined for chromium, vanadium and PGE (platinum-group elements) in the mafic part and for Sn (as cassiterite) and F (as fluorite) in the granitic part (see Chapter 14).

The Bushveld (or Lebowa) granites are typically A-type granites with two lithologies: the medium-grained Nebo granite is the main petrographic type, with a minor fine-grained variant, the Klipkloof granite. The granite roof penetrates the Rooiberg rhyolites of nearly identical age (2061 Ma) and chemical composition. These volcanic rocks are more than 3 km in thickness and may have initially reached a volume of 300,000 km^2, thus representing the largest volume of Precambrian rhyolites ever erupted. The Rashoop granophyre is regarded as a possible subvolcanic equivalent of the Rooiberg rhyolites. A few rocks with granodioritic composition can be observed at the granite base in contact with the underlying mafic rocks, but interactions between mafic and felsic magmas seem to have been restricted.

The Bushveld granitic rocks are characterized by hypersolvus alkali feldspar and interstitial iron-rich silicates (hornblende or biotite). Silica contents are high, ranging from 69 to 76 wt % for the Nebo granite sheet, respectively from base to top. The original granite magma is thought to have been hot (>900 °C) and relatively anhydrous (initial H_2O content: 2.2 wt %). Nevertheless, the upper part of this granite, the Bobbejaankop granite, has been heavily altered by fluid interaction, that has overprinted the original magmatic signatures.

Bulk composition supports a crustal origin for the Nebo granite (Fig. 12.2c). The initial strontium ratio, $Sr(i)$, amounts to 0.715 in average, but is highly variable, possibly due to the addition of Rb after crystallization. Heat advection by the mafic magma was very likely responsible for crustal melting. The Bushveld mafic rocks were mantle-derived, but their relatively high $Sr(i)$ of 0.708 is evidence of some crustal contamination. In common with most mixed mafic–felsic large igneous provinces, there is still uncertainty as to the origin of the felsic magmas (granites, granophyres and rhyolites) either by crustal melting or by prolonged fractional crystallization accompanied by crustal assimilation.

12.2.2 Jurassic granites of Nigeria

These granites, which are also known as 'Younger Granites', are indeed much younger that the granitic and gneissic Precambrian rocks of the Benin–Nigerian shield. Their ages range from 191 Ma in the north to 141 Ma in the south, first suggesting a systematic decrease due to the northward motion of the African plate. However, Begg et al. (2009) noticed that African alkaline magmatism, including Nigerian Younger Granites, has been concentrated in Neoproterozoic to Phanerozoic metacratonic areas with higher geotherms and more fertile mantle with respect to older Archaean–Paleoproterozoic cratonic areas corresponding to cool depleted mantle (Fig. 12.3a). Indeed craton margins likely channelled magmas from the depths. In addition, they were prone to tectonic

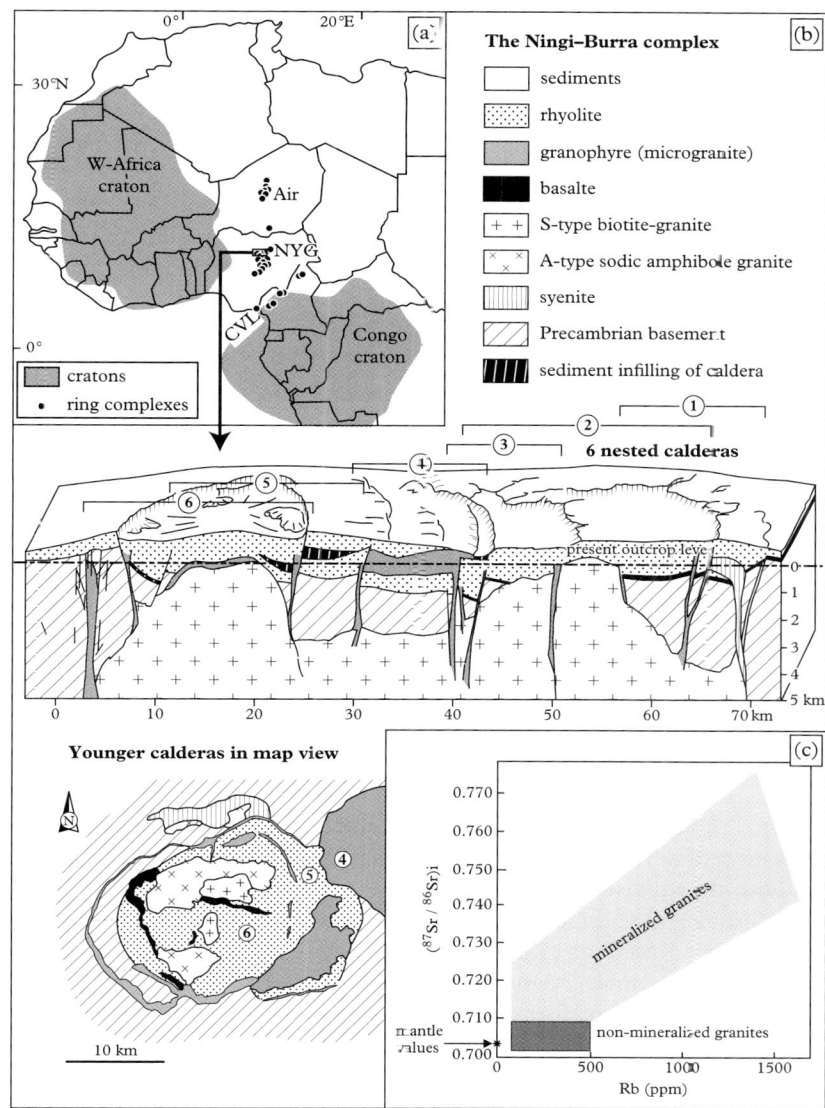

Figure 12.3 *Jurassic ('Younger') granites of Nigeria. (a) Location of ring complexes of West Africa; NYG: Nigerian Younger Granites, CVL: Cameroon volcanic line; craton margins after tomographic image at 100–175 km depth profile (in Begg et al., 2009). (b) Three-dimensional diagram and detailed map of the Ningi–Burra annular complex (ca. 183 Ma) showing six nested calderas, after Turner and Bowden (1979) and Bowden and Kinnaird (1984). (c) Initial Sr isotopic ratios vs. Rb contents after Bonin et al. (1979).*

reactivation such as in the Benue aborted rift. Therefore, anorogenic magmatism can be viewed as representing zones of incipient rifting, which occurred periodically until the opening of the Atlantic Ocean.

The present-day outcrop level is rather shallow, as both volcanic and plutonic rocks can be observed inside ring complexes (Fig. 12.3b). Volcanic sequences are dominated by rhyolites, but also preserve basaltic rocks. Granophyric textures, typical of fast cooling at shallow levels, are common in these granites. From the petrographic and geochemical point of view, it is possible to recognize peraluminous (S-type) biotite-bearing granites and peralkaline (A-type) granites containing sodic ferromagnesian silicates, such as arfvedsonite (a sodic amphibole). These granites are of economic interest because of associated Sn–Zn–REE placers deposits.

The wide range of initial Sr isotopic ratios is striking, from mantle-like values in some syenites to very high values in the mineralized granites (Fig. 12.3c). The highest ratios required an enrichment mechanism by fluids. The volatiles were likely responsible for geochemical and textural variations, such as the development of microcline. Thus, Martin and Bowden (1981) questioned the origin of S-type granites by partial melting of basement rocks and suggested that peraluminous compositions resulted from the mobility of alkalis (first Na, and then K at a lower temperature) as a result of subsolidus interaction with fluids. Besides, micas also departed from their annite (Fe-rich) composition towards Li-, Al- and F-rich compositions.

12.3 Granites and oceanic subduction

12.3.1 Thermotectonic setting

Subduction, i.e. sinking in the mantle, of oceanic plates is a major process compensating for the formation of new crust along the oceanic ridges. It occurs worldwide, associated with an intense magmatism producing rocks of various compositions. Volcanic eruptions are numerous and spectacular (hence the so-called 'ring of fire' around the Pacific Ocean); their less conspicuous plutonic counterparts are actually much more important in terms of magma volumes. Many granitoids are produced above subduction zones, which are the very places of formation of new (juvenile) continental crust. It is now accepted that subduction-related magmatism is the consequence of dehydration of the subducted oceanic plate at depth. Namely, the oceanic crust locally contains mineral assemblages (or parageneses) typical of an early hydration under greenschist facies conditions. These assemblages are characterized by the occurrence of hydroxylated minerals, such as chlorite which contains 14 wt % water. During plate subduction, evolution from greenschist to amphibolite facies conditions in the crust correspond to an important (but incomplete) loss of water, because amphiboles still retain 2 wt % water. Dehydration of the oceanic plate is effective at the amphibolite–eclogite transition, because eclogites are characterized by anhydrous parageneses. Water produced by these metamorphic reactions is responsible for partial melting of the overlying mantle wedge, because it considerably lowers the solidus temperature of mantle peridotites (Fig. 12.4a).

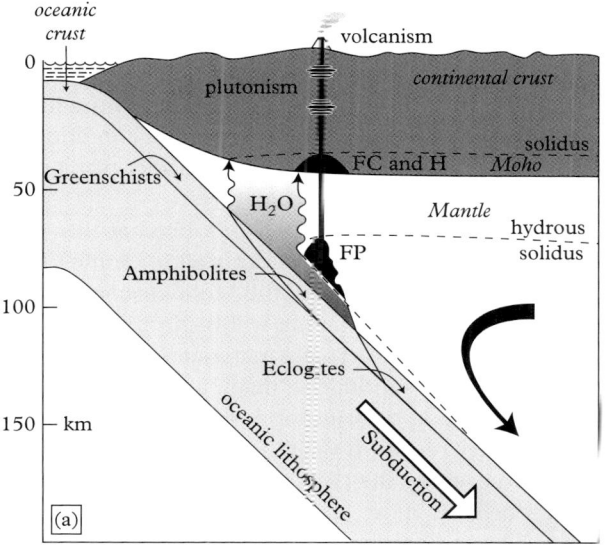

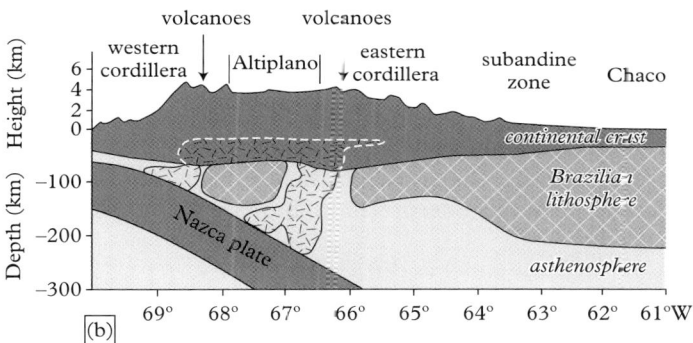

Figure 12.4 *Origin of magmas in a subduction geodynamic setting. (a) Interpretative section of a subduction zone after Wyllie (1984): the hydrated mantle (due to metasomatism) is in grey; magmas are in black. FC: fractional crystallization, PM: partial melting, H: hybridization of magmas. (b) Interpretation of the seismic profile from Central Andes between 18 and 20°S (after McQuarrie et al., 2005); the low velocity zones (dashed areas) indicate fluids and/or magmas. Magma ascent in the continental crust appears to stop at depths of about 14 to 20 km: this is the place where plutons crystallize presently. Note that the altitude scale above sea level is different from the depth scale.*

Convection of the mantle wedge may also occur and will contribute in dragging downward peridotites recently hydrated at the contact of the oceanic plate, hence facilitating their entrainment toward deeper zones, where partial melting temperatures (at least 1000 °C) could be more easily reached. Inside continental margins, melting of the continental crust may also occur under appropriate conditions (thick crust, high geothermal gradient), as suggested by seismic profiles under the central Andes (see Fig. 12.4b), and witnessed by magmatic isotopic signatures.

Experimental data prove that the partial melt produced at the relevant depth in the mantle is of basaltic composition. This magma can stall for some time during its ascent toward the surface, especially when it arrives at the Moho discontinuity beneath an island arc or a continental margin. Then, the magma will begin to differentiate due to crystal fractionation, mixing or assimilation of crustal material. These mechanisms can explain the large range of compositions of subduction magmas, as well as the fact that the most abundant compositions sampled at the outcrop level are andesitic (or tonalitic for their plutonic counterparts), namely rocks whose average silica contents is close to 60 wt %.

In addition, sediments covering the oceanic plate may contribute to magma production, leading to compositions more silicic than the basaltic compositions obtained in melting experiments. The sedimentary contribution has been demonstrated, for instance, in the Antilles. A recent compilation of the compositions of subduction granites from Japan suggests that the more or less reduced character of the granitic magmas may have been controlled by the nature and volume of sediments arriving at depth. The sediment volume depends on the convergence rate. C- and S-rich terrigenous sediments, deposited during the Cretaceous oceanic anoxic event and then subducted with the Pacific plate, probably contributed to the predominant formation of ilmenite-bearing (reduced: see Fig. 10.6) granites in the Cretaceous batholith of the Asiatic margin (Fig. 12.5). Sulfur isotope data confirm this interpretation. Indeed, S isotopes do not fractionate at

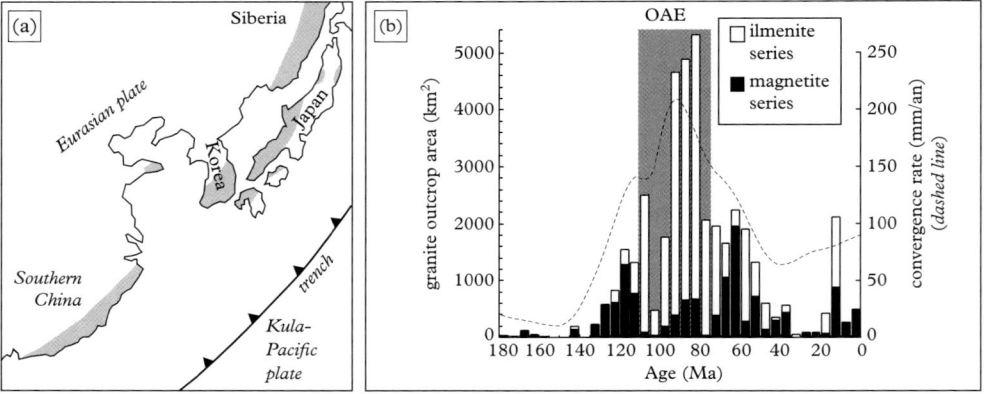

Figure 12.5 *Cenozoic granites in Japan. (a) Map of granitic batholith at the Asian margin for Cretaceous times (grey shaded area). (b) Evolution of granite compositions with time; OAE: oceanic anoxic event. After Takagi (2004).*

magmatic temperatures and, consequently, any difference with respect to mantle values is inherited from the source of the magma. Thus, the lighter S-signatures of ilmenite-bearing granites witness a contribution from sedimentary sulfides, whereas the signatures of magnetite-bearing granites are inherited from hydrothermalized mafic oceanic crust, i.e. they reflect the heavier S-isotope ratios of seawater sulfate.

12.3.2 Island arc granitoids

Crustal nature and thickness of an island arc change with time. Compositions of arc magmas also change with the age of the arc. As a rule, in a young ocean–ocean subduction, magmas are arc-tholeiites. They become more and more potassic (and eventually more and more felsic) when the arc crust becomes thicker. In this case, island arc magmas are calc-alkaline *sensu stricto*, a composition regarded as typical of subduction settings (Fig. 12.6). Few young island arcs display plutonic rocks at the outcrop level. In the French Antilles, there are mainly volcanic rocks, but occasionally a small pluton can be also observed. The Kohistan island arc was first built upon oceanic crust and then tectonically accreted to the Himalayan belt and tilted. It displays a full section from the mantle–crust transition at the floor to volcano–sedimentary sequences at the roof (Fig. 12.7a). The arc crustal section contains a lower complex of gabbros, ultramafic cumulates and amphibolites, overlain by a huge calc-alkaline batholith made of diorites, tonalites, granodiorites and granites (Fig. 12.7c). Mafic to intermediate rocks represent a large volume of the whole magma spectrum. The gabbroic complex is regarded as the root of the arc. The building of the Kohistan granite batholith was induced

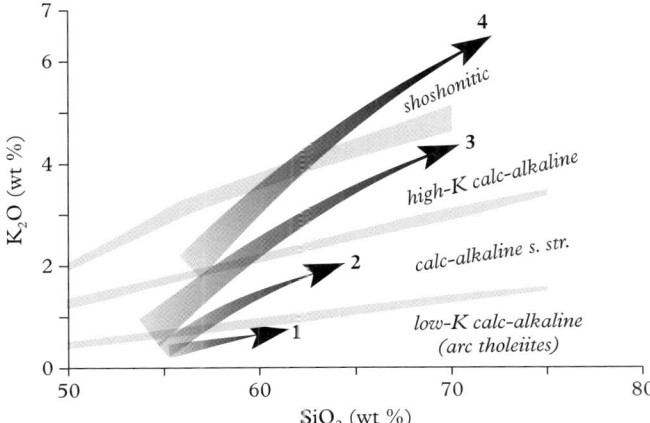

Figure 12.6 *Classification diagram of K_2O versus SiO_2 showing the spatio-temporal evolution of subduction magmatism: magmas become increasingly silica- and potassium-rich with time (series 1 to 4) as well as when the distance to the trench increases. Field limits after Rickwood (1989).*

236　Granites and plate tectonics

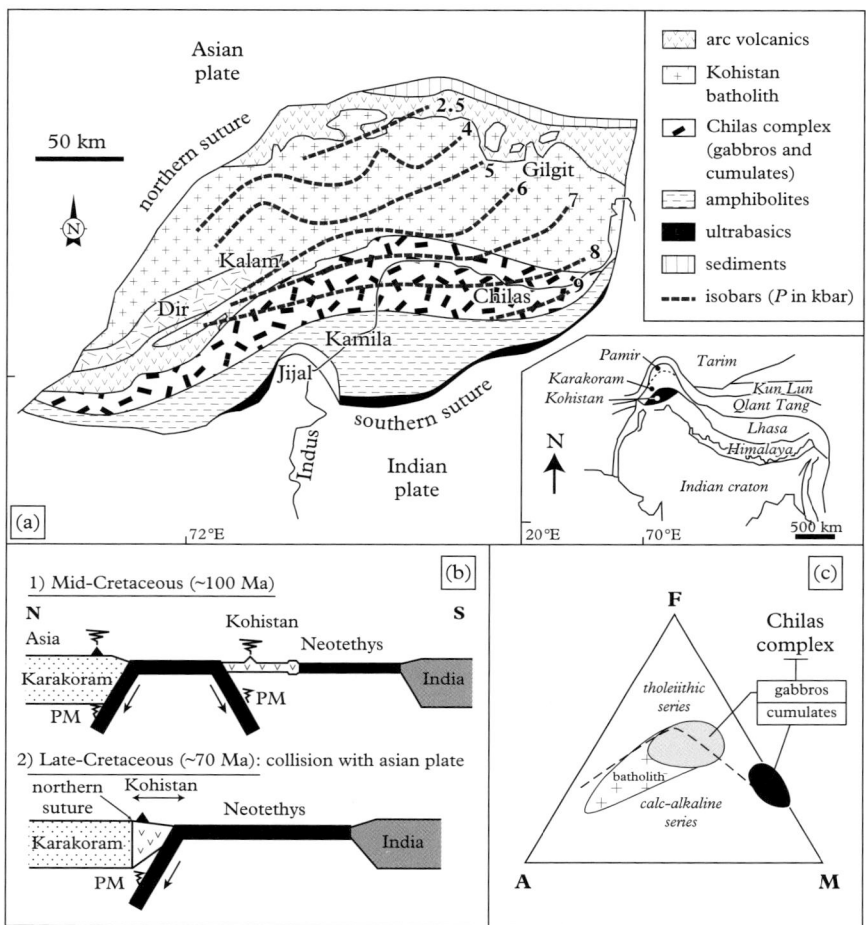

Figure 12.7 *The Kohistan arc (see location in Fig. 12.11). (a) Simplified geological map after Bard (1983) and emplacement pressure of granitoids after Al-in-hornblende barometry by Jagoutz et al. (2013). (b) Geodynamic history, after Khan et al. (1997) and Danishwar et al. (2001); PM: partial melting. (c) Geochemical properties of plutonic rocks, after Takahashi et al. (2007).*

by the subduction of a northern oceanic basin and occurred from 110 to 40 Ma ago (Fig. 12.7b). After the collage of the Kohistan arc to the Eurasian continent along the northern Shyok suture, the Kohistan batholith represented the Eurasian continental margin, under which the Neotethys Ocean was being subducted.

Calc-alkaline magmatism generally comprises different series with small to high potassium contents. Strontium initial ratios (Sr_i) are in the range 0.704–0.705, in agreement with the mainly juvenile nature of the batholith. Hornblende-controlled fractionation of a hydrous mafic melt was responsible for the generation of large volumes of granitoids.

The efficiency of this fractionation process is explained by the silica-poor nature of amphibole and the relatively low viscosity of hydrous melts.

The youngest intrusive rocks (*ca.* 30 Ma) are muscovite (± garnet)-bearing peraluminous leucogranites related to the India–Eurasia collision. They only represent but a very small volume. Further East, the Ladakh batholith has exactly the same history as the Kohistan batholith, because both of them belonged to the same island arc, but its root cannot be observed.

12.3.3 Granitoids in active continental margins

The North American cordillera and the Andes display wonderful granitic outcrops of Meso- to Cenozoic ages due to the existence of many active subduction zones along the western margin of North and South America. These granitoids form multiple plutons arranged as elongate batholiths, parallel to the west coast.

The Sierra Nevada batholith of California (Fig. 12.8) consists of hundreds of plutons whose average volume is 30 km^3 and whose ages range from Triassic to Cretaceous. This batholith is related to the subduction of the now-disappeared Farallon plate. It represents a total volume of *ca.* 100 km^3 of magmas of granodiorite composition. The mixed origin of these granitoids has already been demonstrated: both mantle and crustal contributions were together confirmed by isotopic studies (see Fig. 2.24).

The Andes cordillera is characterized by two batholiths of more than 1000 km in length: the North-Patagonian batholith in Chile and the Coastal batholith in Peru. The latter is 1600 km long and 50 km (or less) wide (Fig. 12.9a). It includes several thousands of plutons gathered as units, and superunits (four from north to south). Ages range from 100 to 30 Ma. Plutons were emplaced at shallow level and are even locally intrusive in the related volcanics. Each pluton displays sharp contacts with the country rocks. Granitoid modal compositions range from tonalites to leucomonzogranites (Fig. 12.9b). Scarce gabbros can also be observed at the outcrop level. The rocks follow a typical calc-alkaline trend. The geochemical features of the most differentiated granites are

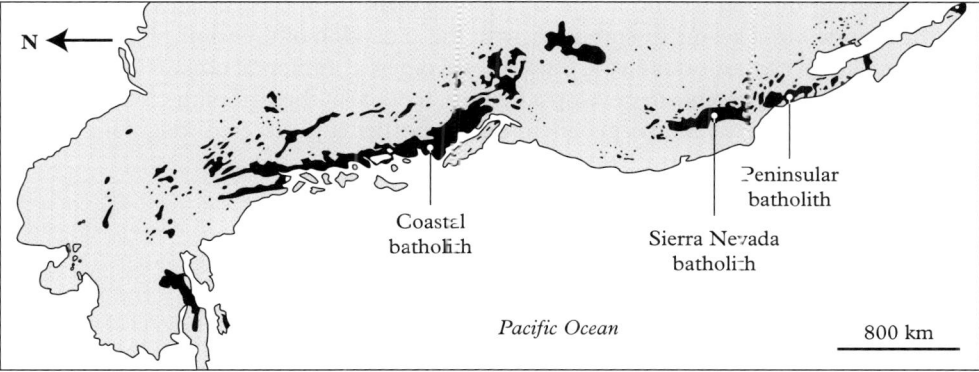

Figure 12.8 *Map showing the granitic plutons along the western coast of North America.*

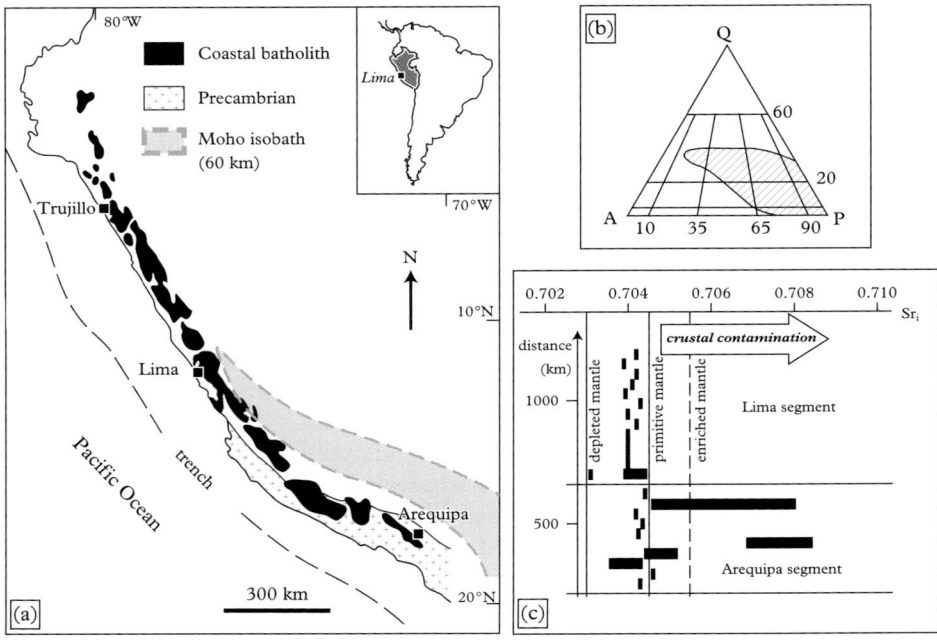

Figure 12.9 *The Peruvian Coastal batholith. (a) Map of the batholith showing the Moho isobath (60 km). (b) Compositional field of all plutonic rocks in the Streckeisen diagram. (c) Variations of Sr_i ratios depending on latitude. After Pitcher et al. (1985).*

consistent with a fractional crystallization process extracting plagioclase and hornblende from a tonalitic magma. Despite the general consensus that granitoids from active margins are hybrid rocks, the relative importance of mantle and crustal contributions is still a matter for discussion. Seismic data from the Central Andes suggest that magmas are present both in the mantle and in the continental crust. In the Peruvian Coastal batholith, the mantle input seems to dominate, with the exception of the southern area around Arequipa, which corresponds to the thickest and oldest continental crust (Fig. 12.9a and c). Besides, it is not possible to develop a definite scheme at the batholithic scale, because each plutonic unit has its own geochemical properties and its own magmatic history despite close similarities between the units.

12.4 Granites and continental collision

12.4.1 Thermotectonic setting

The genesis of a collisional mountain belt results from the closure of an ocean, locally identified by ophiolitic relics along the suture zone. Collision of two continental margins produces an orogenesis with a thickened crust due to stacking of crustal nappes or, more precisely, the underthrusting of crustal units beneath each other. Later, the underthrusted units can be exhumed by tectonics and/or erosion-driven isostatic rebound. The fate of these units and especially their thermal evolution will control the possibility of partial melting and the production of granitic magmas.

Thermal evolution of a thickened continental crust has been numerically modelled by England and Thompson (1984) and the result is shown in Fig. 12.10a. For isostatic reasons, uplift of an orogenesis begins just after the thickening event, hence the decompression of the underthrusted lower unit. At the same time, but more slowly, because rocks are not efficient heat conductors, the temperature of the underthrusted unit increases. This process is called thermal relaxation. It continues until the interplay of erosion and uplift brings the rock close enough to the surface to begin to cool. Finally, the resulting P-T-t (pressure–temperature–time) path of a rock located in the underthrusted unit displays a significant curvature, passing through a thermal peak several tens of millions of years after the crustal thickening event. This is the most favourable time to reach the conditions of an important anatexis, which will therefore always be delayed after nappe tectonics (Fig. 12.10b and c). It is also worth noting that, during this evolution, magma formation always occurs during decompression. These considerations render obsolete former distinctions between 'syn-collisional' and 'post-collisional' granites. Actually, most granites appear to be post-collisional in the sense that they were formed after the beginning of collision, i.e. after the crustal thickening event.

This model can be applied to any collisional orogenesis. However, the volume of magmas produced depends on the amount of crustal thickening and on the rate of crustal uplift. The described model is based on slow erosion-driven isostatic uplift. If tectonically exhumed, the underthrusted crustal units are uplifted faster and may not have reached temperatures high enough for anatexis. This is the case in the French–Italian Alps. 'Cold' orogeneses, i.e. with small volumes of granites, are therefore distinguished from 'hot' orogeneses characterized by multiple granite intrusions.

Moreover, the tectonic history of a collisional orogenesis often takes place as follows. After the crustal thickening stage due to nappe stacking, further plate convergence results in the formation of transcurrent mega-shear zones at variable distances from the suture zone. In these tectonic structures, up to thousands of kilometres long, that involve the whole lithosphere thickness, partial melting of the lithospheric mantle and of the continental crust may occur. The resulting magmas are felsic or intermediate in composition, with possible mantle inputs as shown by their isotopic signatures. The corresponding granitic compositions are diverse: peraluminous, calc-alkaline potassic, shoshonitic or even alkaline.

Meanwhile, the thickened orogenesis follows its own thermal evolution leading to low pressure–high temperature (LP-HT) conditions. In the waning orogenic stages, thinning and extension of the previously thickened crust may occur and will be discussed in Section 12.5. Granites formed in this case are very similar to other granites produced during anorogenic extension, because of similar (LP-HT) conditions and because of the possible involvement of mantle-derived magmas, contributing at least as heat vectors if not as parental magmas to the genesis of granitic magmas.

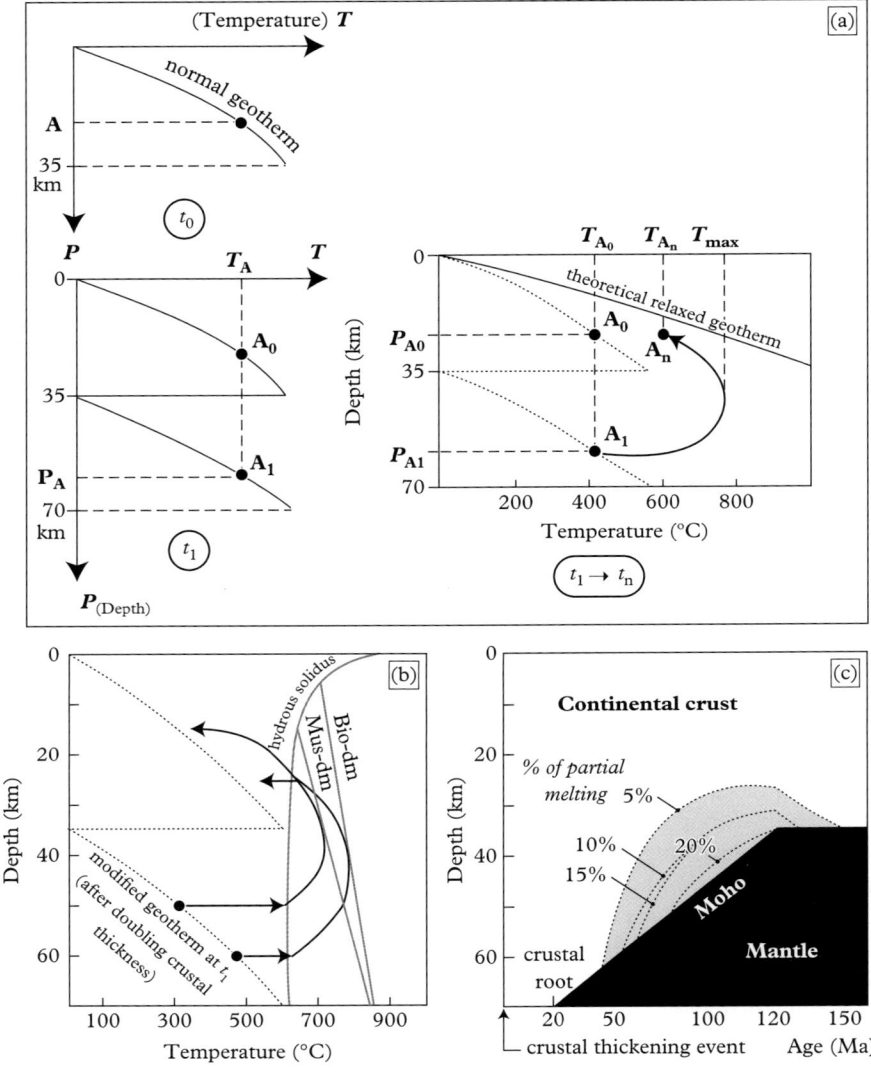

Figure 12.10 *Numerical modelling of collision and crustal melting occurrences. (a) Model of England and Thompson (1984): evolution of (P,T) conditions for rock A, after instantaneous doubling of crustal thickness ($T_{A1} = T_{A0}$, $P_{A1} = P_{A0} + 35$ km); evolution from A_1 to A_n takes into account the heating of the underthrusted crustal unit (thermal relaxation), as well as the erosion-driven isostatic rebound. Fine dotted lines: geotherms at time t; thick curve: P-T-t path for rock A; note that the peak temperature is reached about 60 Myrs after crustal thickening. (b) P-T-t paths for different loci of the underthrusted unit: a minimum crustal thickness of 50 km is required to intersect the muscovite dehydration–melting curve (Mus-dm) and a minimum thickness of 60 km to intersect the biotite dehydration–melting curve (Bio-dm). (c) Percentage of melting in the lower crust with time. Be aware that these magmas do not stay in the lower crust, but generally migrate toward the surface; in addition, magmas modify crustal rheology. (b, c) After Thompson and Conolly (1995).*

Two examples of collisional belts will be addressed in detail: the Cenozoic Himalayan belt and the older (Palaeozoic) Hercynian (or Variscan) belt of western Europe.

12.4.2 Himalayan leucogranites

The Himalayan belt results from the collision of India and the Tibetan Asiatic margin during early Eocene times at around 55 Ma, after subduction of the Tethys Ocean. A few ophiolitic remnants indicate the location of the suture zone (Fig. 12.11a). Before collision, the Asian margin was an active Andean-type continental margin where several calc-alkaline granitoids compose the Trans-Himalayan batholith. As a consequence of the collision, the passive margin of India has been noticeably thickened due to major southward thrusts that were successively active: first the Main Central Thrust (MCT), then the Main Boundary Thrust (MBT) and finally the Main Frontal Thrust (MFT). The MCT is characterized by inverse metamorphism of tectonic origin. Indeed, the high grade gneisses of the Higher Himalaya were thrusted upon lower grade metasediments of the Lower Himalaya (Fig. 12.11b). Anatexis of the gneisses yielded leucogranitic magmas that were emplaced at the base of the overlying sediments (see Figs 2.23 and 5.12a). Leucogranites form a linear succession along the belt (Fig. 12.11a and b). All of them have a Miocene age of about 20 Ma.

They are light coloured S-types, containing minerals typical of their peraluminous composition (muscovite, garnet or tourmaline). Their genesis was first explained by a water-present reaction. Water necessary for melting would have been produced by prograde metamorphic reactions in the sediments underlying the MCT: this was the once fashionable 'hot-iron' model of Le Fort (1975) (Fig. 12.11c), which is consistent with inverse metamorphic zonation observed on both sides of the MCT. However, water-present melting is dismissed by the whole-rock geochemistry of the Himalayan leucogranites, and a water-absent dehydration-melting of the muscovite-bearing High Himalaya gneisses is now preferred (Harrison et al., 1999). The final partial melting stage occurred at low pressure as indicated by the crystallization of andalusite and cordierite in some leucosomes. The P-T-t path of the crustal unit corresponding to the High Himalaya is rather steep, suggesting a relatively fast exhumation, partly of tectonic origin. The detachment fault on top of the exhumed unit is therefore a particular sort of normal fault, active at the same time as the MCT, which opened preferential sites for the emplacement of granitic magma. The most recent magmatism is not observed in the Himalayas, but further north, especially in relation to major transcurrent faults in Tibet. It is a mafic to intermediate volcanism, very rich in potassium (shoshonitic), but whose plutonic equivalents crystallized at depth and therefore cannot be observed at the current outcrop level.

12.4.3 Hercynian granites of western Europe

The Hercynian orogenesis of western Europe comprises two belts with opposed vergencies. The northern belt resulted from the closure of the Rheic Ocean and the collision of two microplates, successively detached from the northern part of the Gondwanan

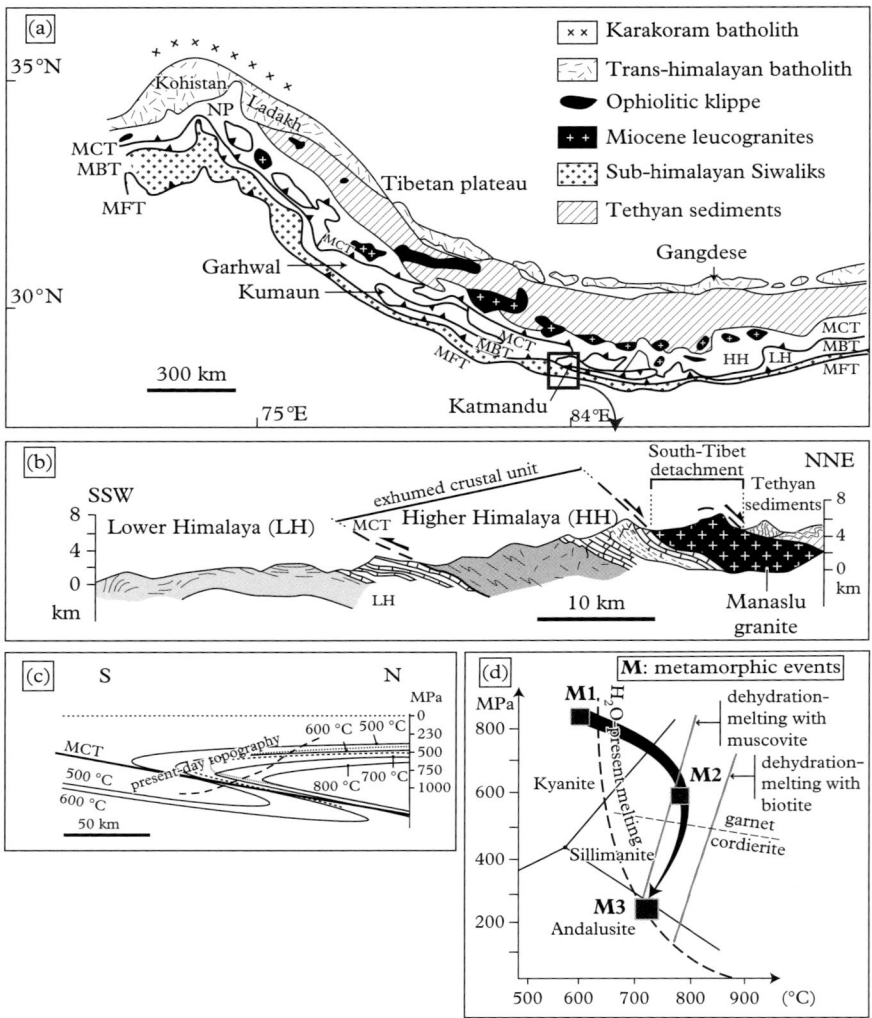

Figure 12.11 *Himalayan leucogranites. (a) Simplified geological map after Johnson (2003). (b) North-south cross-section at the Manaslu longitude, after Guillot (1999). (c) Le Fort (1975) petrogenetic model. (d) P-T-t path of the Higher Himalaya gneisses, after Pognante and Benna (1993).*

supercontinent, namely the Avalon (or Avalonia) microplate (accreted to the future Devonian Old Red Sandstone continent, or Laurasia, during Silurian times) and the Armorican microplate. The southern belt resulted from the collision of Armorica and Gondwana. Other microplates, also detached from Gondwana, are sometimes considered, e.g. for the basement of the External Crystalline Massifs of the western Alps. Two

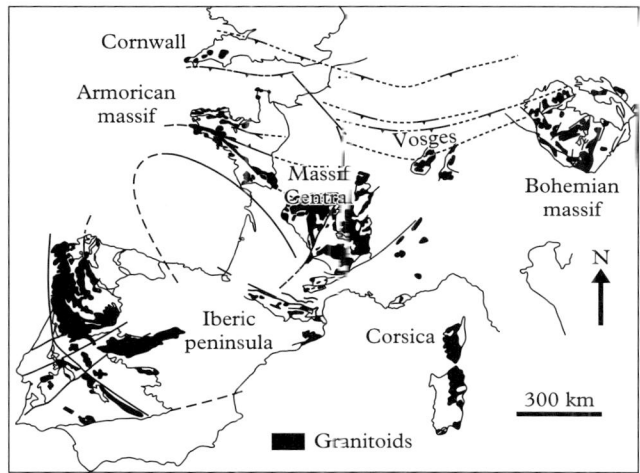

Figure 12.12 *The Hercynian orogenesis of western Europe: location of the main tectonic structures and granitic plutons (black).*

suture zones can be recognized: the northern one due to the closure of the Rheic Ocean, whose remnants correspond to the Lizard ophiolites in Cornwall and possibly to the Channel magnetic anomaly, and a southern one delineated by a succession of blue schists and eclogitic relics in the French Armorican and Central massifs, as well as in the Vosges massif (Fig. 12.12). The total width of the subducted oceans is a matter of debate, but it is worth noting that only a few granites can be related to a continental margin active before collision. This may indicate that the subducted oceans were relatively narrow, thus preventing the establishment of subduction-related magmatism of long duration and the production of large magma volumes. The Limousin tonalites, now completely transposed by post-collisional nappe tectonics, the Marche granites in northern Limousin, and the 'Champ du Feu' granodiorite in the northern Vosges are likely examples of syn-subduction ante-collision granitoids. All of them are Devonian in age. Nappe emplacement resulting from continental collision occurred at the end of Devonian times (at about 360 Ma). In the southern Hercynian belt, nappe tectonics are responsible for the stacking of two gneissic units upon domains regarded as autochtonous (e.g. the mica schists of Millevaches plateau).

Most of the granitic magmatism is Carboniferous in age. Despite its small volume, the S-type Chanteix granite in the Limousin provides convincing evidence for post-nappe occurrence of crustal anatexis and granite emplacement (Fig. 12.13). Its age is Lower Carboniferous.

Granitic magmatism became more and more important during mid-Carboniferous times in the whole Hercynian orogenesis. At that time, it was coeval with transcurrent tectonics and locally with post-collisional extension. Indeed, mid-Carboniferous S-type granites in the French Central massif were emplaced in an extensional setting (see Section 9.3 and Fig. 9.18). The Margeride granite, the largest granitic massif in France, is

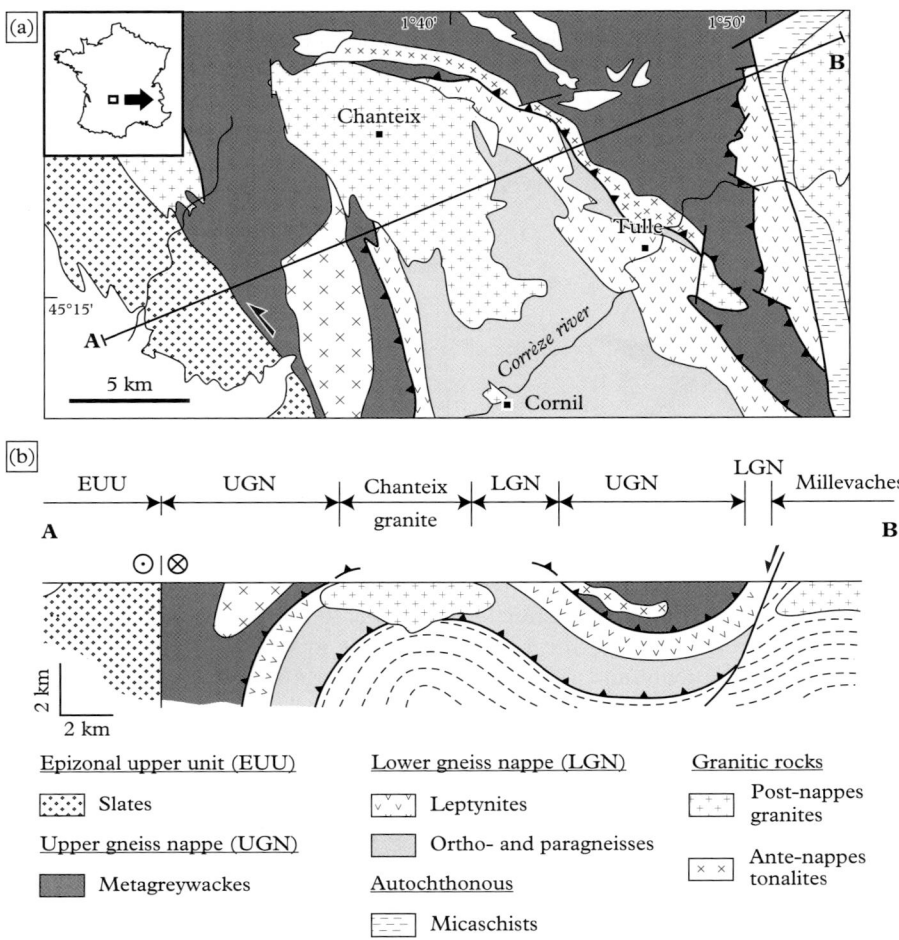

Figure 12.13 *The Chanteix granite. (a) Geological map of Limousin. Note that the granite is cutting the thrusted contact between the lower and upper gneiss units, whereas tonalites emplaced before nappe tectonics are only observed as deformed rocks in the upper gneiss unit. (b) Cross-section showing post-nappe folds and Argentat normal fault. After Roig et al. (1998).*

a thick sill characterized by magmatic lineations striking NW–SE witnessing the extension direction of the orogen (Fig. 12.14). Extensional tectonics finally prevailed in Late-Carboniferous times (see also Section 12.5.2).

In the Pyrenees, Hercynian granitoids were emplaced during mid-Carboniferous times (310 ± 5 Ma) in relation to a major transpressive dextral shear zone (see Fig. 10.22). Whereas all post-collisional granites cited so far are S-types, the Pyrenean granitoids are I-type and follow a typical high-K calc-alkaline differentiation trend

Granites and continental collision 245

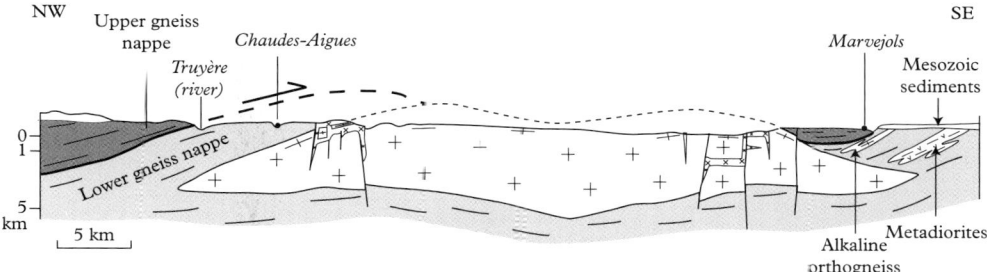

Figure 12.14 *Cross-section of the Margeride granite. After Talbot et al. (2005).*

(Fig. 12.6). They contain numerous mafic enclaves. Magma sources are therefore different from those of the S-type granites from the Armorican and Central massifs. Isotopic data enable the identification of their source protoliths and point to the existence of significant mantle contributions in the more mafic rocks.

In the Armorican massif, S-type two-mica leucogranites were synkinematically emplaced along the south Armorican shear zone at around 315 Ma. Their '*cornue*' (or tear) shapes are kinematic indicators of a dextral sense of motion (Fig. 12.15a). Structures

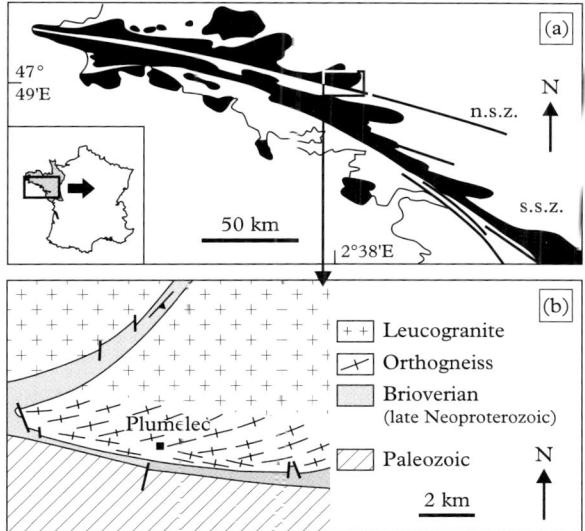

Figure 12.15 *The synkinematic leucogranites of southern Brittany. (a) Location of the granites (n.b.: northern branch of the south Armorican shear zone; s.b.: southern branch). (b) Detailed map with foliation strikes in the synkinematically emplaced granites close to the shear zone. After Berthé et al. (1979) and Jégouzo (1980).*

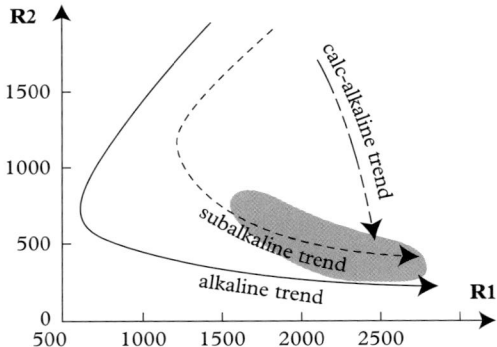

Figure 12.16 *Plot of the Hercynian red granites from the Alps (grey shaded area) in the R1–R2 diagram of La Roche et al. (1980), with R1 = 4Si − 11(Na + K) − 2(Fe + Ti) and R2 = 6Ca + 2Mg + Al, all in millications. After Debon and Lemmet (1999).*

and microstructures witness their progressive deformation as orthogneisses during emplacement (Fig. 12.15b). Finally, late-orogenic transcurrent tectonics favoured the emplacement of high-K calc-alkaline to subalkaline granites, e.g. the bimodal Ploumanac'h pluton (Fig. 11.8). This pluton belongs to a series of late pink granites, such as the Flamanville granite in Cotentin (Normandy). The Ploumanac'h pluton is a zoned pluton of 302 Ma age. It comprises three centripetally-nested granites and a gabbro-noritic stock. The nature of the contacts suggests three nearly-coeval separate magmatic pulses. The outer rim is composed of a reddish porphyritic alkali feldspar granite. The fine-grained granite in the inner rim has similar affinities, whereas the core consists of a two-mica- and cordierite-bearing S-type granite derived from a different source, as can be deduced from the mineralogy and geochemistry (see the O isotope data in Fig. 11.8).

The red granites of the external crystalline massifs in the Alps are coeval, but their tectonic setting has not been studied in detail. However, they might have been emplaced along a late-Hercynian shear zone. From the geochemical point of view, they are subalkaline potassic granites (Fig. 12.16).

It is important to note the diversity of granitoids formed during late-orogenic transcurrent tectonics. Such a diversity results from different possible crustal protoliths as well as from the eventual contribution of mantle magmas.

12.5 Granites and continental extension

12.5.1 Thermotectonic setting

Continental extension may occur in different ways: it may be localized along a rift or distributed in a much a larger area (e.g. the Basin and Range province in the western

USA). It is worth noting that extension may also occur in an orogenic setting. In this case, extension may result from the thermal evolution of the crustal root of the mountain belt, sometimes associated with the detachment (delamination) of the lithospheric root. Extension may also happen as an indirect consequence of continental collision and orogenesis. This is exemplified by rift opening in the foreland, immediately at the front (or even at some distance) from the mountain belt. Finally, continental extension may also occur without any relation to orogenesis. Indeed, extension occurs at the beginning of continental rifting, and will possibly continue until drifting and oceanic crust formation. Such an extension is sometimes improperly called 'anorogenic', namely without any relation to orogenesis. It may or not be heralded by a mantle plume and its typical voluminous magmatism of mantle derivation.

In any case, continental extension results in crustal thinning and isotherm tightening (Fig. 12.17a), hence a geothermal gradient warmer than the average. In this setting, LP-HT (low pressure and high temperature) metamorphism is observed, but crustal melting will not necessarily happen if the Moho temperature does not increase. On the other hand, if lithospheric thinning and formation of mantle-derived mafic melts occur together with crustal thinning, crustal melting will be favoured. Indeed, mafic melts en route to the surface will sometimes stop and remain at the base of the crust (*underplating*) or in the lower crust (*intraplating*) inducing a significant temperature increase (Fig. 12.17b). This latter situation is very similar to the case of granite formation in relation with 'hot spot' activity (see Section 12.2).

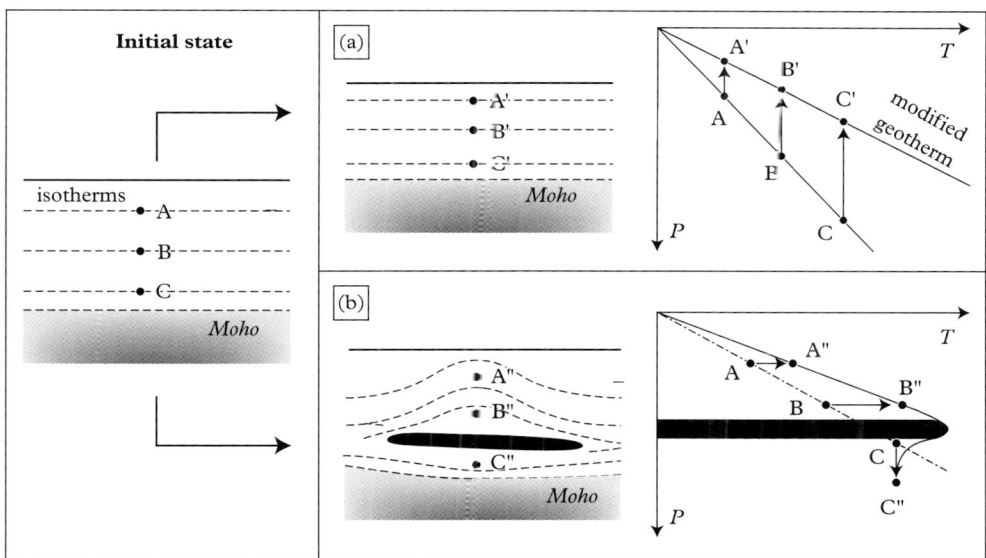

Figure 12.17 *P-T conditions during continental extension. (a) Modification of isotherms and geotherm after crustal thinning; resulting P-T paths for A, B and C. (b) Same after crustal thinning and emplacement of a mafic sill in the lower crust.*

To obtain a complete overview of granitic plutonism associated with the Hercynian belt of western Europa, the following sections are dedicated to the late-orogenic extension in the Hercynian belt and then to the Oslo rift that opened at some distance in the foreland of this belt.

12.5.2   Granites related to Hercynian late-orogenic extension

Late-Hercynian extensional tectonics occurred at the end of the Upper Carboniferous system during the so-called 'Stephanian' stage (304–299 Ma). The name of this stage is derived from the city of Saint-Etienne (i.e. Saint-Stephen) in the eastern French 'massif Central'; it corresponds to the upper part of the Pennsylvanian epoch. Extensional tectonics produced grabens or half-grabens (for instance, the sedimentary coal basin at Saint-Etienne) in the upper brittle crust, whereas the geothermal gradient was still increasing in the ductile lower crust. Actually, the conditions of partial melting had already been reached in the lower crust because of the phenomenon of thermal relaxation (see Section 12.3.1). The degree of partial melting was thus higher and the crustal viscosity decreased by several orders of magnitude. The mountain belt could not be sustained in these conditions and would eventually collapse (Vanderhaeghe and Teyssier, 2001).

As soon as the divergent gravity forces passed over the external convergent forces at plate boundaries, extensional tectonics began in the innermost part of the belt, despite continuing plate convergence! The partially-molten lower crust was exhumed as migmatitic and granitic domes, e.g. the Velay dome (Fig. 5.5). The Velay granite is a very heterogeneous S-type cordierite-bearing granite of crustal derivation. Numerous mica-rich enclaves represent restitic material (i.e. crystals inherited from the granite protoliths). Some granite types, typically the poorest in SiO_2 and the richest in Al_2O_3, FeO and TiO_2, may contain up to 35% restite. A few coeval mafic magmas, the so-called 'vaugnerites' (high-K diorites), are indicative of lithospheric thinning and mantle melting, possibly due to the sinking of the cold and dense lithospheric root (a process also called 'lithospheric delamination'). The cold root was then replaced by hot asthenosphere, which might have been subjected to decompression melting.

The contribution of mantle magmas is more easily recognized in crustal sections, where recent tectonics tilted the lower crust and where mafic sills originally emplaced close to the Moho crop out at the surface. These magmatic intrusions were metamorphosed as mafic granulites. Due to the high P and T conditions at the base of the crust, these rocks consist of garnet-pyroxenites (containing Na-poor clinopyroxene) characterized by a conspicuous annealed texture. The gabbroic texture is sometimes well preserved in mafic intrusions emplaced at higher crustal levels, namely at the granulite-amphibolite transition. The Ivrea and Lake (*Laghi*) zones in northern Italy (Fig. 1.7) or the Calabria in southern Italy (Fig. 7.14) provide excellent sections of the late-Hercynian continental crust. Granites from the *Serie dei Laghi* are either I- or A-type (Baveno red granite). In Calabria, both S- and I-type granites were formed in this thermal and tectonic event.

Finally, the A-type Permian granites in Corsica correspond to the transition toward an anorogenic extensional setting, which will trigger the rupture of Pangea in the Mesozoic. These granites have already been discussed at the end of Chapter 2 (see also Fig. 2.27).

12.5.3 Granites related to the Oslo rift in Norway

The Oslo rift opened in the foreland of the Hercyian belt at the southern margin of the Baltica craton (Fig. 12.18a). Its axis has a north–south orientation that is nearly perpendicular to the orogenesis. Extension began at the end of the Carboniferous and

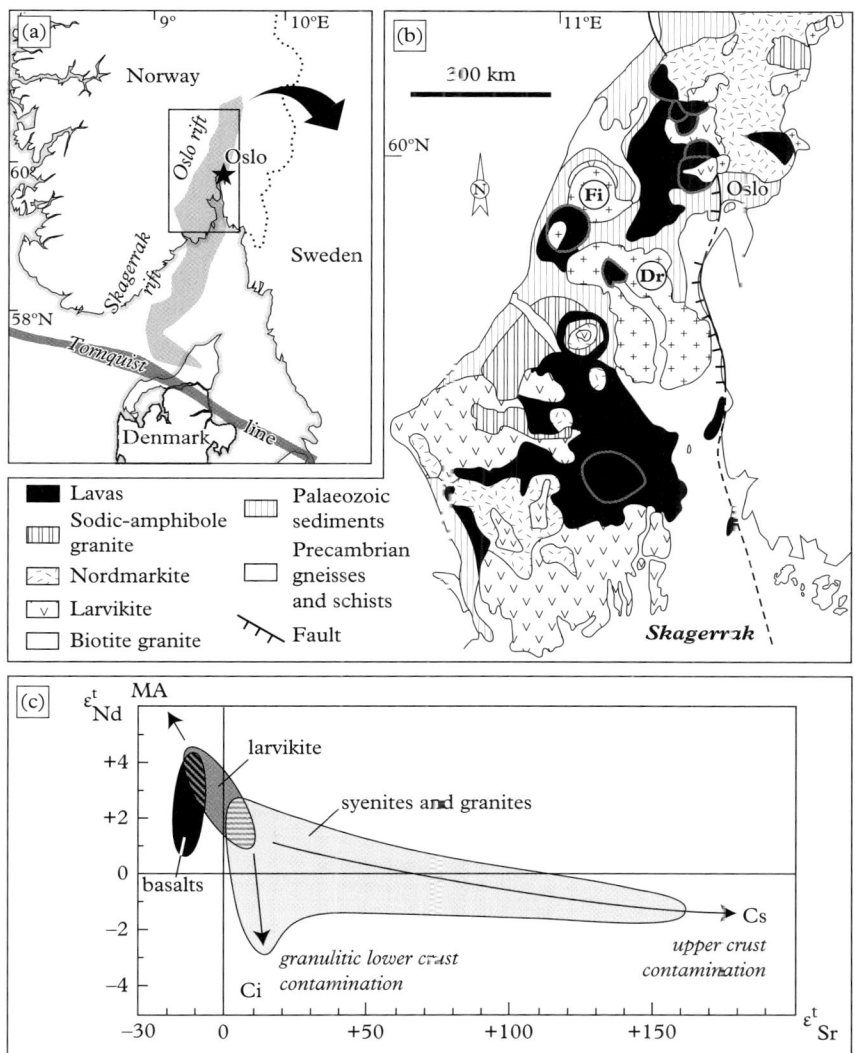

Figure 12.18 *The Oslo rift. (a) Location. (b) Intrusion map. (c) Isotope data; LC: lower crust, UC: upper crust, Dr: Drammen, Fi: Finnemarka. After Tronnes and Brandon (1992) and Neumann et al. (1995).*

Box 12.1 What are the links between geochemistry and tectonics?

Figure 12.19 shows all the previously studied magmatic series. It is difficult to correlate each of them to a unique tectonic setting, e.g.

- A-type granites can be formed in anorogenic settings (in relation to hot spots or continental rifting) or during late- to post-orogenic extensional conditions;
- S-type granites are abundant in collision belts, but are also present in anorogenic or extensional settings;
- I-type granites are typical of subduction zones and also occur in continental collisional belts.

Warning! The so-called 'tectonic discrimination' diagrams (e.g. Pearce et al., 1984) must be used with caution. These diagrams are excessively used, and even misused, to determine geodynamic settings of emplacement, whereas the relevant geochemical indicators are controlled by the nature of the protoliths. For instance, immature sedimentary protoliths (e.g. greywackes) are abundant in active margins, but may be also present in collisional belts. No tectonic conclusions should be derived without a structural study of the intrusion and its country rocks, associated with the determination of the regional metamorphic conditions.

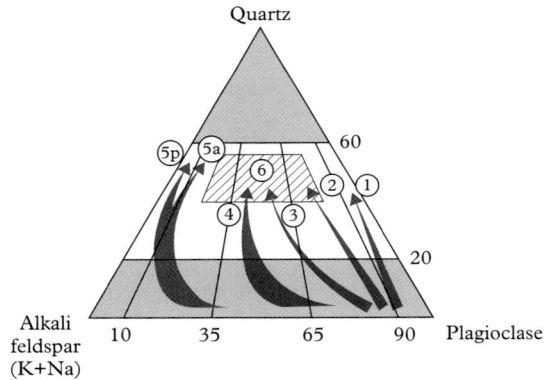

Granite types	Magmatic series	Geodynamic contexts						
		hot spot	oceanic accretion	subduction		continental collision		continental extension
				insular arc	continental margin	post-nappes	transcurrent s.z.	
I	1 tholeiitic		●	●				
	2 calc-alkaline			●	●			
	3 high-K calc-alkaline				●	●		
	4 shoshonitic				●	●		
A	5a aluminous	●						●
	5p peralkaline	●						●
S	6 peraluminous	●				●	●	●

Figure 12.19 *The different granitic series and their occurrence in the main geodynamic settings.*

was on-going during the Permian. Magmatism began with flows of alkali basalt. Several annular complexes were emplaced during the main extensional phase. Intrusive rocks comprise monzonites (the well-known larvikites quarried for ornamental use), quartz-syenites (or nordmarkites) and granites (Fig. 12.18b). The most extensive granite is the Drammen biotite-bearing granite that displays aluminous to peraluminous characteristics (S-type). In contrast, the Finnemarka granite displays alkaline characteristics (A-type). These granites are rich in silica (SiO_2 = 70–78%). Their isotopic signatures are consistent with a fractional crystallization process of mantle-derived magmas at depth, followed by contamination of the most evolved magmas mainly by the upper crust, as shown by high Sr initial ratios (Fig. 12.18c). In contrast, the contribution of the lower crust was limited, because of the granulitic and refractory nature of the Norwegian lower crust. The parental mantle magmas were derived from the asthenosphere or from the lithosphere. Large dense intrusions are substantiated by geophysical data from the lower crust in the rift zone. These intrusions most likely correspond to gabbroic cumulates. Their estimated volumes compared to those of the most differentiated rocks (syenites and granites) are consistent with a fractional crystallization process. It must be emphasized that the continental crust acted as a filter retaining mafic magmas at depth, whereas the evolved more felsic magmas were able to ascend to shallow levels: this is the crustal filter concept first suggested by Bonin (1996).

12.6 Conclusions

Formation of granitic magmas requires distinctive geodynamic settings, either intraplate conditions or at plate boundaries, especially in convergent settings. In intraplate conditions, heat must be provided by mantle-derived magmas. At plate boundaries, crustal melting may result from similar causes or from geothermal perturbations following the thickening of the continental crust.

Juvenile granitoids responsible for the growth of the continental crust are typically produced above subduction zones. This is the most suitable context for the formation of hybrid granitoids (with mixed isotopic signatures pointing to both crustal and mantle inputs). In other geodynamic settings, granites are mainly derived from the recycling of an older continental crust, resulting in an increased crustal differentiation, i.e. an increased compositional contrast between the upper and the lower continental crust.

Continental collision generally favours the production of large volumes of granitic magmas. However, granitic plutonism occurs after a delay in the tectonic history, typically a few tens of million years after crustal thickening.

Finally, it must be emphasized that no single granite-type is a specific indicator of a unique geodynamic setting.

13

Precambrian granitic rocks

The duration of the Precambrian supereon spans approximately four billion years (4000 Myr). It encompasses a large part of Earth's history. The Precambrian supereon is divided into three eons: Hadean (4.6–4 Ga), Archaean (4–2.5 Ga) and Proterozoic (2.5 Ga to 541 Ma). It is worth noting that the Archaean and Proterozoic eons are divided, respectively, into four and three eras. Precambrian granitic rocks are common and represent the dominant part of the continental crust as in Phanerozoic times, but they are quite different from younger granites.

After preliminary discussion of dating methods used for assessing the age of granitic rocks, this chapter will review the dominant types of Precambrian granites *sensu lato*. The last section in this chapter is dedicated to the special case of granitic rocks derived from important high velocity bolide impacts, likely common during the formation of early Earth and of sporadic occurrence in recent times.

13.1   Age determination of granitic rocks

13.1.1   Crystallization ages

To determine the age of a granite, or of any magmatic rock, is to assess when the magma crystallized. The best method to achieve this goal is by dating zircon, an early crystallizing mineral, that is also quite resistant to metamorphism and weathering. Zircon ($ZrSiO_4$) contains uranium atoms substituting for zirconium. Uranium has two radioactive isotopes, ^{238}U and ^{235}U, whose radioactive decay yield radiogenic lead, respectively, ^{206}Pb and ^{207}Pb. Owing to these two isotopic pairs, a reference curve can be calibrated: it is known as 'concordia' (Fig. 13.1a). In theory, measurements of isotopic ratios of zircons should directly provide the granite age on the concordia curve.

Zircon crystals are often zoned (Fig. 13.1b). Zoning may be of magmatic origin (without any effect on the age determination) or, in other cases, an inherited older core can be identified, eventually witnessing the crustal origin of the granitic host and providing the opportunity to determine the age of the protolith. Besides, a younger metamorphic event may have been responsible for some loss of radiogenic lead from the zircon crystal, or for the crystallization of an outermost rim, leading to a younger age. Thus, the measured

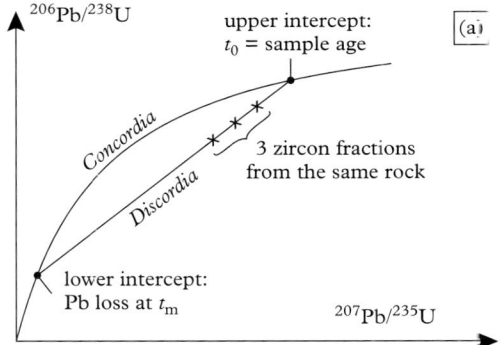 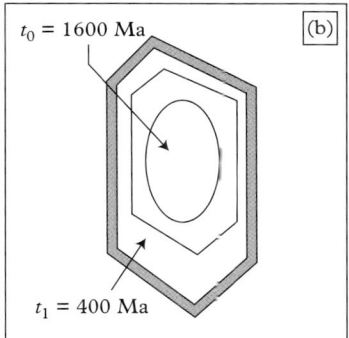

Figure 13.1 *Principle of zircon isotopic dating (a) Isotopic evolution diagram with Concordia and Discordia. (b) Zircon with magmatic zoning enclosing a relict core (these observations can be made using cathodoluminescence).*

isotopic ratios may plot away from the concordia curve, but are aligned along a straight line called 'discordia', which cross-cuts twice the concordia curve at the respective upper (older) and lower (younger) intercepts, therefore yielding two different ages that remain to be interpreted depending on the geological setting. To avoid this problem, Krogh (1973) recommended abrading zircon crystals to remove any rejuvenated outer rim and obtain more concordant zircons. The measured isotopic ratios should then plot closer to the upper intercept, hence yielding a better determination of the corresponding age.

The traditional method (ID-TIMS: isotope dilution-thermal ionization mass spectrometry) uses several zircon grain fractions, each fraction being gathered on the basis of crystal external features (shapes, sizes, colours . . .) and measured for its own isotopes ratios. This method is very fastidious, but provides very good results for rocks with a rather simple history (two magmatic and/or metamorphic events at the maximum), although the risk of mixing ages cannot be completely excluded. Recently, the *in situ* dating of individual grains has become more and more popular (Fig. 13.2). Two instruments are currently used: the sensitive high-resolution ion microprobe (SHRIMP), first built in Australia, and the laser ablation microprobe (LA-ICP-MS), enabling one to decipher a complex history corresponding to several different ages, if the zircon grains and their internal zones are large enough (>500 μm).

Several granites, especially S-types, contain monazite, another accessory mineral that can also be analysed for isotopic dating. Monazite does not lose any radiogenic lead after crystallization and its age would therefore theoretically provide the crystallization age of its host rock. Nevertheless, this mineral is prone to recrystallization when interacting with hydrothermal fluids. Thus, monazite ages must be regarded with caution.

13.1.2 Cooling ages

Other methods yield ages younger than the crystallization age of the granitic rock, which can be of interest (Fig. 13.3). They are cooling ages corresponding to the closure age of

254 Precambrian granitic rocks

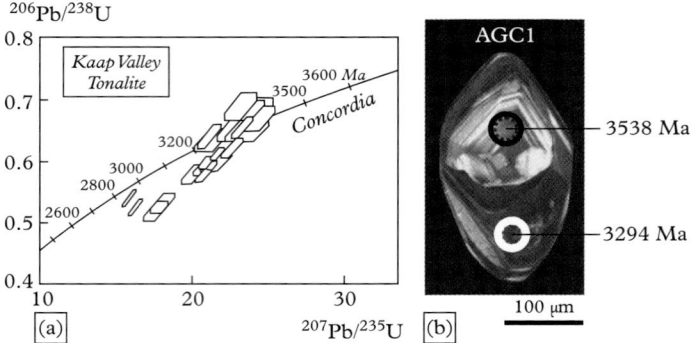

Figure 13.2 *Examples of zircon dating. (a) Age determination of the Archaean Kaap Valley tonalite from South Africa (location in Fig. 13.5b) using the sensitive high-resolution microprobe (SHRIMP); after Armstrong et al. (1990). (b) Cathodoluminescence image of a zoned zircon crystal from a tonalitic sample of the Ancient Gneiss Complex near Barberton; circles correspond to the laser spots from which the U–Pb ages and Hf initial ratios were obtained by Zeh et al. (2009).*

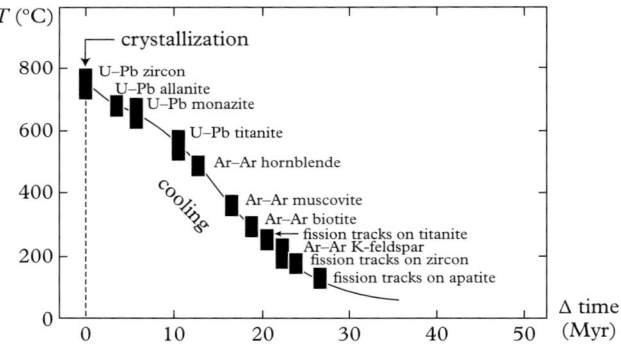

Figure 13.3 *Closure temperatures of the different isotopic systems used in thermochronology: all ages, with the exception of the U–Pb zircon ages, are cooling ages enabling a thermochronological curve to be drawn.*

the studied isotopic system below a specific isotherm. With all these ages (crystallization and cooling ages), it is possible to draw a thermochronological curve following the whole cooling history of the granite sample. The shape of this curve provides information on the evolution of the geological context after the emplacement of the granitic rock. Each method uses a mineral whose closure (or blocking) temperature for an isotopic system is well known. With the help of different minerals, it is possible to derive the thermal evolution of a rock since its initial crystallization.

Box 13.1 The mystery of the Hadean crust

The oldest continental crust identified so far is made of tonalitic to granodioritic gneiss cropping out over 40 km² in the Acasta complex of north-western Canada. These rocks contain magmatic zircons, whose ages range from 4002 (±4) to 4031 (±3) Ma (Bowring and Williams, 1999). An older continental crust likely existed during Hadean times (4.5–4.0 Ga), but it has been never been preserved as whole-rock samples. However, many detrital zircons with Hadean ages (4.3–4.0 Ga) have been found, especially in the Jack Hills quartzites from

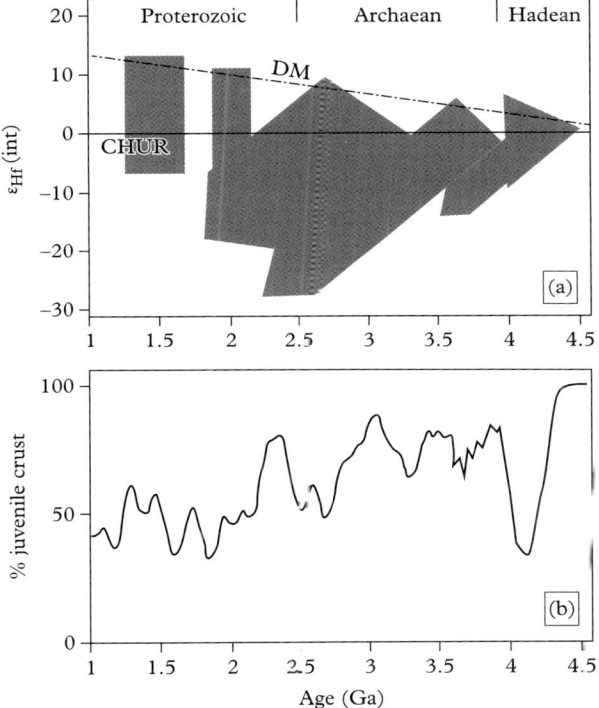

Figure 13.4 *(a) Initial ε_{Hf} values vs. age from magmatic and inherited zircons worldwide after Zeh et al. (2009): both superchondritic and subchondritic values were analysed in Hadean zircons. CHUR: bulk silicate earth as a chondritic uniform reservoir; DM: depleted mantle. (b) Proportion of juvenile (mantle-derived) vs. crustal components after Belousova et al. (2010); recycling processes already occurred during the Hadean, suggesting the existence of some sort of continental crust; the juvenile proportion was 70% in average from 3.9 to 2.2 Ga and decreased to 50% from 2.2 to 1 Ga.*

continued

> **Box 13.1** *continued*
>
> Western Australia (Compston and Pidgeon, 1986). One crystal from these quartzites gave an age as old as 4.404 Ga (Wilde et al., 2001). Mineral inclusions, $\delta^{18}O$ temperatures and significant negative $\varepsilon_{Hf}(t)$ values suggest that the Hadean zircons crystallized in granitoid-like melts (Cavosie et al., 2005; Harrison et al., 2008).
>
> Zeh et al. (2009) compiled a U–Pb and Hf isotope dataset of magmatic and inherited zircons from the Hadean to the Proterozoic worldwide (Fig. 13.4). The dataset suggests that continental crust and a complementary depleted mantle reservoir formed as early as the Hadean eon. Nevertheless, Hadean detrital zircons tell us nothing about the amount of created crust or about the size of the depleted mantle reservoir. The Hadean protocrust seems to have been completely recycled and/or replaced between 3.9 and 3.3 Ga.

13.2 Archaean granitoids: tonalites, trondhjemites and granodiorites (TTGs)

13.2.1 Lithology and structure

Where it is still recognizable (Fig. 13.5a), the Archaean crust comprises many granitic rocks corresponding to the first continental nuclei. These rocks are always associated with so-called 'greenstone belts', or 'greenstones', including low-grade metavolcanics and subordinate metasediments (Fig. 13.5b). These volcanic rocks are mainly tholeiitic basalts and locally komatiites (ultramafic Mg-rich lavas without any 'modern', i.e. post-Archaean, equivalent), that result from a high degree of partial melting of the mantle under a high geothermal heat flow.

All Archaean granitoids worldwide are very similar. First of all, they are not granites strictly speaking, but rocks poorer in potassium: tonalites, trondhjemites and granodiorites, collectively named by the initials 'TTGs' (Fig. 13.6). These rocks contain plagioclase (typically oligoclase of average composition An_{27}, or more sodic) as the only or dominant feldspar. They are rich in quartz. Ferromagnesian minerals are biotite and hornblende (± epidote, ± pyroxene). A few TTGs display enderbitic mineral assemblages, i.e. they are orthopyroxene-bearing tonalites.

Many TTGs were orthogneissified after their crystallization and are described as 'Archaean grey gneisses' in the field. The TTG plutons (or orthogneisses) are often dome-shaped. They are more or less coeval with adjacent greenstones. However, the contact zones between TTGs and greenstones have often been deformed. In the Archaean Dharwar craton (southern India), Bouhallier et al. (1993) demonstrated steeply plunging lineations and triple points resulting from the downward dragging of the greenstones with respect to the TTGs (Fig. 13.7). These Archaean domains seem to have been affected by gravity-driven vertical tectonics due to the density contrasts between TTGs and greenstones. The interpretation of such a structural pattern has been a matter of discussion. Recent studies by van Kranendonk et al. (2004) in the Pilbara craton

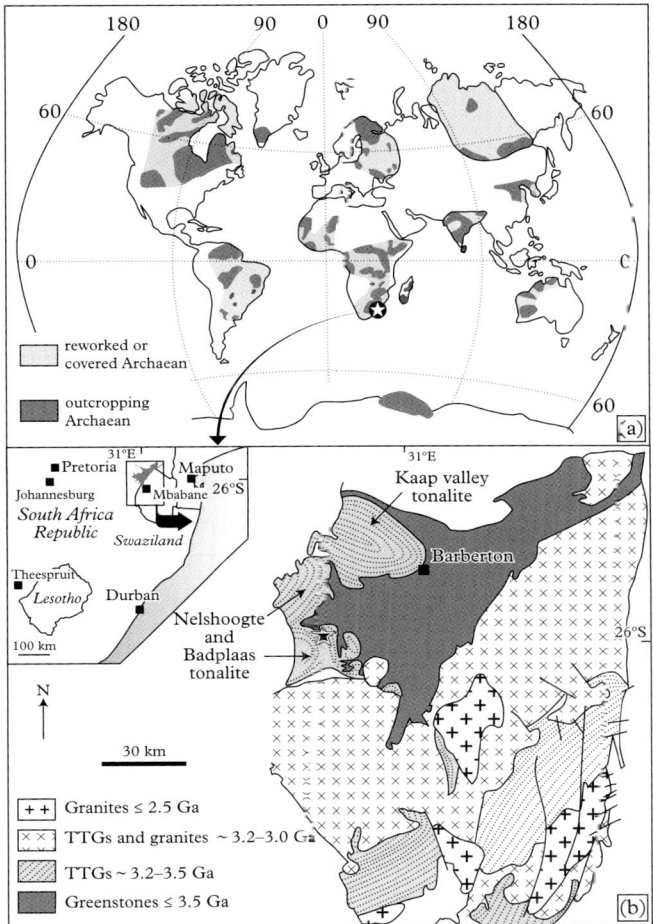

Figure 13.5 (a) Locations of rocks of Archaean age worldwide. (b) Geological map of granitoids and greenstone belts near Barberton, South Africa; star: location of migmatitic amphibolites studied by Nédélec et al., 2012.

of western Australia suggested that these old TTG plutons display structural features expected for diapirs (Fig. 13.8), and confirmed the existence of gravity-driven tectonics during Palaeoarchaean times (>3.2 Ga). The vertical motions took the form of rising TTG magmas, eventually followed by solid-state uplift of the crystallized TTG domes, and of downward sagging greenstones. This emplacement mechanism has no modern analogue.

258 Precambrian granitic rocks

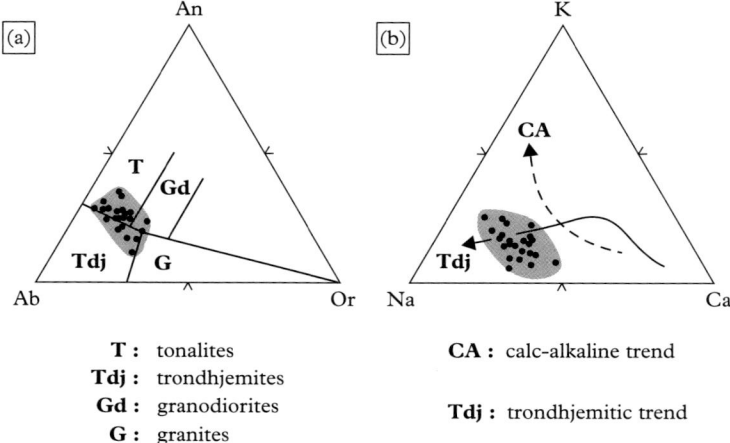

T : tonalites
Tdj : trondhjemites
Gd : granodiorites
G : granites

CA : calc-alkaline trend

Tdj : trondhjemitic trend

Figure 13.6 *Location of Archean TTGs in the Ab–An–Or normative diagram (a) and in the K–Na–Ca diagram (b). After Martin (1993).*

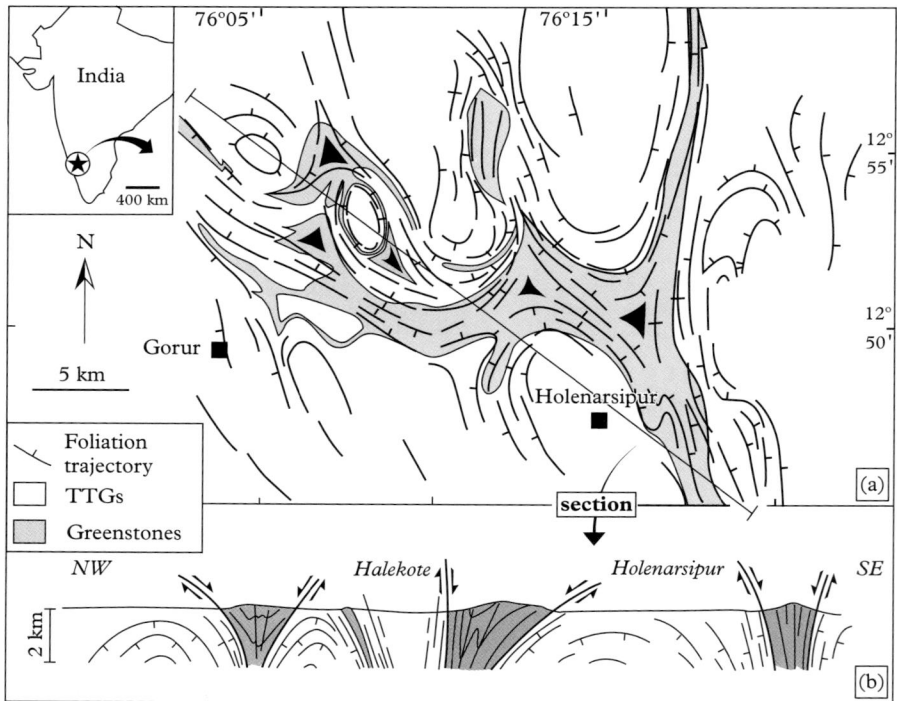

Figure 13.7 *(a) Structural map of Holenarsipur area (Karnataka, India). (b) Interpretative cross-section. After Bouhallier et al. (1993).*

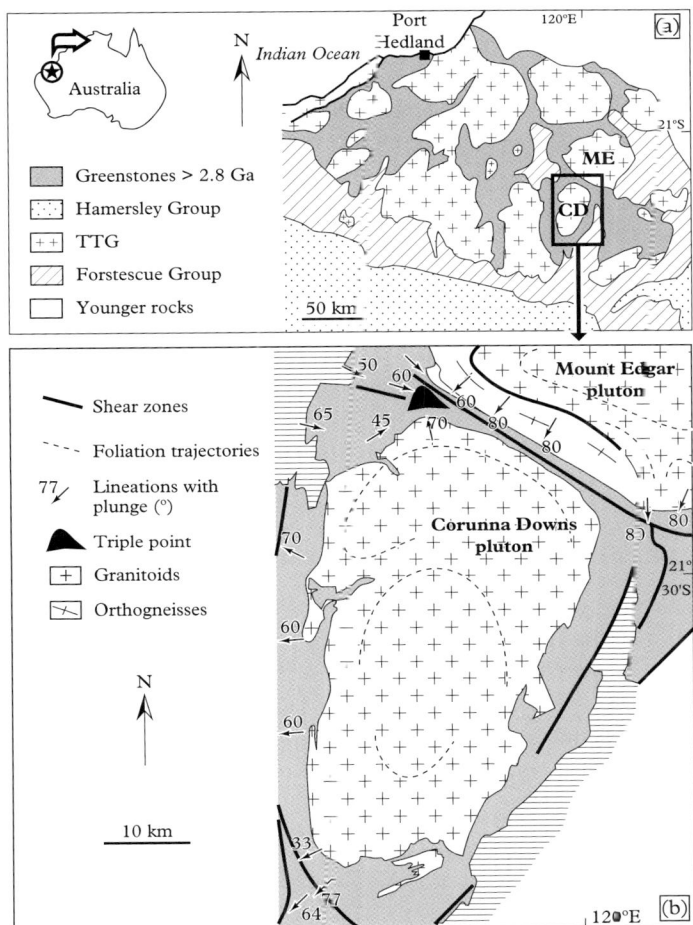

Figure 13.8 *(a) Pilbara craton (Western Australia): Corunna Downs (CD) and Mount Edgar (ME) TTG 'plutons' (actually gneissic domes). (b) Detail map of structures around the Archaean (3.3 Ga) Corunna Downs and Mount Edgar 'plutons' (gneiss domes): note the radiating lineations trending away from the TTGs and their steep plunges in the greenstones, consistent with dome-shaped TTGs, whose rims were deformed after crystallization. A triple point (black area) corresponds to the highest plunging lineations in the greenstones trapped between the two domes. After van Kranendonk et al. (2004).*

13.2.2 Geochemistry

TTGs are I-type metaluminous (calc-alkaline) rocks. They follow the trondhjemitic differentiation trend and not the classical calc-alkaline trend, i.e. they become richer in sodium and not in potassium with increasing magmatic differentiation (Fig. 13.6b and Table 13.1).

Their compatible element contents are low, excluding a direct derivation from the mantle. Their rare earth element (REE) distribution patterns are typical, and differ from most patterns of younger granites. TTG patterns are fractionated with a rather high average $(La/Yb)_N$ of *ca.* 24; their shapes are concave upwards due to the relatively low heavy REE contents, typically less than eight times chondritic values (Fig. 13.9a and c). They require the occurrence of a fractionation process *sensu lato* involving hornblende and/or garnet during the genesis of TTGs (Fig. 13.9b). By contrast, plagioclase is unlikely to have fractionated, because there is no europium (Eu) negative anomaly. Younger granites (<2 Ga) generally display very different REE patterns, i.e. less fractionated and often characterized by a more or less pronounced Eu negative anomaly (Fig. 13.9a and c).

13.2.3 Petrogenesis: geochemical and experimental constraints

Major and trace element contents of TTGs are not consistent with a juvenile origin by partial melting of the mantle. Besides, TTG isotopic signatures, e.g. their strontium initial ratios, point either to a mantle protolith or to a protolith recently extracted from

Table 13.1 *Average composition of 1439 TTG plutonic rocks in Moyen and Martin (2012). TTGs are felsic rocks characterized by an average K_2O/Na_2O ratio of 0.4 (see average granite compositions in Table 1.1 for comparison).*

Oxides	% (wt)
SiO_2	69.15
TiO_2	0.36
Al_2O_3	15.53
FeO (Fe total)	2.73
MnO	0.05
MgO	1.16
CaO	3.14
Na_2O	4.84
K_2O	1.70
P_2O_5	0.12

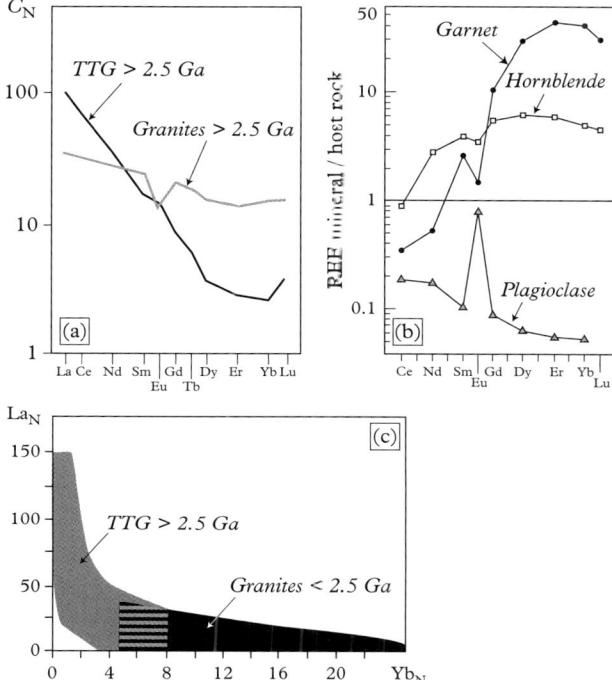

Figure 13.9 *(a) Typical REE distribution patterns of Archaean TTGs and of younger granites. (b) REE ratios of hornblende, garnet and plagioclase with respect to their host granitic rock: the high heavy REE partition coefficients of hornblende and especially of garnet are conspicuous. Plagioclase is characterized by a strong europium affinity explaining that plagioclase fractionation will be responsible for a strong Eu negative anomaly in the cogenetic magma. Conversely, hornblende will induce a small negative anomaly in the magma. Therefore, any process fractionating garnet, hornblende or plagioclase would yield a mirror image in the REE distribution pattern of the resulting magma. (c) $(La/Yb)_N$ vs. $(Yb)_N$ diagram. After Martin (1986, 1993).*

the mantle, i.e. with a short crustal residence time. The oceanic crust would meet these requirements and would also generate a K-poor magma by partial melting. It is made of tholeiitic basalts, or their deeper analogues, that have been variably hydrated by hydrothermal circulation close to an oceanic ridge. Moreover, partial melting of a hydrous basalt can begin at $T \leq 800\,°C$, as shown by numerous experiments. The hypothesis of a hydrous tholeiitic protolith has been tested using a numerical model based on rare earth elements (Fig. 13.10). Using the distribution coefficients of REEs between several minerals (hornblende, plagioclase, garnet ...) and a tonalitic melt, it is possible to calculate

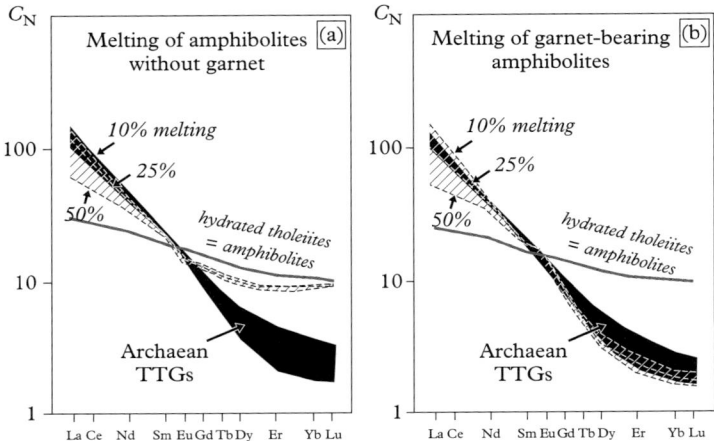

Figure 13.10 *Numerical petrogenetic model based on rare earth elements (after Martin, 1987). (a) Partial melting of amphibolites (primitive hydrous tholeiites, resembling Archaean tholeiites) does not yield enough fractionated REE patterns. (b) 25% partial melting of garnet-bearing amphibolites yield the best fit between calculated and observed REE patterns.*

the REE contents of a liquid derived from partial melting of a protolith resembling an Archaean tholeiite metamorphosed to amphibolite. The best fit between the calculated REE distribution patterns and the observed patterns (in TTGs) was obtained with a melt in equilibrium with hornblende and garnet.

Among others, Clemens et al. (2006) presented experimental results that clarified this point. Their experiments were conducted at 1.6 GPa and at temperatures ranging from 875 to 1000 °C, using amphibolites from the Barberton area. Within this temperature range, the amount of melt increased from 5 to 30%; garnet and clinopyroxene were always present; the amount of orthopyroxene increased with temperature; plagioclase was entirely consumed in the melt; complete disappearance of hornblende only happened at 1000 °C. All taken together, the suggested melting reaction was:

hornblende + plagioclase → garnet + clinopyroxene ± orthopyroxene + melt

The melt composition was granodioritic to trondhjemitic. The melt was also characterized by an Mg/Mg + Fe ratio, or Mg# (Mg number) of 33 and contained a plagioclase component of composition An_{23}. It is therefore possible to form TTG-like magmas from a hydrous tholeiitic protolith under a geothermal gradient of about 18–20 °C/km and at a minimum depth of 50 km.

13.2.4 Evidence from the Barberton migmatitic amphibolites

Despite the general consensus that TTGs are the main constituents of the Archaean continental crust worldwide, originated by partial melting of garnet-bearing amphibolites, natural evidence was lacking until recently. Nédélec et al. (2012) studied migmatitic garnet-amphibolites exhumed as a tectonic melange in a shear zone along the border of the Badplaas pluton near Barberton (Fig. 13.5b). The migmatitic amphibolites are characterized by quartz-plagioclase leucosomes in equilibrium with garnet, amphibole, titanite ± epidote. Thermobarometric estimates cluster in the range 720–800 °C and 1.1–1.2 GPa for the melting reaction and the derived geothermal gradient is 17–22 °C/km. These conditions are consistent with water-absent epidote-dehydration melting (Fig. 13.11). This reaction is unlikely to have produced a large volume of magmas. Nevertheless, the same geothermal gradient is assigned to the genesis of nearby TTG plutons, whose magmas were formed at greater depth at the hornblende-out solidus.

Hornblende cumulates were also exhumed by the same shear zone. The hornblende crystals have the same composition as hornblende crystals in the nearby TTG pluton. These magmatic cumulates crystallized at ca. 0.6 GPa and provide evidence that the

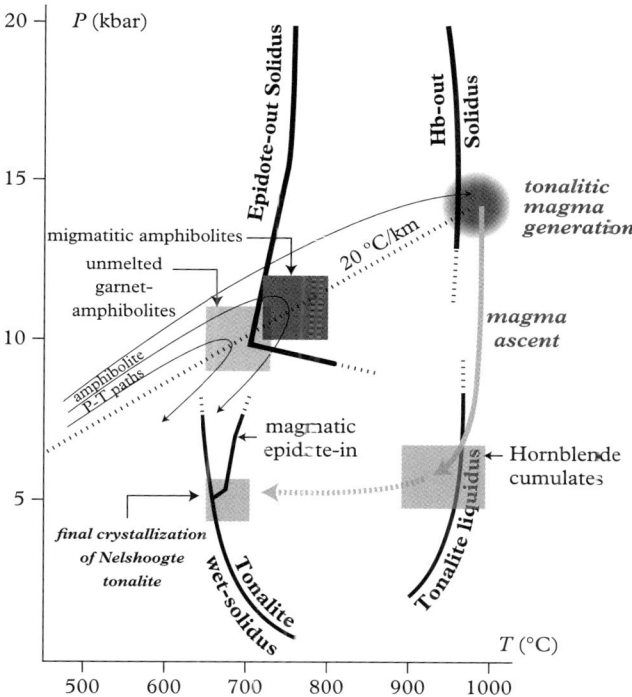

Figure 13.11 *P–T diagram with compilation of thermobarometric results and P–T paths for Barberton amphibolites and TTG melts; amphibolite solidus after Vielzeuf and Schmidt (2001); tonalite liquidus after Prouteau and Scaillet (2003); tonalite solidus and magmatic epidote stability after Schmidt and Thompson (1996).*

TTG parental magmas en route to the surface crossed their liquidus at this depth (Fig. 13.11) and experienced fractional crystallization of a large amount of amphibole before crystallizing as tonalite–trondhjemite plutons at slightly shallower depths (*ca.* 0.5 GPa).

13.2.5 Geodynamic context

Martin (1986) suggested a geodynamic model of Archaean subduction characterized by a warmer geothermal gradient than today (Fig. 13.12a). The latter condition is a sensible hypothesis for Archaean times and is confirmed by the common occurrence of Archaean komatiitic lavas resulting from a high degree of mantle partial melting. In this context of high terrestrial heat flow, a greater length of oceanic ridges (hence smaller plates) is likely, together with a greater number and a hotter nature of mantle plumes, all of them (ridges and plumes) contributing to dissipate the internal heat. In this hypothesis, a subducted oceanic plate would cross the hydrous basalt solidus (or amphibolite solidus) and begin to melt (Fig. 13.12b), whereas today it is fully dehydrated by prograde metamorphic reactions and will not melt.

In present-day conditions, the subducted oceanic plate is dehydrated and the released water may trigger partial melting of the mantle wedge; magmas are mainly granodioritic in active continental margins (see Chapter 12). Rarely, subduction of a young and still warm oceanic plate below a volcanic arc may induce partial melting of the hydrated basalts in the oceanic crust. This happened near the Taitao triple junction in Chile, where peculiar rocks are observed, mainly intermediate to felsic volcanic rocks, the so-called 'adakites', and even a 'granite' (in Cabo Raper), whose chemical compositions share many similarities with those of the Archaean TTGs (Fig. 13.13).

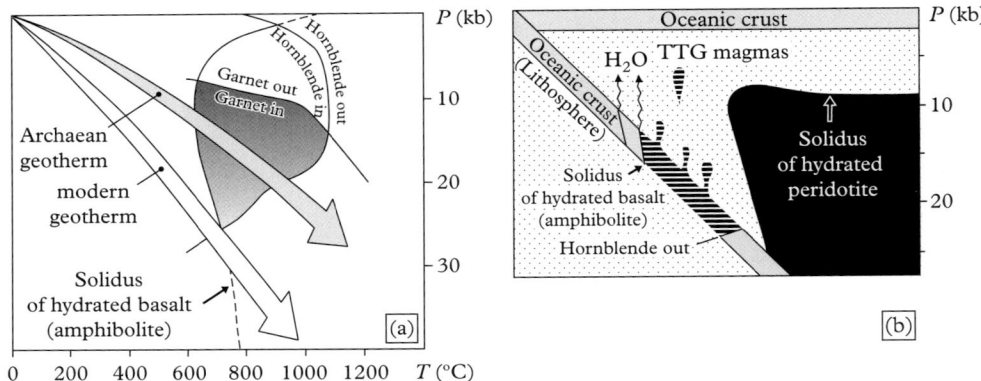

Figure 13.12 *(a) P–T diagram with the key metamorphic reactions and partial melting conditions of a hydrous tholeiitic metabasalt (amphibolite). (b) Hypothetical section of an Archaean subduction zone. After Martin (1986).*

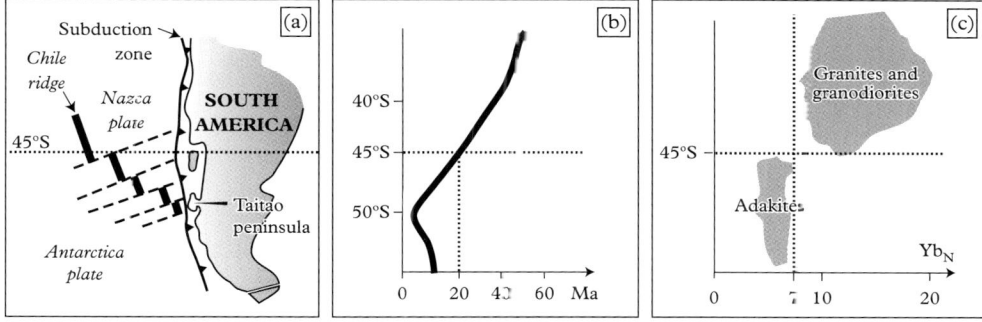

Figure 13.13 *(a) Location of Chile triple junction. (b) Age of the subducted oceanic crust with respect to latitude: the age is younger than 20 Ma south of latitude 45°. (c) Adakitic magmas observed south of latitude 45° are Yb-poor like TTGs. After Martin (1999).*

Another context was suggested by Bédard (2006), among others. This author only considered a very thick basaltic crust, such as can be found in an oceanic plateau. The lowermost part of this crust could have reached partial melting conditions, providing TTG magmas, eventually followed by granitic magmas, themselves derived from partial melting of the TTGs. This model requires either hydrothermal circulation at greater depths or gravity-driven downsagging of hydrated basalts to reach water-present melting conditions of garnet-bearing amphibolites at the base of the plateau. The interpretation (subduction model versus no-subduction model) has remained controversial for many years.

Most studies discussing the origin of Archaean TTG series regard them as one single, unique rock type corresponding to only one tectonic site of formation. Yet the detailed review by Moyen (2011) indicates three distinct sub-types (high-, medium- and low-pressure: HP-, MP- and LP-TTGs), reflecting specific depths of melting. Elements regarded as pressure indicators behave as follows: Al, Na and Sr are sequestrated preferentially into plagioclase (Fig 13.14a and b), Y and Yb are compatible in garnet and Nb and Ta are controlled by rutile (Fig 13.14c). Indeed, experimental melting of the same source over a range of pressure (i.e. at different depths) shows distinct residual assemblages corresponding to the different TTG sub-types:

- at 10 kbar: amphibole–plagioclase–pyroxene residue (corresponding to the case of LP-TTGs);
- at 15 kbar: no plagioclase, but clinopyroxene, amphibole and garnet (case of MP-TTGs, actually the most common case);
- at 20 kbar: eclogitic residue: omphacite, i.e. jadeitic clinopyroxene, garnet and rutile (case of HP-TTGs).

Therefore, the three TTG sub-types are interpreted as reflecting primarily the various depths of melting (Fig. 13.15). The respective geotherms are 10, 15–20 and

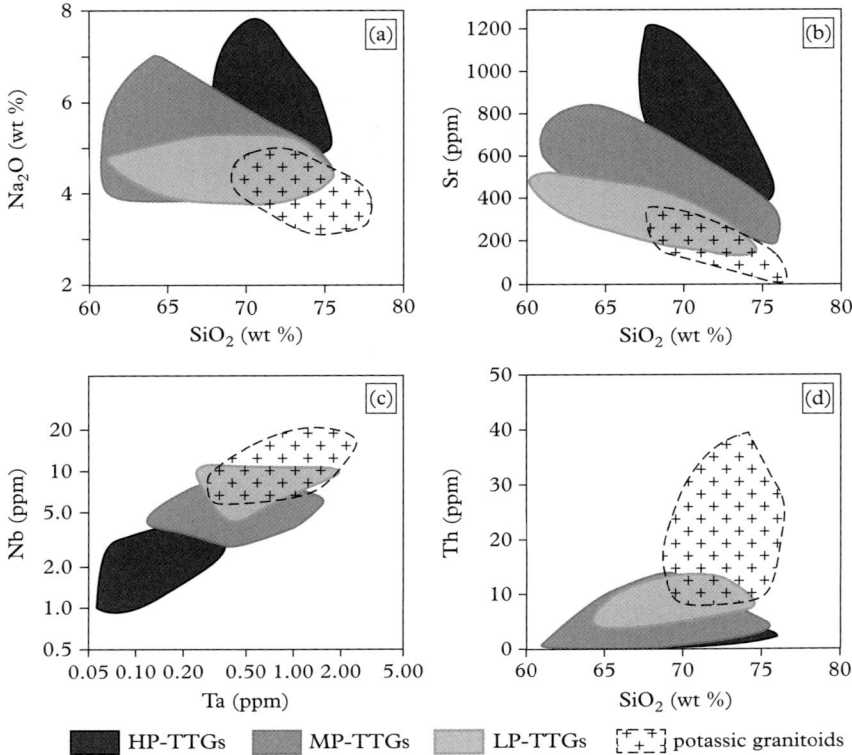

Figure 13.14 *Discrimination diagrams for the three TTGs sub-types and the younger potassic granitoids after Moyen (2011): (a) Na_2O vs. SiO_2; (b) Sr vs. Y; (c) Nb vs. Ta; (d) Th vs. SiO_2.*

20–30 °C/km. Notice that the geotherm required for the genesis of the MP-TTGs corresponds precisely to the geotherm established from migmatitic amphibolites near Barberton. Geotherms from 10 to 30 °C/km cannot occur in one unique tectonic environment. Hence, there is growing recognition that both slab melting and intraplate crustal sinking (possibly in a non-plate tectonic context) are probable valid models.

In addition, a recent study of mineral inclusions in diamonds by Shirey and Richardson (2011) provided important constraints on the beginning of plate tectonics. Indeed, before 3.2 Ga, sulfide inclusions in diamonds are never of an eclogitic-type, i.e. with Re/Os ratios in the range 2–30; rather, they are all of peridotitic-type with Re/Os of *ca.* 0.02. The opposite situation is observed in younger diamonds. Actually, the absence of eclogitic inclusions in older diamonds does not rule out other downwelling processes dragging appropriate lithologies at depth, such as gravity-driven

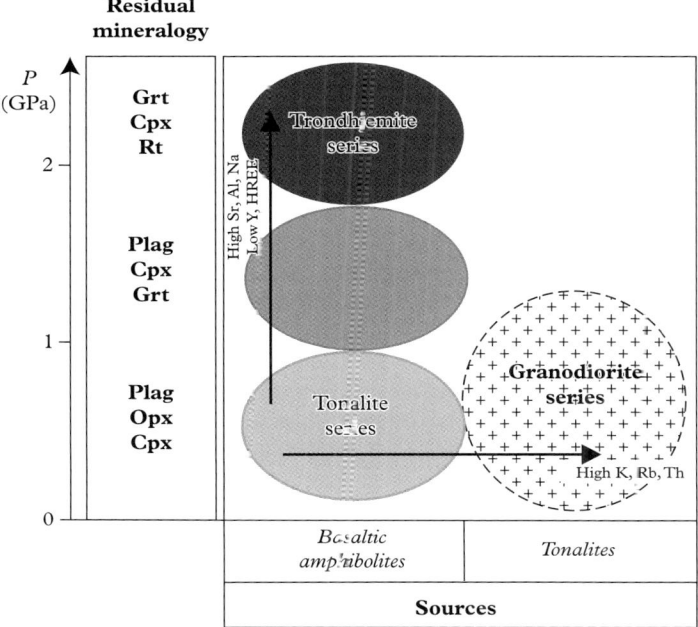

Figure 13.15 *Summary of interplay between source compositions and depths of melting (hence modal composition of residues) for the three TTG sub-types and the late-Archaean (or younger) granodiorite series. After Moyen (2011).*

vertical tectonics or convective counterflows (downwellings) associated with plume-like instabilities. Hence, a cold sink (no-subduction) model appears more appropriate for TTGs older than 3.2 Ga and a plate-tectonic model with subduction zones for younger TTGs.

13.3 Late-Archaean granitoids

13.3.1 Temporal evolution of Archaean TTGs

The TTG chemical composition changes during the Archaean: compatible element contents (Ni, Cr, etc.) increase, as well as the Mg content (Fig. 13.16). Martin and Moyen (2002) explained these evolutions by an increased interaction of TTG magmas with mantle peridotites, possibly due to the steepening of the subducted plates. Indeed, the subduction angle depends on the geodynamic context. In the case of a warm and less dense oceanic plate, subduction occurs with a low angle and magmas derived from partial melting of the oceanic crust do not ascend through a developed mantle wedge

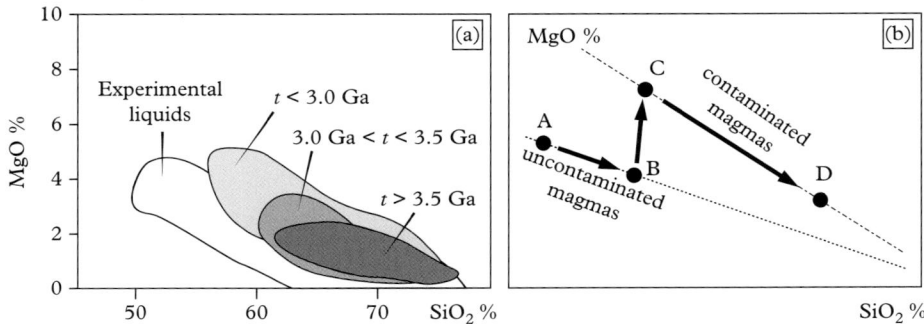

Figure 13.16 *(a) MgO vs. SiO$_2$ diagram for TTGs and experimental melts after Martin and Moyen (2002). (b) Interpretation by mantle–melt interactions: A → B, fractional crystallization trend of TTG parental melts; B → C, MgO increases at a given SiO$_2$ content due to interaction with the mantle peridotites; C → D, shifted fractional crystallization trend for contaminated TTG melts. After Moyen and Martin (2012).*

(a situation actually different from the oversimplified representation of Fig. 13.12b). When the geothermal gradient decreases, the subducted oceanic plate has a higher density and the subduction angle increases. Then, magmas ascend to the surface through the mantle wedge, where they may interact with mantle peridotites. Such an interaction will result in the increase of Mg and other elements due to the dissolution of small quantities of mantle olivine in the magmas.

13.3.2 Potassic granitoids of continental origin

All TTGs are juvenile granitoids that contributed to the net growth of continental crust through time. Granitoids produced by crustal recycling appeared as early as Mesoarchaean times (3.2–2.8 Ga), for instance the voluminous 3.1 Ga potassic granites near Barberton (Fig. 13.5b). They were formed by partial melting of an existing continental crust, probably in a (micro)plate collision context. They belong to the granodiorite series, whose modal compositions and geochemical features are easy to distinguish from those of the TTGs (Figs 13.14 and 13.15). They became more and more abundant through time.

13.3.3 Sanukitoids: granitoids of the Archaean–Proterozoic transition

Towards the end of Archaean times, but diachronously depending on the considered craton, special plutonic rocks are observed, namely magnesian diorites and their differentiation products. These rocks are called 'sanukitoids' because of their chemical compositions resembling those of the sanukites, which are Mg-rich andesites from Japan. Sanukitoids were first defined in the Superior Province of Canada by Stern et al. (1989):

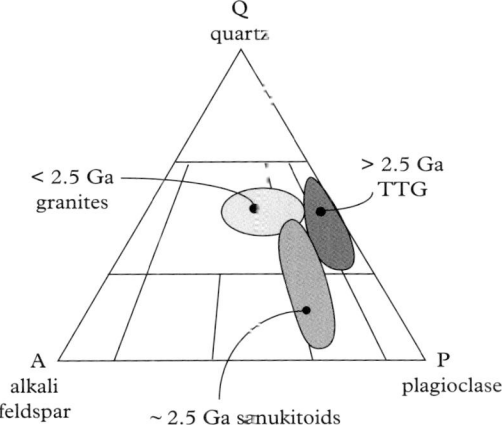

Figure 13.17 *Composition of sanukitoids (after Martin et al., 2005).*

they consist of monzodiorites and granodiorites, with light REE-enriched and fractionated REE distribution patterns. They are also characterized by high Sr, Ba, Ni and Cr contents and their Mg# is about 50–55. Hence, they are less silicic, but more Mg rich than typical TTGs (Fig. 13.17). They are regarded as hybridized slab melts or rather products of the interaction between slab melts (TTG-like magmas) and magmas derived from water-present partial melting of mantle peridotites (Fig. 13.18). They correspond to the transition from Archaean-type subduction to modern-type subduction, in a broader perspective of secular mantle cooling.

The oldest sanukitoids were identified in the Pilbara craton of Western Australia: they are 2.95 Ga old, hence much younger that their nearby TTGs, whose ages range from to 3.47 to 3.30 Ga. The Closepet batholith of India (Fig. 5.13) is an example, partly made of sanukitoid-type plutonic rocks (clinopyroxene-bearing monzonites), whose ages are about 2.52 Ga. Actually, this batholith is rather complex, because it resulted from different proportions of mixing of the sanukitoids with a felsic end-member derived from the anatexis of the Archaean TTGs (Moyen et al., 2001).

13.4 Proterozoic granitoids

13.4.1 Eburnian/Transamazonian granitoids (*ca.* 2 Ga)

Section 13.3.2 suggests a continuous continental crustal growth, with a steady-state granitic magma production. Actually, U/Pb zircon age histograms call for a discontinuous process with peaks of crustal production, notably around 2.7, 2 and 1.2 Ga (Fig. 13.19).

270 Precambrian granitic rocks

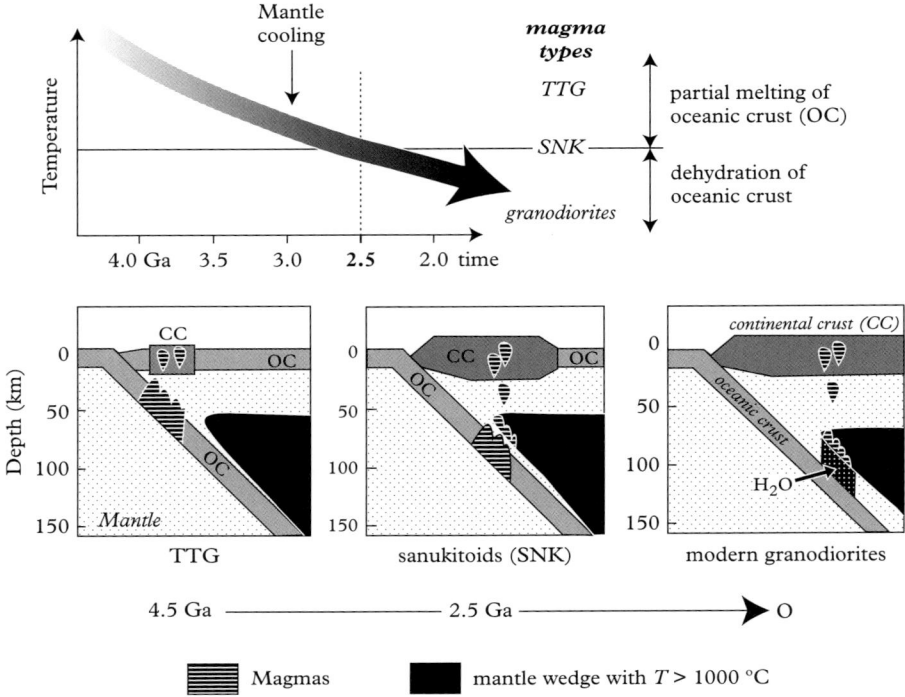

Figure 13.18 *Different petrogenetic models for granitoids in an active margin context depending on the mantle thermal evolution (here constrained by the decrease of radiogenic heat production). After Martin and Moyen (2002) and Martin et al. (2009).*

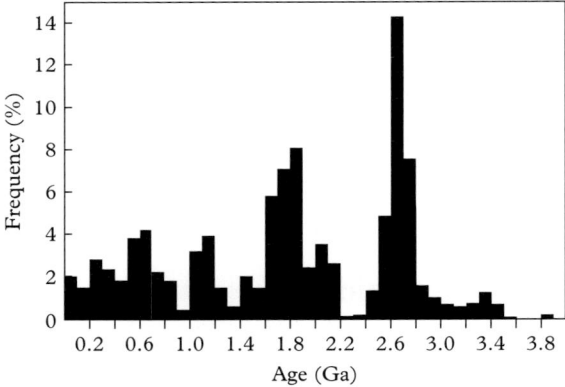

Figure 13.19 *Histogram of zircon U/Pb ages for juvenile granitoids in the continental crust. Despite probable sampling bias, magmatic production appears to have been discontinuous through time. After Condie and Aster (2010).*

Crustal growth at about 2 Ga was related to the Eburnian orogeny (defined in the Ivory Coast, hence the adjective 'eburnean' meaning similar to ivory) or to the Transamazonian orogeny (its equivalent in Brazil). The last peak corresponds to the Kibaran or Grenvillian orogeny that was responsible for the formation of the Rodinia supercontinent. The Eburnian orogeny was also responsible for the formation of another supercontinent, Columbia (sometimes called Nuna), whose paleogeography is still poorly constrained. These Proterozoic tectonic events provide evidence for some modern-type active Wilson cycles. The Eburnian/Transamazonian orogeny was heralded by an intense magmatic activity in the mantle, apparently resulting in the formation of several oceanic plateaux, and was even associated with a new burst of TTG-like granitoids (Fig. 13.20). This magmatic activity looked like a return to Archaean thermal conditions, whose cause remains unclear, although melting the base of a thick plateau crust might have generated LP-TTG magmas. Some authors consider temporary perturbations of mantle convection, such as the initiation of superplumes by slab avalanches, or even a complete reorganization of mantle convection. These processes may not have modern analogues.

13.4.2 AMCG (anorthosite–mangerite–charnockite–granite) associations

Associations of anorthosites, mangerites, charnockites and granites of Late-Paleoproterozoic to Mesoproterozoic ages, namely from 1.9 to 0.9 Ga, are commonly observed and referred to as AMCG rocks. Despite their spatial and temporal links, these rocks are not necessarily comagmatic (i.e. derived from the same parental magma) or even cogenetic (i.e. derived from the same process). However, anorthositic magmas were at least responsible for heat advection in the continental crust and, thus, indirectly responsible for crustal anatexis. Indeed, there is a consensus to discriminate anorthosites (A) of mantle origin from the monzonitic suite (MCG) of partly crustal origin. More mafic rocks may also be present: gabbros in relation with the anorthosites on one hand, norites and jotunites (orthopyroxene-bearing gabbros and monzodiorites: Fig. 13.21) in relation to the monzonitic suites on the other hand. The terminology used for the monzonitic suite reflects the ubiquitous presence of hypersthene (a ferromagnesian orthopyroxene). These rocks are sometimes called 'charnockitic' *sensu lato*, but it is better to use the name 'charnockite' only for hypersthene-bearing granite (Fig. 13.21).

Granites from AMCG associations are often A-type and display a special texture, the so-called 'rapakivi' texture, where alkali feldspars are rimmed by plagioclase (Fig. 13.22b). If these feldspar crystals form ovoid phenocrysts, the rock is called 'wiborgite' (from the Wiborg batholith in Finland: Fig. 13.22a). The adjective 'rapakivi' is a Finnish word meaning 'easily broken apart'. Southern Finland is the homeland of rapakivi granites: it is the place where they were first described by Sederholm (1891). Several massifs crop out on both sides of the Finland gulf (Fig. 13.22a). Their ages range from 1.6 to 1.5 Ga. Here, the mafic to intermediate rocks of the

272 *Precambrian granitic rocks*

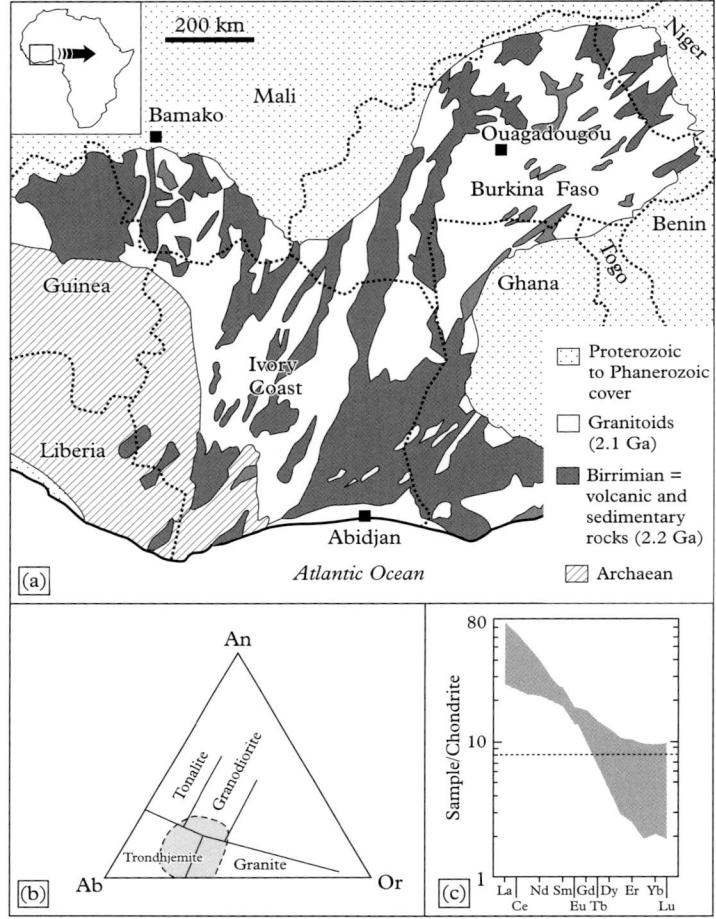

Figure 13.20 *Eburnian granitoids from the Baoulé–Mossi domain in West Africa. (a) Location map. (b) Composition. (c) REE distribution patterns. After Dioh et al. (2006).*

AMCG association are much less abundant than in other AMCG complexes worldwide. Nevertheless, geophysical data suggest that mafic rocks do exist at depth (Fig. 13.22c and d).

The Wiborg batholith (Fig. 13.22a) of 1.6 Ga age comprises mainly granites of rapakivi-type with scarce anorthosites and leucogabbronorites at the present outcrop level. Conventional isotopic studies have not been able to provide an answer to the crust–mantle controversy over their source, because both silicic and mafic rocks yield rather similar Nd initial isotope ratios. The new Hf isotope data of Heinonen et al. (2010) argue for the truly bimodal nature of rapakivi magmatism, involving at least two

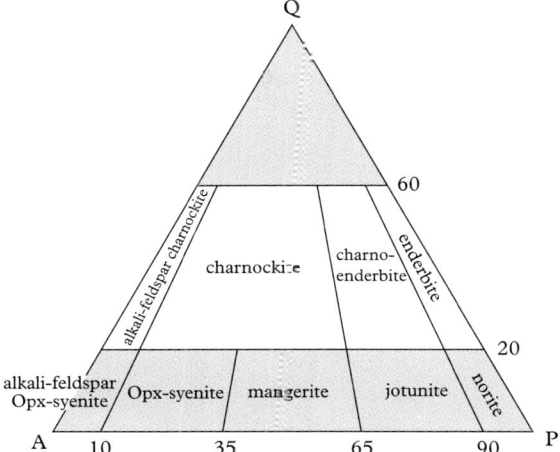

Figure 13.21 *Streckeisen classification diagram for hypersthene-bearing rocks. Compare with Figure 1.1 for a reminder of the usual terminology.*

distinct magma sources: a Paleoproterozoic crustal component and a depleted mantle component (Fig. 13.23). The initial Hf values in the rapakivi granites imply that these rocks were mainly derived from a crustal source. By contrast, the mafic rocks display less homogeneous values and their within-sample variation in zircons is considerably greater than in granites.

The Bjerkreim–Sokndal (BKSK) intrusion from southern Norway belongs to the much younger Rogaland anorthositic complex (930 Ma). It was studied in detail for petrology and structure. It includes a lower 7 km-thick layered part, of noritic cumulate composition, passing upward to mangerites, then to charnockites that form the synclinal core of the massif (Fig. 13.24a). Charnockites look isotropic to the naked-eye, but systematic AMS measurements revealed a centripetal deeply-plunging lineation pattern (Fig. 13.24b and c). Kinematic indicators in the cumulate confirmed a downward magmatic shearing of the BKSK intrusion together with an upward relative motion of the nearby anorthosites. These gravity-driven tectonics were permitted by hot granulitic conditions in the country rocks, which enhanced ductile deformation. The layered base of the BKSK intrusion resulted from fractional crystallization of a jotunitic magma (SiO_2 = 55%) leaving a residual mangeritic liquid. Isotopic data also confirmed that this residual liquid was contaminated by a leucogranitic magma derived from partial melting of the intrusion roof (Nielsen et al., 1996). The upper charnockites resulted from this hybridization. As the contaminant was a melt of eutectic granite composition (i.e. similar to the final product of fractional crystallization alone), it could not be identified using major element geochemistry, but only with the help of trace element and isotope studies. Several pillow-like leucogranitic enclaves in the upper part of

274 *Precambrian granitic rocks*

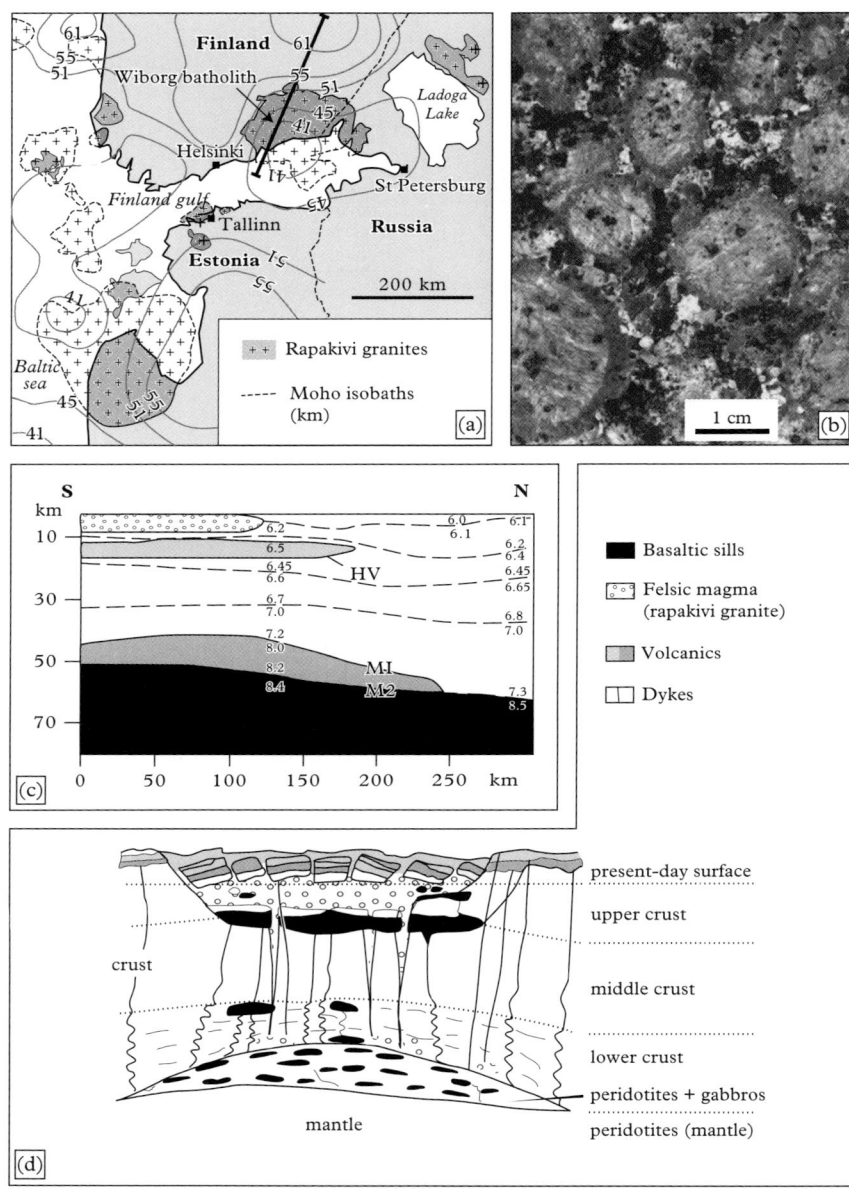

Figure 13.22 *Rapakivi granites of southern Finland. (a) Location map. (b) Rapakivi texture (wiborgite-type). (c) Interpretative seismic profile through Wiborg pluton: note the mafic sill verified by high seismic velocities (HV) at ca. 12–13 km depth and the mantle shallowing at the base of the crust; seismic reflectors M1 and M2 indicate the limits of a mafic magmatic underplate made of alternating gabbros and peridotites. (d) Geological reconstruction. After Rämö and Haapala (1996).*

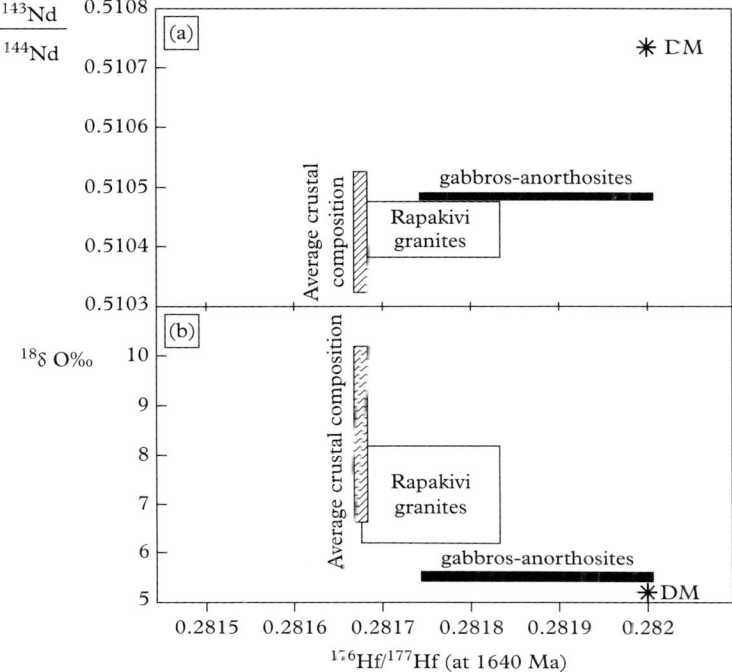

Figure 13.23 *Comparison of single-zircon Hf isotope compositions from rocks of the Wiborg batholith with (a) whole-rock Nd and (b) zircon O isotope compositions. After Heinonen et al (2010).*

the mangerites witness the coexistence of two different magmas with variable degrees of mingling or mixing.

Taken together, 1.6–1.3 Ga granitoids from the Mesoproterozoic AMCG complexes represent the largest volume of intraplate granitic magmatism in the Earth's history. Although there is a general agreement about their anorogenic nature, the geodynamic context of their formation and emplacement remains unclear. Vigneresse (2005) suggested that they are an indirect consequence of the formation of the first large supercontinent (Columbia) after the Eburnian orogeny (2.1–1.8 Ga). The thick continental crust of this supercontinent would have had a thermal blanketing effect, leading to a temperature increase in the mantle and in the lower crust, responsible for the formation of AMCG magmas. In addition, these magmatic complexes seem to have been emplaced along a longlive 12,000 km long linear structure through the supercontinent, preferentially in areas of juvenile crust. These complexes do not show any consistent change in age, hence they were possibly formed by several discrete small mantle plumes rather than by one unique hot spot under a moving plate.

276 Precambrian granitic rocks

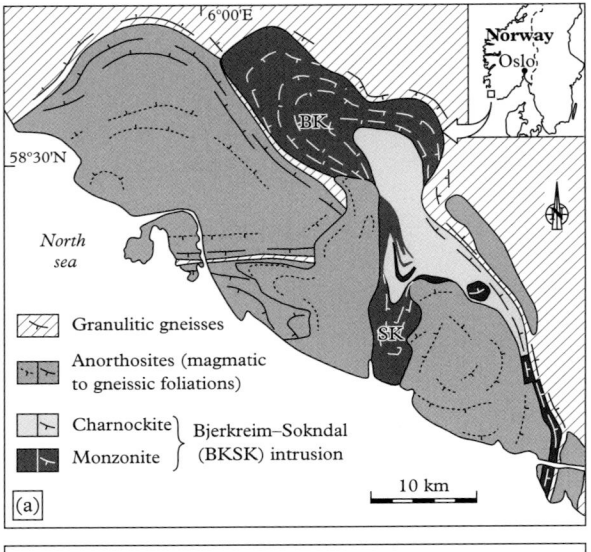

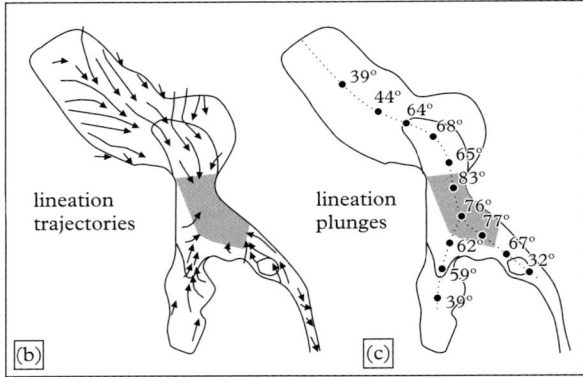

Figure 13.24 *Bjerkreim–Sokndal intrusion (Rogaland, Norway). (a) Map after Duchesne and Wilmart (1997). (b) Trend and plunge of magnetic/magmatic lineations. (c) Average lineation plunge inside the intrusion. After Bolle et al. (2000, 2002).*

13.5 Impact granites

This section deals with the famous Sudbury granite in Canada, which belongs to a magmatic complex of the same name formed 1.85 Gyrs ago, due to a meteoritic impact. The size of the impactor was estimated at *ca.*12 km in diameter and its speed at 25 km/s. Conversion of its kinetic energy into heat resulted in the production of about 30,000 km³ of magmas from the various crustal rocks at the impact site: Archaean granites, gneisses and greenstones to the north and Paleoproterozoic metasediments and metavolcanics to the south (Fig. 13.25a). The transient crater would have been excavated from the continental crust down to the Moho. Nevertheless, the magma isotopic signatures exclude any significant mantle input and point to a crustal origin. All crustal rocks would have been molten as temperatures may have reached at least 1700 °C, namely a temperature higher than the liquidus of all crustal rocks (granite, basalt, etc.). The duration of magma formation was very short. The transient crater was modified immediately after impact: its rim collapsed inward and its core was uplifted, and then eventually collapsed outward. The melts were collected as a continuous sheet on top of the impact structure. The melt sheet is covered by a layer of ejecta, whose thickness may have reached 1.5 km in the centre and whose area exceeded the limit of the impact structure. These ejecta are known as the Onaping formation breccias. The Onaping breccias contain clasts of various sizes embedded in a glassy matrix. The glassy component increases downward. A long time after its formation, the impact structure was deformed into a syncline during the Mesoproterozoic Grenville orogeny. Its incomplete erosion enables observations to be made at different levels. The Sudbury complex is well-known for its economic potential, because it contains huge ore deposits mined for nickel, copper and PGE (platinum-group elements) (Fig. 13.25a).

The magmatic layer is less than 3 km thick and displays a conspicuous bimodal composition, with a lower gabbroic layer and an upper granitic layer. The Sudbury 'granite' has a granophyric to fine-grained texture (Fig. 13.25b). Its silica content is 69 wt %. It represents one half of the magmatic complex. The other half is made of norites (with a silica content of 57 wt %) and quartz-gabbros (silica content: 60 wt %) in the transition zone. The equal proportions of mafic and felsic rocks and the lack of intermediate compositions (Fig. 13.25c) apparently exclude that the whole Sudbury magmatic complex would have been derived from fractional crystallization of a unique parental magma. However, as mafic and felsic magmas formed at the same time from the continental crust, they may not represent two distinct magmas. Zieg and Marsh (2005) suggested an original differentiation process from a heterogeneous liquid (Fig. 13.25d). Indeed, the very short melting duration is consistent with the formation of a heterogeneous magmatic layer, where different magmas issued from melting of each rock type and even each mineral might have been juxtaposed but not well mixed. Nevertheless, the calculated average silica content for the whole Sudbury magmatic complex is 64 wt %, namely a granodioritic composition corresponding to the average composition of the crust at the impact site. The different silicate melts were theoretically miscible, but actually they behaved as an emulsion because of their different viscosities. Zieg and Marsh (2005) considered that the melts might have been segregated according to their densities: globules of equal-density liquids being finally merged to achieve the collection of two main magma types: the lower gabbroic melt and the upper granitic melt. These super-liquidus melts have cooled inwards from the roof and from the floor, until they reached their liquidus and began to crystallize. Because of its mafic composition, the noritic magma behaved as a rigid crystalline network before full crystallization of the granitic magma (see Chapter 4), thus limiting the interactions at the transition zone.

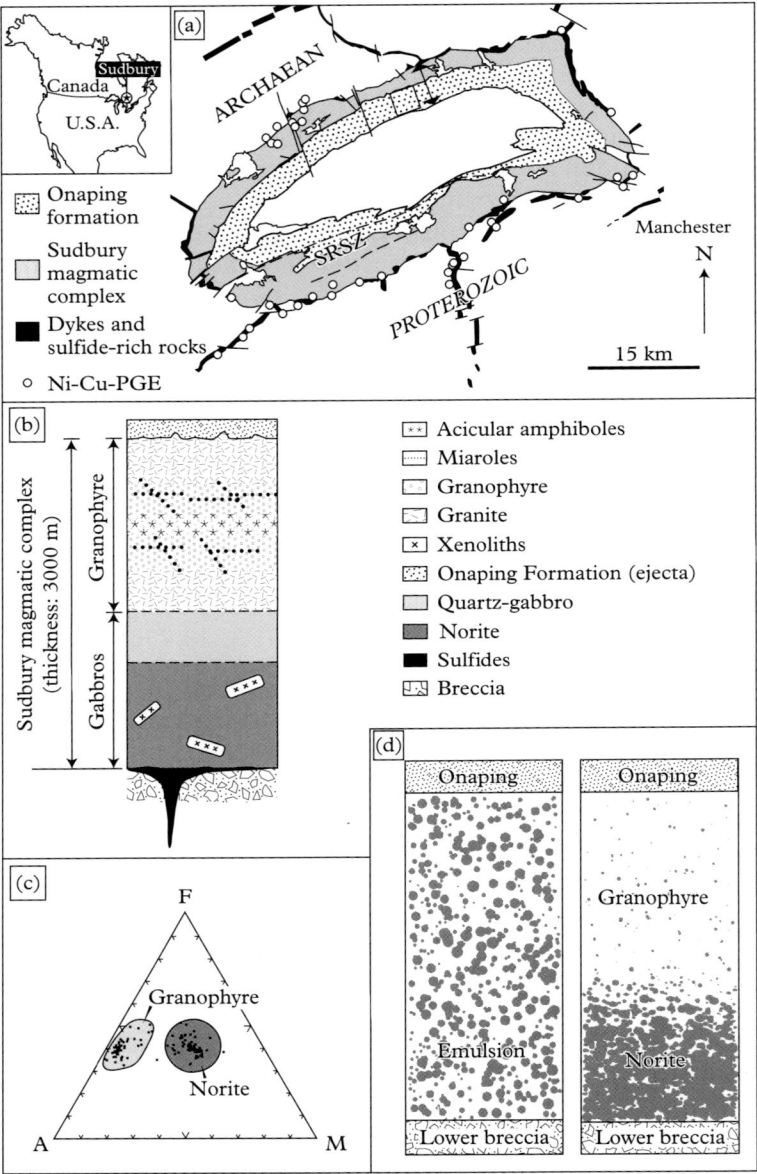

Figure 13.25 *Sudbury complex. (a) Map of the impact structure. (b) Vertical section of the magmatic complex. (c) Composition of granophyres and norites in the AFM diagram. (d) Interpretative sketch. After Ames (2002) and Zieg and Marsh (2005).*

All multi-kilometric sized meteoritic impacts contain molten products, but the Sudbury complex is an outstanding example because of its size and of the good preservation of the melt sheet. It seems rather unique, but similar complexes may have existed at the surface of the early Earth, and even formed more recently. In western Europe, two large meteoritic impact structures, Ries in southern Germany (14 Ma) and Rochechouart in France (201 Ma), both slightly more than 20 km in diameter, only display impact breccias with a glassy component, but no proper granitic melt sheet. The Manicouagan structure (215 Ma) in Quebec, Canada, has a larger diameter (90 km) and contains a monzodioritic sill-like melt sheet a few hundred metres thick, recently studied by Spray et al. (2010). The largest impact structure identified so far, the Vredefort impact (2.1 Ga) near Johannesburg (South Africa), is more than 200 km in diameter, but it is now too deeply eroded to preserve any melt sheet. Nevertheless, the corresponding Vredefort dome contains a suite of homogeneous granophyric dykes with an average silica content of 67%, regarded as downward injection products from an overlying melt sheet into the uplifted Archaean basement. Therriault et al. (2004) calculated that the Vredefort granophyre could represent a mixture of five molten rock types: basalt (40%), quartzites (30%), gneisses (25%), shales (3%) and carbonates (2%). In the Hadean to Early Archaean times, impact cratering and melting may have been an important mechanism of crustal differentiation and recycling.

13.6 Conclusions

A continental crust of granitic composition *sensu lato* may have existed as early as the very beginning of Precambrian time. Archaean granitoids display specific characteristics related to various geodynamic settings, possibly different from more recent settings in Earth's history. These rocks are mainly TTGs (tonalites–trondhjemites–granodiorites), i.e. plutonic rocks more sodic than potassic, unlike common younger granites. Their major and trace element contents are consistent with partial melting of a hydrous basaltic protolith at depth, possibly a subducted oceanic crust, at least after 3.2 Ga. Older TTGs may have formed by gravity-driven vertical tectonics. The secular decrease of mantle temperatures led to a change of petrogenetic processes that formed juvenile continental crust. For instance, late-Archaean sanukitoid plutons and monzonitic suites in Mesoproterozoic AMCG complexes correspond to transitional stages from the hot Archaean conditions to colder 'modern' geotherms. Granitoids derived from the recycling of an older continental crust were already formed during the Archaean.

14

Granite metallogeny

Granites are a source of mineral ores of economic interest. The average crustal abundance of an element was named 'clarke' after the American geochemist F.W. Clarke (1847–1931): see Table 14.1 for a few examples. The clarke of concentration of an element is its amount in a rock relative to its crustal abundance. An ore concentration can also be called its 'grade'. It is a common observation that valuable ores with high-grade concentrations are scarce and difficult to find. Their origin often seems enigmatic a priori. Metallogeny studies the process resulting in an increased concentration of an ore element, and eventually in the genesis of a valuable ore body, depending on its grade and volume. The economic value may vary over time depending on the market price, which is governed by demand and supply, both related to historical and technological evolutions, to operating costs and to eventual discovery of new high-grade deposits and the opening of new mines.

Granite-related deposits of interest result from their magmatic and post-magmatic history. Therefore this chapter will often refer to concepts discussed in Chapter 8. After a review of the main processes of mineral concentration, a few examples will be explained in detail.

14.1 Development of ore mineral concentrations

A mineral concentration is generally obtained after a number of successive stages. Depending of the nature and efficiency of each concentration mechanism, an economic deposit may or may not be formed. Being aware that each concentration stage has a low probability of occurrence, the scarcity of mineral deposits is understandable and their probability of occurrence can even be regarded as unlikely. The chemistry of the source plays an important role, as S-, I- and A-type granites are respectively specialized in lithophile or chalcophile elements, or in rare metals. But the involvement of fluids, by late- to post-magmatic processes, is of major importance for the formation of most ore deposits, both chemically (dissolution and precipitation of elements) and physically (element transfer). The unequal distribution of an element between two phases (either liquid or fluid) is the preferential mechanism of concentration of an ore. Therefore, different

Table 14.1 *Average concentrations for some elements of economic interest in the continental crust (clarke) and in I-, S- and A-type granites for comparison (data in Whalen et al., 1987, Chappell and White, 1992, Wedepohl, 1995, and Konopelko et al., 2009). The very low values show the extent of the concentration processes required to obtain a valuable ore body.*

Elements (ppm)	Continental crust	Granites (I)	Granites (S)	Granites (A)
Copper (Cu)	25	9	11	2
Tin (Sn)	2	6	10	22
Niobium (Nb)	19	11	12	37
Lead (Pb)	15	19	27	24
Thorium (Th)	9	18	18	23
Uranium (U)	2	4	4	5
Zinc (Zn)	65	49	62	120

phase transitions and the physical and chemical parameters that influence the distribution of an element between two phases in equilibrium will be examined. Eventually, the tectonic setting may also favour the formation of mineral deposits.

14.1.1 Importance of magmatic source and granite type

14.1.1.1 S-type granites

S-type granites are derived from partial melting of the continental crust, hence from a material that has been previously enriched in lithophile elements (Sn, W and U, as well as Nb, Ta, Be and Li). However, the enrichment factors are variable. For instance, the beryllium (Be) content of a magma is indirectly related to the pressure conditions. Indeed, this element can only enter into cordierite or beryl mineral structures. If cordierite, a mineral stable at low pressure, is already present in the protolith or is generated by the melting reaction, it will retain Be in the crystal lattice and the magma will no longer contain Be (Evensen and London, 2002).

More generally, a crustal protolith is water-rich as it contains an amount of hydroxylated minerals, such as micas. Thus, crustal-derived granitic magmas will inherit a water content and will easily become water-saturated during magma ascent toward the surface. Generally speaking, the water content of a magma is an important parameter in metallogeny. Dissolved water decreases the viscosity of the magma, making crystal fractionation easier, thus enhancing the consequential enrichment of incompatible elements in the residual liquid. In this way, a few S-type leucogranites, or their related pegmatites, can reach rather high contents in fluorine and metals, such as tantalum (Ta) and tin (Sn). Due to low viscosity, the residual magma may ascend to shallow crustal levels, where water-saturation will be easily reached, thus liberating large volumes of hydrous fluids that are the carriers of potential ores.

In addition, as pelitic sediments often contain a significant amount of organic (carbonaceous) matter that is changed into graphite by metamorphism, derived magmas will have a reduced character. In fact, the oxygen fugacity of a granite magma influences the elements with variable valencies. It was shown previously that a low oxygen fugacity does not allow the crystallization of magnetite, resulting in the formation of ilmenite-bearing granites (see Fig. 2.14 and Section 10.2). Indeed, all S-type granites as well as a few I-type granites are ilmenite-bearing. In these reduced granites, tin is present as Sn^{2+}, which behaves incompatibly and will remain in the evolved liquid, the first step to obtain an Sn-enriched magma and eventually an Sn-ore deposit (Fig. 14.1). By contrast, in

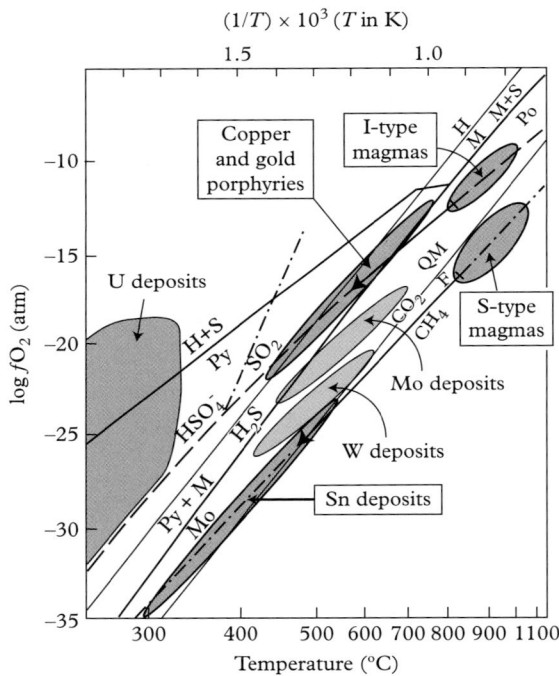

Figure 14.1 *Conditions of temperature and oxygen fugacity in S- and I-type granitic magmas and in hydrothermal ore deposits of tin, tungsten (W), molybdenum, copper and uranium. Conditions of ore deposits were deduced from the study of fluid inclusions. Arrows show the direct relationships (with buffered fO_2) between S-type magmas and Sn deposits, and between I-type magmas and porphyry copper and gold deposits. Note that formation of uranium deposits requires a high oxygen fugacity and a rather low temperature. After Ohmoto and Goldhaber (1997).*

magmas with a high oxygen fugacity, tin is present as Sn^{4+}, which will easily enter into the crystalline network of some early magmatic minerals, for instance replacing Ti^{4+}, so these oxidized (magnetite-bearing) granites will never be associated with tin deposits.

14.1.1.2 I-type granites

Many chalcophile elements, namely elements with an affinity for sulfur, including copper, molybdenum, silver and gold, are often found in ore bodies associated with I-type granites. The example of copper is the most important in terms of volumes, because the so-called 'porphyry copper' deposits, the main ores derived from I-type granites, represent about 20% from the 500 most large ore concentrations worldwide (Laznicka, 1999). These deposits are formed above subduction zones along convergent plate margins (Fig. 14.2).

The origin of copper can be traced using the isotopic ratio of osmium, another chalcophile element (Fig. 14.3) that is found together with copper in sulfides (pyrite, chalcopyrite). The ^{188}Os isotope is stable, whereas ^{187}Os results from the decay of rhenium ^{187}Re. Unlike osmium, rhenium is highly incompatible: it will therefore leave the mantle protolith more easily than osmium during partial melting events. The mantle $^{187}Os/^{188}Os$ ratio is presently equal to 0.13. The osmium crustal ratio is higher because of a greater production of radiogenic osmium due to the decay of radioactive rhenium. Sulfide mineralization can be dated by Re/Os using the slope of the isochron $^{187}Os/^{188}Os$ vs. $^{187}Re/^{188}Os$, in the same way than Rb/Sr isochron dating is used (see Section 14.2.4). The y-intercept of the isochron line yields the $^{187}Os/^{188}Os$ initial ratio. In Chile, this method was applied to several ore deposits, demonstrating that copper was mainly of

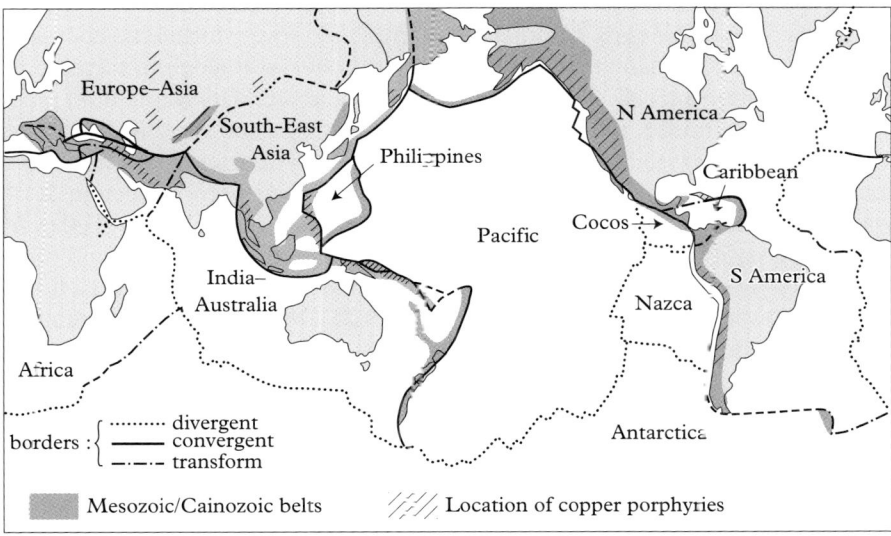

Figure 14.2 *Global distribution of porphyry copper deposits.*

mantle origin. Moreover, the largest deposits (Chuquicamata, El Teniente) are characterized by the lowest $^{187}Os/^{188}Os$ initial ratios, namely ratios that are closest to the mantle ratio (see Fig. 14.16c). Conversely, smaller copper deposits display higher $^{187}Os/^{188}Os$ initial ratios indicating a greater crustal input in the formation of magmas or during the post-magmatic hydrothermal history (Mathur et al., 2000).

14.1.1.3 A-type granites

A-type granites and their cogenetic pegmatites are characterized by high contents of fluorine, lanthanides and high field strength elements (HFSE), and thus could be the source of zirconium (Zr), niobium (Nb), rare earth element (REE) and fluorine (F) mineralization. The related deposits may display high grades and be of great economic interest despite their generally small volumes. The high HFSE contents of A-type granites have been discussed earlier (Chapter 2).

Fluorine often substitutes for OH in hydroxylated minerals (biotite, phlogopite, amphibole) in the lower continental crust. The releasing of fluorine requires high temperature anatexis of an igneous crustal source. In this case, magmas will be water under-saturated, but relatively halogen-rich (F and/or Cl). Hereafter, we will see that F significantly decreases the viscosity of granitic magmas and may also form stable complexes with some elements (Zr, Nb, REE and possibly Sn, Zn and W) in hydrothermal fluids. Thus, fluorine can influence processes that play a role in the genesis of mineralizations in two ways.

14.1.2 Orthomagmatic processes

Metallogenic processes occurring in magmas and only involving the melt and its crystal products are called 'orthomagmatic'. They represent an intermediate concentration stage for the incompatible elements of potential interest. However, their efficiency in felsic magmas is less than in mafic magmas, and they cannot result alone in forming potential ores of economic interest. Other concentration stages in relation to a fluid phase are generally required. Indeed, the efficiency of the orthomagmatic stage is directly related to the importance of crystal fractionation. Actually, two parameters make it easier: a low magmatic viscosity and a long duration of crystallization. As seen in Section 8.4.1, the viscosity of a residual granitic melt is decreased by water and/or halogens (F, Cl). A low-viscosity melt is easily extracted from the already-crystallized solid fraction and can result in a magma highly enriched in incompatible elements. In addition, fluorine is a fluxing element, which lowers the solidus temperature (Fig. 8.25). Phosphorus plays a similar role. These fluxes, or fluxing elements, increase the total duration of the magma crystallization, thus enhancing the efficiency of crystal fractionation and favouring the formation of a highly evolved melt, which will be noticeably enriched in most incompatible elements.

14.1.3 The pegmatitic stage

The question of the origin of pegmatites (a supercritical aqueous fluid phase after Jahns and Burnham, 1969, or a water-rich residual magma after London, 1986) was addressed

Box 14.1 The periodic classification of chemical elements

At the beginning of the nineteenth century, chemists were looking for a way to classify chemical elements. An arrangement according to their atomic weights was their first idea, but, at the same time, chemists were also looking at how to show the common properties of some elements. In 1869, the Russian chemist Mendeleev published a classification table containing six columns corresponding to recurring properties of the elements. These columns included the 63 known elements at that time according to their increasing atomic weights and to similar properties. The initial columns correspond to the periods of the present-day table (also called Mendeleev's Periodic Table: Fig. 14.3), including elements discovered later, such as the noble gases, the lanthanides (or rare earths) and the actinides. From the beginning, Mendeleev recognized the importance of grouping elements with common properties, which led him to propose the substitution of iodine for tellurium in his Table despite the apparent order suggested by the respective atomic weights of both elements.

In 1888, Rydberg proposed to use the concept of atomic number rather than atomic weight and in 1913 the English physicist Moseley established an empirical link between these numbers and the characteristic X-rays, which are regularly offset following the elemental order.

The actinide series has been extended with the discovery, eventually by synthesis, of elements with atomic numbers greater than 94.

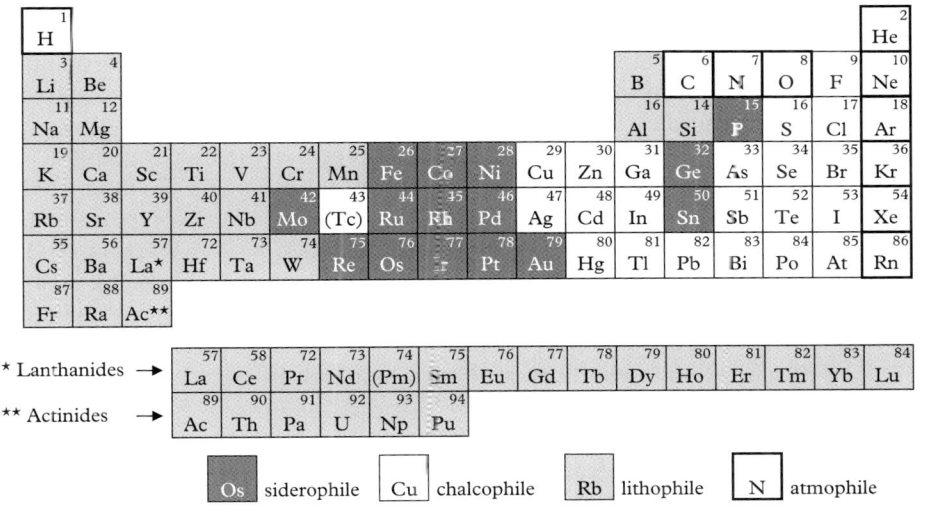

Figure 14.3 *Classification of natural chemical elements according to their atomic numbers (Mendeleev's Periodic Table). The atomic number (Z) is the number of protons included in the atomic nucleus, whereas the atomic mass corresponds to protons plus neutrons.*

in Section 8.4.3. The key point is to determine when exsolution of an aqueous fluid phase occurred with respect to pegmatite formation. Studies of melt and/or fluid inclusions in plutonic rocks brought new evidence to this debate. The first identification of melt inclusions in minerals was due to Sorby (1858) at the University of Sheffield. However, the study of melt inclusions did not significantly develop until approximately twenty years ago, whereas the study of fluid inclusions began earlier (see the books by Roedder, 1984, and Shepherd et al., 1985). The study of melt inclusions was mainly focussed on volcanic rocks at the beginning, because the melt is preserved as glass. Thus, it is easy to discriminate these melt inclusions (MI) from fluid inclusions (FI), which contain liquid and gas (usually CO_2), and sometimes crystals of salts and occasionally economic minerals (Fig. 8.29).

Melt inclusions in plutonic rocks are smaller (10 μm of average diameter) than in volcanic rocks (diameter >50 μm), and always appear fully crystallized. To be studied, they need to be heated in order to be homogenized by complete melting. The homogenization temperature is noted and the melt is analysed. A whole population of similar inclusions is required in order to derive credible information on the magmatic evolution. The method is described in detail by Bodnar and Student (2006).

Thomas et al. (2000) identified melt inclusions in pegmatites from Ehrenfriedersdorf (Germany), strengthening the hypothesis that pegmatites could represent a water-rich, but not yet unmixed, magma (David London hypothesis). Indeed, experiments by Sowerby and Keppler (2002) show that some elements, e.g. F or B, when dissolved in a water-rich albitic melt, postpone water saturation, hence forming a free aqueous fluid phase, even at relatively low pressure. However, natural examples appear slightly more complex. Thomas et al. (2006) finally identified two types of magmatic inclusions in Ehrenfriedersdorf pegmatites, both of them containing a silicate melt, but one type (MI-type A) with a moderate water content and the other (MI-type B) with a much higher water content (Fig. 8.26). More specifically, type A inclusions contain a melt with about 70–75% silica and 5–20% water (wt %), whereas type B inclusions contain 65–25% silica and 25–55% water. Is this latter case still a melt when SiO_2 is not the main constituent? Actually, the important point is the coexistence of two phases, both of them containing silica and water in variable proportions, that will form pegmatites and eventually another water-rich phase when reaching water saturation.

These water-rich melts contain other elements. Cerny (1991) published a four-fold classification of pegmatites depending on crystallization conditions (T and P, hence depth) and on their compositions (Table 14.2). The recent book by London (2008) provides a discussion of this classification and emphasizes the importance of the two pegmatite families: LCT (Li, Cs, Ta) pegmatites and NYF (Nb, Y, F) pegmatites.

Indeed, only the rare-element pegmatite class has an economic interest, because these pegmatites result from an extended fractional crystallization process, which led to a noticeable enrichment in these elements. The LCT pegmatites correspond to peraluminous liquids and are mostly associated with S-type granites and some highly fractionated I-type granites, whereas the NYF pegmatites correspond to peralkaline liquids and are associated with A-type granites.

Table 14.2 *Classification of granitic pegmatites. After Cerny (1991).*

Classes		Typical elements	Metamorphic context	Associations with granites
Abyssal pegmatites		U, Th, Zr ...	amphibolite to granulite facies (700–800 °C; 400–900 MPa)	no, rather associated with migmatite leucosomes
Muscovite pegmatites		Li, Be ...	amphibolite facies (650–580 °C; 500–800 MPa)	yes (outer veins from pluton) or no (migmatites)
Rare-element pegmatites	LCT	Li, Rb, Cs, Sn, Ta, B, P	LP amphibolite facies (500–650 °C; 200–400 MPa)	yes (outer veins)
	NYF	Y, REE, Nb, F	Shallow depth (100–200 MPa)	yes (inner or marginal veins)

14.1.4 Concentration processes in the aqueous fluid phase

These processes occur during the pneumatolytic stage (with a supercritical fluid phase) and the hydrothermal stage (with a liquid and/or a vapour phase). In both cases, the coexistence of two different phases, either a silicate melt and a fluid phase, or a hydrous liquid (or brine) and a vapour phase, provides the opportunity to concentrate dissolved elements in one of these phases depending of the distribution coefficients of these elements (Fig. 8.28). Dissolved elements in water are free ionic species or complexes. Alkaline (group I of Mendeleev's Periodic Classification: Fig. 14.3. and alkaline earth (group II) elements are easily dissolved as ions, whereas elements from the following columns (columns 3–12 between groups II and III), or transition elements, preferentially enter in complexes with other ions called 'ligands'. The reaction is:

$$M(\text{metal cation or nucleus}) + n\, L\, (\text{ligand}) = C(\text{polyatomic complex})$$

Cations are divided into two classes: 'hard' (a) and 'soft' (b), in the same way as ligands. Thus, for instance, F^- is a hard ligand and $S_2O_3^{2-}$ a soft ligand. Hard cations, such as the REEs and HFSEs, form very stable complexes with hard ligands. Soft cations, such as Cu^+, Ag^+, Au^+, Zn^{2+}, Sn^{2+}, rather form complexes with sulfur: they are chalcophile elements.

Formation of complexes is a key point in the understanding of ore genesis. The cases of copper and gold are illustrative. These metals sometimes reach very high concentrations in the hydrothermal vapour phase, whereas several other metals get preferentially concentrated in the brine (Fig. 14.4). The experimental work of Pokrovski et al. (2008) demonstrated that copper and gold become preferentially concentrated in the vapour phase where they form volatile sulfide complexes (Fig. 14.5).

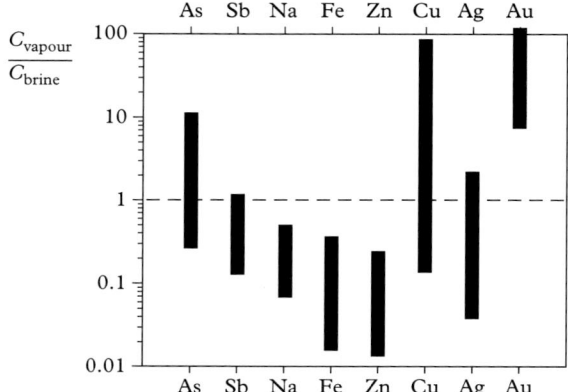

Figure 14.4 *Concentration ratios in the vapour phase and in the brine of some metals from hydrothermal deposits (compare with Fig. 8.28b). After Pokrovski et al. (2008).*

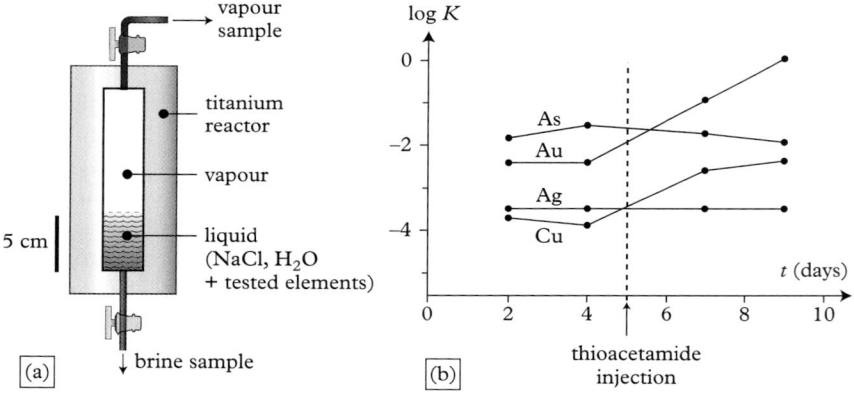

Figure 14.5 *Experimental study of the complexing effects of gold and copper. (a) The experimental device. (b) Consequences of thioacetamide injection on the distribution of metals; thioacetamide (C_2NH_2S) is decomposed in water, yielding H_2S which is important for the formation of sulfide volatile complexes with gold and copper, hence an increase of the relevant element contents in the vapour phase; K is the ratio of vapour to liquid concentrations. After Pokrovski et al. (2008).*

During the pneumatolytic and hydrothermal stages, the free water phase reacts with the granite and its country rocks. Some elements are precipitated and others are leached: these open-system interactions are called 'metasomatism' or pneumatolytic/hydrothermal alteration, and were summarized by Halter and Webster (2004). When the granite itself suffers many changes due to the influence of hydrothermal fluids, it becomes a 'greisen'. The beginning of greisenization is characterized by the development of muscovite from the breakdown of K-feldspar:

$$3KAlSi_3O_8 + 2H^+ \rightarrow KAl_3Si_3O_{10}(OH)_2 + 6SiO_2 + 2K^+,$$

but the initial texture and the primary minerals of the granite can still be recognized. More advanced transformations erase the initial texture and result in a rock mainly made of quartz, muscovite and topaz, whose formula is $Al_2SiO_4(F,OH)_2$, and which is sometimes of gem quality. Fluorite, beryl and cassiterite (SnO_2) may also be present, contributing to the economic interest of greisens. More subtle modifications due to the interaction of evolved magmas with late-magmatic fluids can be verified by their influence on REE distribution patterns. In this case, REE distribution patterns display a typical shape, the so-called 'tetrad', because they can be split into four convex or concave segments in addition to a large negative Eu anomaly (Bau, 1996; Jahn et al., 2001). An example from the Ore Mountains (German: *Erzgebirge*, Czech: *Krušné Hory*) at the border between Germany and the Czech Republic is provided in Fig. 14.6. This Variscan metallogenic province is characterized by Sn–W enriched, fractionated S-type and A-type granites emplaced between 330 and 290 Ma (Breiter et al., 1999). The tetrad effect is observed in the altered A-type leucogranite of Cinovec (Zinnwald). The modifications confirm an input of REEs during the percolation of an F-rich fluid, also responsible for the crystallization of accessory minerals, such as fluorite (CaF_2), bastnaesite (($Ce,La)CO_3F$) and synchysite ($Ca(Ce,La)CO_3F$). Transportation of Sn and W likely occurred also as F-complexes in this fluid phase (Johan and Johan, 2001). Then, Sn

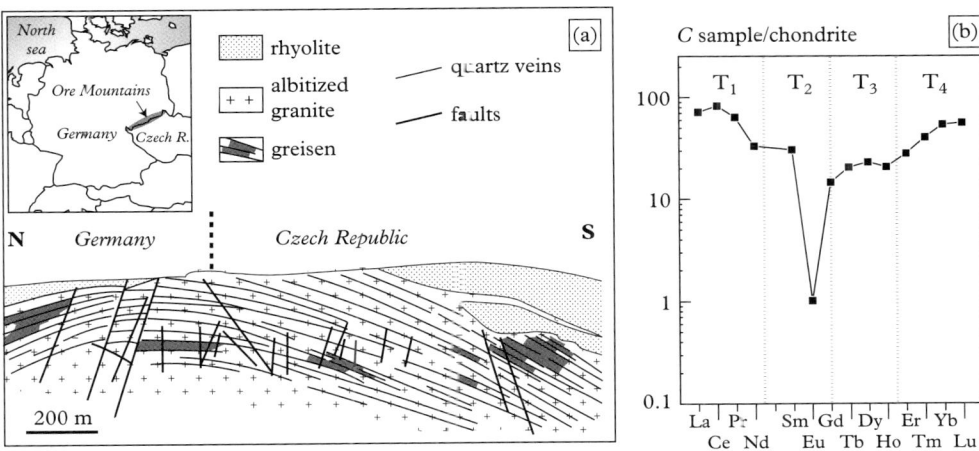

Figure 14.6 *(a) Section of the Sn–W (cassiterite and wolframite) ore body of Zinnwald (Cinovec) near the German–Czech border. (b) Tetrad-style REE distribution pattern from the albitized and greisenized granite, where four more or less convex successive segments (T_1 to T_4) can be recognized witnessing HREE enrichment by fluids. After Monecke et al. (2007).*

precipitated as cassiterite in the greisenized granitic cupola, whereas W was transported in the uppermost quartz veins due to its higher volatility.

Modifications of the granite country rocks are variable both in nature and intensity. Some of these modifications were already described in Chapter 6. From an economic point of view, the more interesting ones occur in carbonates, where the formation of skarns may be associated with increasing metal concentrations. The case of tungsten-bearing skarns is the most common example (see, for instance, Fig. 7.12).

14.1.5 Transfer of metals and deposition of ore minerals

Pegmatitic liquids and hydrous fluids (here regarded as 'late-magmatic fluids') are able to percolate through the granite along local anisotropies (foliations, shear zones or other planes of weakness), and then are channelled via extensional fractures, which may become mineralized lodes (Fig. 14.7).

At the regional scale, quartz veins with tourmaline and cassiterite in the country rocks of the Shah Kuh pluton (Iran) are systematically orthogonal to the finite extension direction registered in granodiorite by magnetic lineations (see Section 10.4.1). This is evidence that these veins were formed at the end of crystallization of the granodioritic magma (Fig. 14.8). At water saturation of the magma, formation of a water-rich, low-density fluid is responsible for a local overpressure, eventually causing brittle fractures. This is the explanation for pegmatites and quartz veins that escaped by fracturing the solidified parts of a pluton and can be traced upwards or laterally outside of the pluton. Low viscosity and low density favour this fluid transfer far away from the granite margin, with the result that the spatial linkage between lodes and their parent granite may not appear directly related.

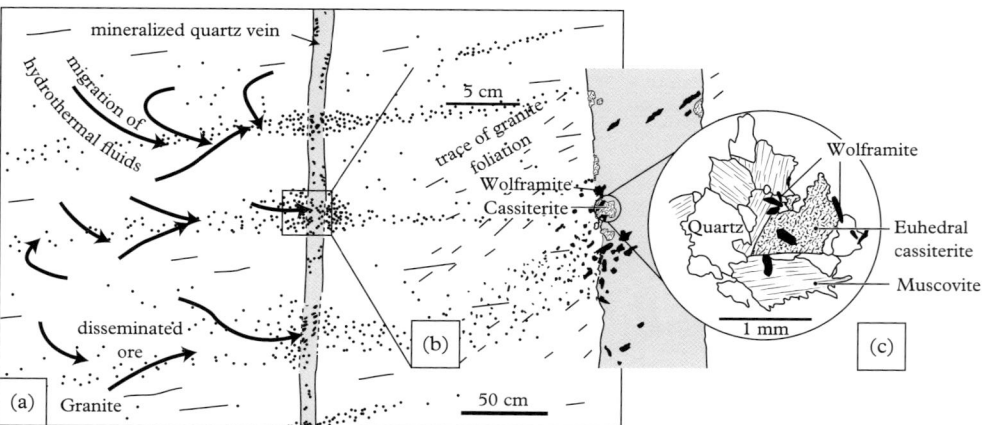

Figure 14.7 *Formation of the Sn–W mineral deposits in the Santa Comba ore body from Galicia (Spain): late-magmatic fluids (arrows) first precipitated scattered oxides in the granite (a), then escaped through fractures to give ore-bearing veins (b); automorphic cassiterite is associated with quartz, muscovite and wolframite in the veins (c). From Nesen (1981).*

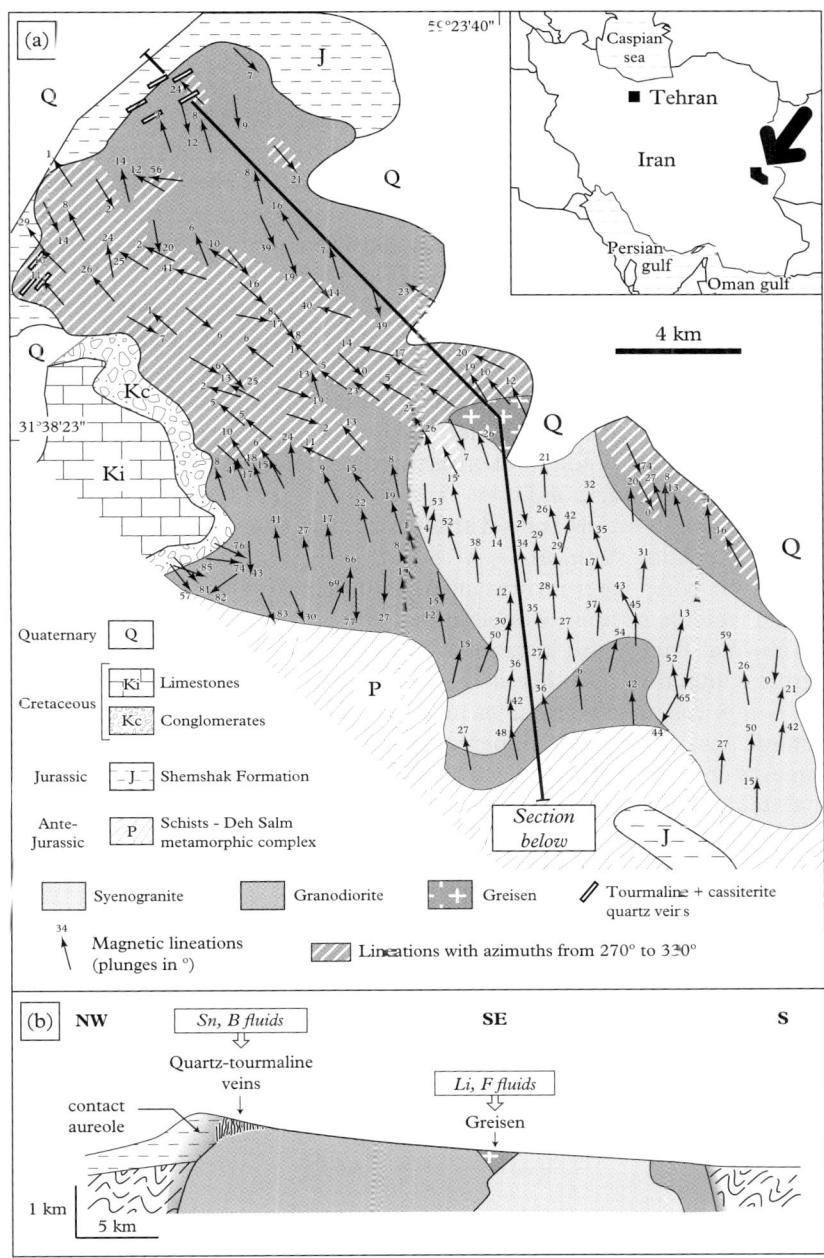

Figure 14.8 *Shah Kuh pluton (Iran) and its related mineralized zones. (a) Pluton map showing two types of granitoids: a reduced (ilmenite-bearing) I-type granodiorite and a cross-cutting S-type. Quartz veins containing tourmaline and cassiterite are always orthogonal to the granodiorite lineations; hence they are coeval with late-magmatic deformation of the granodiorite. (b) Interpretative section showing granitoid-derived fluids; the greisen is due to fluids expelled from the S-type granite. After Esmaeily et al. (2005, 2007).*

Late-magmatic fluids react with their country rocks and modify them as long as the water/rock ratio is high (open system). Consequently, the fluid physical and chemical properties are also modified, providing new opportunities of element concentrations, possibly leading to precipitation of elements that have become insoluble in these new conditions. A decrease of pressure and especially of temperature influences the element solubility and explains the well-known zonal distribution of mineral deposits in the roof cupola or in the vicinity of a granitic pluton (Fig. 14.9).

Precipitation is also controlled by the evolution of redox conditions, for instance in the case of tin, soluble as Sn^{2+}, but precipitating to form cassiterite (SnO_2) after oxidation to Sn^{4+}. Fluid chemical properties are influenced by the composition of the country rocks or by mixing with meteoritic waters, which are often oxidized. As shown in Fig. 14.1, formation of uranium ores requires fluids that are more oxidized than the usual magmatic fluids. Indeed, uranium may have two different valencies, U^{4+} and U^{6+}, and is only soluble in oxidizing conditions (contrary to tin!). It precipitates if conditions become more reduced (for instance in sediments containing organic matter or in hydrothermal veins containing sulfides).

The role of meteoric water is most efficient in the brittle upper crust and, especially, in extensional tectonic settings, where it can enter deep into the crust due to the development of normal faults. This external fluid contributes to enhancing the duration of hydrothermal convection cells that were active in the roof zone of shallow plutons. The external (i.e. non-magmatic) input of meteoric hydrothermal fluids can be measured by the isotopic signatures of the mineral deposit and its trapped fluid inclusions in associated quartz or other minerals. Elemental (re)mobilization and concentration by hydrothermal fluids sometimes occur well after granite crystallization. The final stage may happen under supergene conditions, when granite becomes exposed to weathering. This last stage may be essential in order to produce an ore body of economic interest (see the example of kaolin in St Austell, Cornwall, UK, in Section 14.2.1).

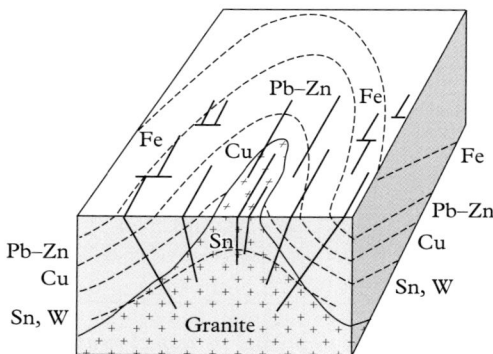

Figure 14.9 *Typical sketch of a granitic cupola with its aureole of radial veins and the zoning of deposits at the roof. The country rocks are barren at some distance from the contact.*

14.2 Examples of granite-related mineral deposits

The selected examples do not provide a thorough panorama of granite metallogeny. They will be presented in relation to the different granite types (S, I and A, and their related mineralizations in this order), and have been selected to illustrate the diversity of geodynamic settings (collision, subduction, intraplate) and of mineral products (base metals, rare metals, non-metals).

14.2.1 Tin and kaolin in the Cornubian batholith, Cornwall

The Cornubian batholith consists of several plutons that crop out in Cornwall as well as in Devon (Fig.14.10), and were emplaced into epizonal (greenschist facies) metamorphic rocks of Devonian and Carboniferous age. The granite mineral assemblages are typically S-type (Manning, 1998). Granite magma formed by partial melting of the lower continental crust, but was emplaced at shallow levels (some plutons contain magmatic andalusite). The most common lithology is a two-mica granite, often with a porphyritic texture. Primary (magmatic) tourmaline of typical Fe,Al-rich schorl composition is abundant (St Austell, Carnmenellis, Bodmin). During the hydrothermal stage, tourmaline (containing less Fe and Al than magmatic tourmaline), fluorite and locally topaz (Tregonning–Godolphin) crystallized in the abundant lodes.

Cornubian granites were synkinematically emplaced at the end of the Hercynian orogeny (280–290 Ma). At that time, the strike of crustal thinning was NNW–SSE, as indicated by recent AMS studies on Carnmenellis and Bodmin granite plutons (Fig. 14.10b and c).

The Cornubian plutons are the source of important concentrations of tin (cassiterite) worked as placers since prehistoric times, then in lodes from Roman times until 1990, when the mines were closed because it became uneconomic to continue exploitation. Cassiterite crystallized in hydrothermal quartz veins. At Carnmenellis, lodes generally strike ENE–WSW near the border of the pluton, which is perpendicular to magmatic lineations (Fig. 14.10b). They also contain wolframite [(Fe, Mn)WO$_4$] and hydrothermal tourmaline. Whereas most tin mines are no longer active, kaolin (from Chinese *kao ling*) deposits at St Austell are still worked.

First recognized in the small Tregonning–Godolphin stock in 1746 by William Cookworthy, kaolinite (Al$_4$Si$_4$O$_{10}$(OH)$_3$) actually occurs in all massifs and especially in St Austell. Kaolin production in Cornwall increased during the nineteenth century to reach 0.7 Mt/yr in the 1950s and 3 Mt/yr more recently. Only 12% of this production is used for making china crockery, whereas 80% are then used as an additive in painting and the paper industry, i.e. as filler clays and/or surface coating (papers).

Kaolinite results from orthoclase (KAlSi$_3$O$_8$) alteration in mildly acidic conditions and at rather low temperature ($T < 200$ °C):

$$2\,\text{orthoclase} + 2\text{H}^+ + \text{H}_2\text{O} \leftrightarrow 1\,\text{kaolinite} + 4\text{SiO}_2 + 2\text{K}^+$$

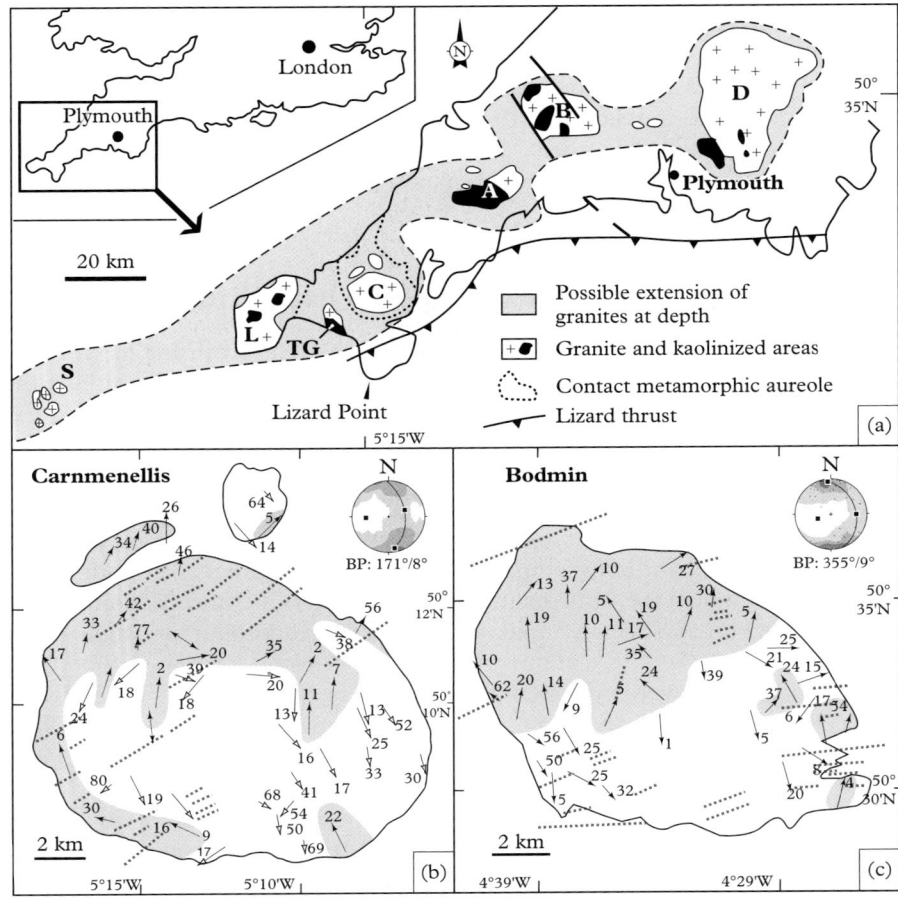

Figure 14.10 *(a) The Cornubian batholith. Plutons: A: St Austell, B: Bodmin, C: Carnmenellis, D: Dartmoor, L: Lizard Cape, S: Scilly Islands, TG: Tregonning–Godolphin. (b) Lineation map in Carnmenellis pluton. (c) Lineation map in Bodmin pluton. Grey shaded areas in (b) and (c): N-plunging lineations suggesting that plutons have an E–W elongate dome shape; dotted lines: lodes (Sn). After Mintsa mi Nguema et al. (2002) and Bouchez et al. (2006).*

Large volumes of granites were bleached and weakened by the kaolinization process and can therefore be worked by water hosing (Fig. 14.11). In other cases, only small rock volumes contain kaolinite, mainly in veins or along fractures (Fig. 14.12).

The origin of kaolin has been debated as either supergene or hydrothermal. Fluid inclusions observed in quartz spatially associated with kaolinite record relatively high homogenization temperatures: 170–200 °C (Charoy, 1979). As quartz is a resistant phase to supergene weathering, the fluid inclusions therefore registered a hydrothermal event before kaolinization. The St Austell massif, the richest in kaolin, is also the

Figure 14.11 *Extraction of kaolin by high pressure water jet at St Austell.*

richest in tourmaline, good evidence for the significant activity of late-magmatic to hydrothermal fluids.

Alternatively, stable isotope data ($\delta^{18}O$ and δD) derived from kaolinite suggest a meteoritic origin (Sheppard, 1977) by surface weathering due to the hot and wet climate that prevailed at times during the Tertiary period. Kaolin deposits in Brittany (France), also associated with S-type Hercynian granites, were the subject of a similar debate. Here, the late-Hercynian hydrothermal event was followed by a younger supergene stage which may have been necessary to obtain a kaolin deposit of economic interest.

14.2.2 Uranium from Limousin province (Massif Central, France)

The Massif Central is located in the inner part of the Hercynian belt of western Europe (Fig. 12.12). It was at the heart of the past mining industry in France, especially for uranium. Uraniferous minerals were first described in Limousin (western Massif Central) in 1804. The first discovery of a pitchblende lode containing globular concretions of uranium oxide, UO_2 (uraninite), occurred in 1948 at la Crouzille (the so-called 'Henriette' lode), in Limousin. Since then, more than 40,000 t of uranium (metal) have been

296 *Granite metallogeny*

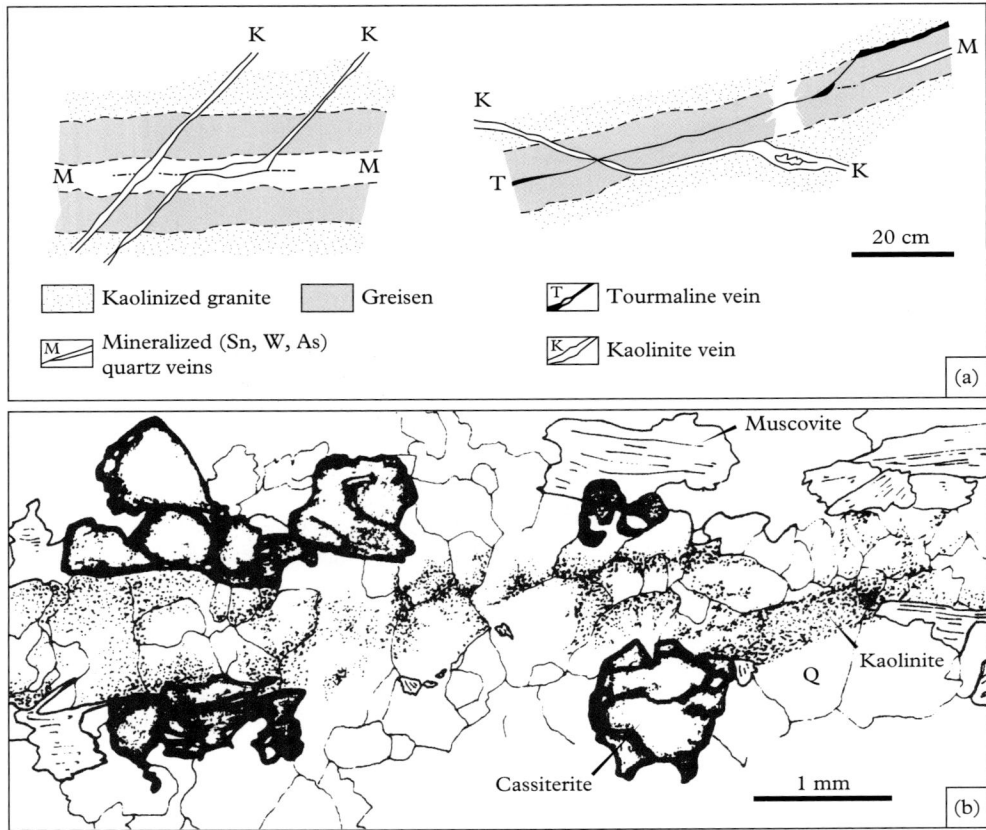

Figure 14.12 *Kaolinization along veinlets (a) or in greisen (b). Observed by Charoy (1979) in W–Sn mines at Cligga Head (Cornwall).*

extracted from the Massif Central granites, with two-thirds of the total volume from la Crouzille area. The last third was mined from other granitic areas in the Massif Central (Marche, Millevaches, Margeride, Forez). Besides, a large amount of uranium has been mined in sedimentary basins (mainly from the Permian sediments in the Lodève basin). Here, we are only concerned with granite-related uranium from Limousin (Fig. 14.13). A general introduction to the Hercynian plutonism in the French Massif Central is provided in Section 12.4.3.

During the so-called 'post-thickening' extensional phase from Namurian to Lower Stephanian times (325–315 Ma), two-mica peraluminous leucogranites were emplaced at about 10 km depth. The WNW–ESE extensional direction has been registered by magmatic lineations in Limousin (Fig. 9.18). Several ore bodies are located in hydrothermal veins. Inclusion planes of trapped fluid have been preserved in their quartz crystals. These ore fluid inclusions suggest relatively high initial temperatures (420–350 °C).

> **Box 14.2** Kaolin and china porcelain
>
> Kaolin is the raw material required to make china, a hard and very fine ceramic invented some 2000 years ago in China, hence its English name (whereas it is called '*porcellana*' in Italian and '*porcelaine*' in French). The Chinese ceramics industry reached its perfection in the twelfth century. The mixture used for making china also contained quartz and feldspar (as a flux) and must be heated up to 1300–1400 °C. China porcelain from China was imported to Europe as early as the fifteenth century, but china production in Europe only began in 1709 in Saxony (Germany), where kaolin deposits were known since the end of the seventeenth century.
>
> In France, a few kaolin samples were brought back in 1712 by Father d'Entrecolles, a Jesuit missionary in China. J.-E. Guettard (1715–1786), the French geologist who was the first to recognize the Auvergne Mountains as volcanoes, also identified white clays used by potters near Alençon (Normandy) as kaolin. He created the first French ceramic factory at Vincennes in 1740. Ceramic manufacturing moved to Sèvres in 1756, to a new building (still visible today) ordered by the Marquess of Pompadour. The first products from the royal manufacturing factory at Sèvres were not pure (hard paste) china (or *porcelaine dure*), but a sort of soft-paste ceramics mimicking china (*porcelaine tendre*). Hard-paste china was only produced in France after the discovery of high quality kaolin at Saint-Yrieix-la-Perche (near Limoges) in 1768. The first china factory in Limoges was founded a few years later. In the mid-nineteenth century, Limoges had about thirty china firms, but only a third of them are still active. Today, the Sèvres factory is a public service controlled by the French Ministry of Culture, and produces small-scale art craft ceramics. The mineralogist Alexandre Brongniart was the head of this factory from 1800 to 1847. He established a ceramic classification still in use worlwide and created the Ceramic Museum at Sèvres.
>
> In Great Britain, the famous Wedgwood pottery firm was founded in 1759 by Josiah Wedgwood in Staffordshire. Bone china, a special type of soft-paste porcelain, was first created in England. It is made from a clay mixture containing bone ash (Ca-phosphate) in addition to kaolin, feldspar and quartz, and has been much appreciated for its white and translucent aspect, as well as a hardness nearly equal to hard-paste china.

They correspond to the depositional succession of Li–F, W–Sn, Au, Sb, Zn–Ge, Pb–Ag and F–Ba ores with decreasing temperatures. An orthomagmatic preconcentration of uranium (± thorium and REE) occurred in the Limousin leucogranites, mainly as monazite (a REE-bearing phosphate with 0.5–3% U) and minor Th-rich uraninite (UO_2). These minerals appear distributed in a heterogeneous way reflecting the source heterogeneity of the leucogranitic magmas. Their preconcentration was especially high (U > 100 ppm) in Brâme and Saint-Sylvestre areas where highly differentiated magmas were intruded into a magmatic megashear zone (Fig. 14.14).

As shown by Ar/Ar ages, the NW Massif Central was still subjected to WNW–ESE directed extensional tectonics and was coevally exhumed from 320 to 315 Ma ago

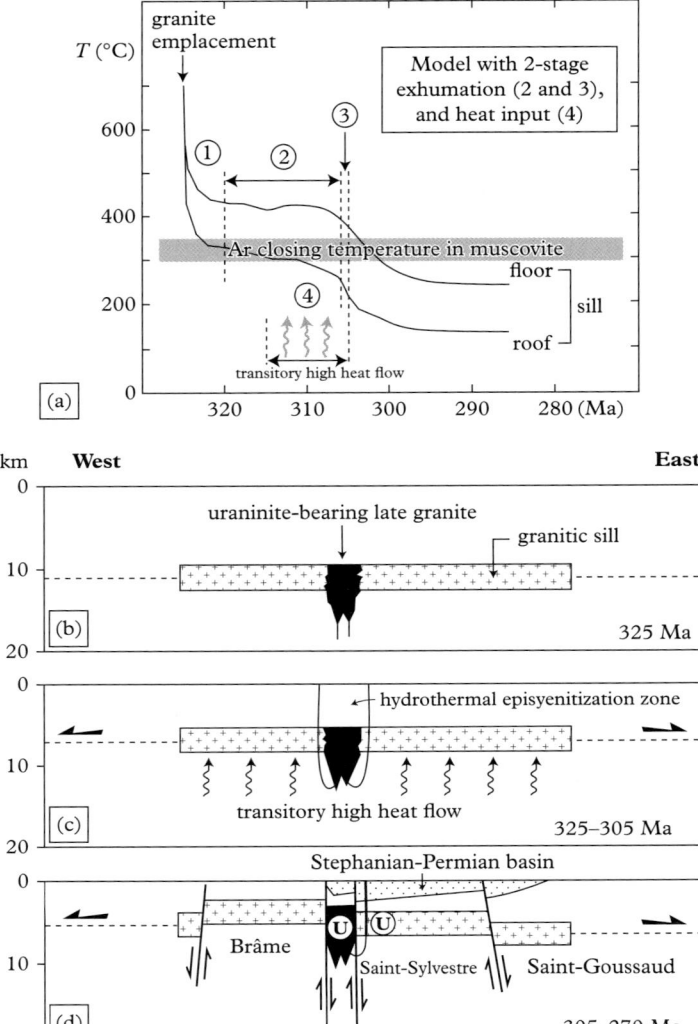

Figure 14.13 *Formation of the uranium ore deposit in Limousin. (a) Cooling model of the Marche–Limousin leucogranitic complex (here regarded as a 3 km thick sill), as deduced from muscovite Ar–Ar ages. The roof and the floor of the sill yield ages differing by 10 to 14 Myrs. The high heat flow between 315 and 305 Ma is correlated with the late-Hercynian stage of lithospheric delamination. (b) Granite emplacement (immediately below the brittle-ductile transition); the late and most differentiated magmas are characterized by some preliminary concentration of U. (c) The exhumation, first slower (2), then faster (3 and see also (a)), carried granites upwards into the brittle zone of the upper crust. It resulted in fractures being formed where convective fluid circulation was favoured, leading to the local formation of hydrothermally altered granites. (d) Final stage: remobilization, concentration and deposition of uranium as pitchblende. After Scaillet et al. (1996a, 1996b).*

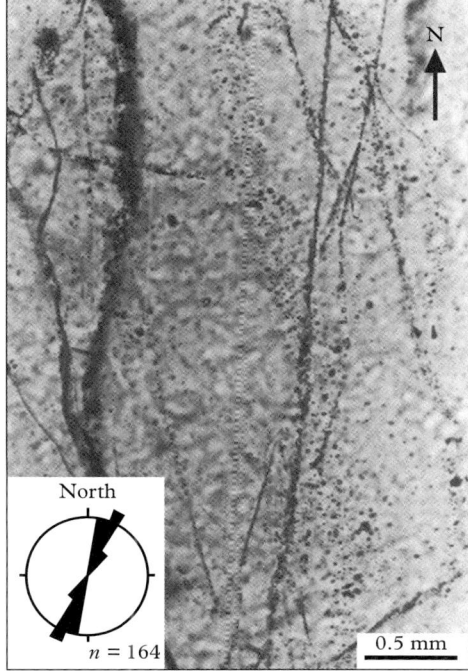

Figure 14.14 *Fluid inclusion planes striking NNE in a quartz grain from Saint-Sylvestre granite, and rose diagram showing the cumulative orientation distribution for a whole thin section. After Lespinasse and Pêcher (1986).*

(Fig. 14.13). Then a new hydrothermal event concentrated fluids in fractures, altered granitic rocks and dissolved the small-sized (a few tens of microns) accessory minerals, that contained uranium. The most altered granites, that will become the most mineralized, look like columnar bodies of so-called '*episyenites*', i.e. hydrothermally altered granites located at the intersection of subvertical fractures where quartz has been dissolved and partly replaced by albite. This was due to warm fluids (370 ± 25 °C), trapped at relatively low pressures (20 MPa) as shown by fluid inclusions in quartz from the altered granites. The aligned inclusions mimic a hydrothermally fractured network with a consistently average strike toward the NNE, thus orthogonal to the WNW–ESE direction of the principal extensive stress coeval with this hydrothermal stage (Fig. 14.14).

During Stephanian and Lower Permian times (270–280 Ma), the whole area was subjected to a new extensional phase in a post-orogenic setting, characterized by an approximate E–W direction, as shown by the submeridian ductile normal faults at Nantiat (Fig. 9.18) and, further south, at Argentat. The first sedimentary basins were formed, namely the Brive, Saint-Affrique and Lodève basins. A final stage of fluid circulation occurred with lower temperatures

(200 ± 20 °C) and a variable origin, either external (oxidizing fluids) or connate (reduced fluids containing CO_2). It was responsible for uranium (re)mobilization, and final concentration during the change from soluble (U^{6+}) to insoluble (U^{4+}). The main uranium deposits mined in the granites of the Hercynian belt were formed at this stage.

This uranium saga shows that formation of uranium ore bodies of economic interest results from a succession of geological processes, each of them corresponding to an increased concentration of this element. Its history has been deciphered through the patient work of many geologists, especially from CREGU (*Centre de Recherche et d'Étude sur la Géologie de l'Uranium*, Nancy, France). The interested reader will find other references in Scaillet et al. (1996a, 1996b) and Marignac and Cuney (1999).

14.2.3 Tantalum from Tanco (Canada)

Tantalum (Ta) is a rare metal, generally associated with niobium (Nb), both entering into oxides of (Fe, Mn)(Ta, Nb)$_2$O$_6$ formula, collectively named 'colombotantalite' (or 'coltan' for miners), that may be either Ta or Nb rich. The main Ta producer is Australia. Recently, many ore bodies in the eastern Congo Democratic Republic have been involved in illegal mining and ore trafficking.

Ta-rich ores are associated with peraluminous (S-type) leucogranites or to LCT pegmatites (as in Tanco). Tantalum is mainly used for the fabrication of electronic hardware and for making cell phones. Niobium is also a rare metal, very much sought-after, that is important for the making of special steels. It is more abundant in NYF pegmatites, as mentioned earlier in Table 14.2.

The pegmatite from 'Tanco' (a name standing for Tantalum Mining Corporation) is located in Manitoba state, near Bernic lake, close to the Ontario border. It is 2640 Ma old. The dyke measures 1.5 km long by 1 km wide, but is only 100 m thick. Pegmatite emplacement was probably guided by the presence of several faults and shear zones. This ore body belongs to a pegmatite field derived from peraluminous leucogranites, which has been well known for its economic potential since 1929. The country rocks were metamorphosed at conditions of low pressure amphibolite facies. The Tanco dyke does not outcrop at the surface. It was discovered and mined in galleries. The mining company was first prospecting for lithium. The lithium mine was closed in 1961, but reopened in 1969 for the exploitation of tantalum.

The Tanco pegmatite displays a complex mineral composition with nine different zones from core to rim (Cerny, 2005). Its bulk composition corresponds to a highly differentiated S-type granite with an average silica content of 76%. The outer zones are granitic in composition and contain abundant albite and quartz, with other minerals indicative of specific elements: tourmaline (B), beryl (Be) and muscovite (Li). The intermediate zones are still granitic and characterized by significant abundance of perthitic microcline, quartz and Li-rich minerals. The outer and intermediate zones crystallized at temperatures ranging from 700 °C to 500 °C and at about 300 MPa pressure. Tantalum is present as different oxides in the intermediate zones and in the Li-rich inner zones. Ta is especially abundant in the albitic aplite lenses of the intermediate zones (Fig. 14.15). The Ta concentration process in the aplite was purely magmatic and probably resulted

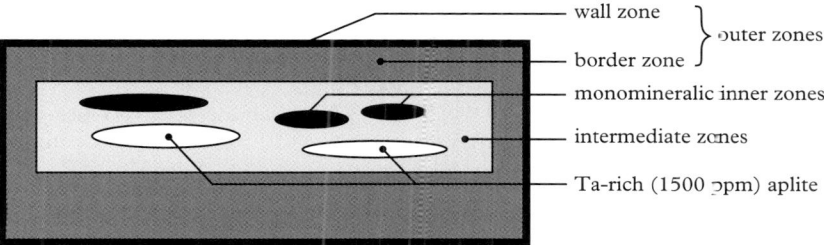

Figure 14.15 *Simplified cross-section of the Tanco pegmatite. After Stilling et al. (2006) and van Lichtervelde (2006).*

from extensive crystal fractionation. Indeed, high contents of water, boron (B_2O_3 = 0.07 wt % on average) and fluorine (F = 0.12 wt % on average) were responsible for a significant decrease of the solidus temperature, which promoted enrichment in several incompatible elements, such as Ta, Cs and Li. Precipitation of Ta oxides occurred when the removal of Li and B in a fluid phase suddenly decreased Ta solubility in the highly evolved and enriched aplitic magma. The inner zones are nearly monomineralic discontinuous lenses, made of quartz or Li-rich minerals, hence no more granitic in composition. These zones likely crystallized at temperatures ranging from 500 °C to 300 °C and at pressures of 300–200 MPa. Some of these zones are rather Ta-rich, but the Ta concentration process is still being debated.

14.2.4 Porphyry copper deposits in Chile

Chile is the world's largest producer of copper and has the largest known reserves in base metals. Copper, as well as other chalcophile elements that may be found together (Mo, Au, Ag), are extracted from the well-known porphyry copper deposits, typical high-tonnage but low-grade deposits (Cu < 1% on average), mined in giant open pits that are several hundreds of metres deep.

These copper porphyries are formed in the roof zone of calc-alkaline plutons of dioritic to I-type monzogranitic composition, emplaced at shallow level in the brittle crust (2–8 km deep at maximum). These conditions favoured water saturation of magmas and led to voluminous hydrothermal degassing, which was noticeably efficient from the metallogenic point of view. Thus, ore bodies were formed both at shallow levels (in a subvolcanic context) and at high elevations in the Cordillera (often between 3000 and 4000 m elevation). Consequently, they may be easily eroded and their age is mostly recent. This calc-alkaline magmatism, explained in Chapter 12, was active along the whole Andean belt, and especially in northern Chile. Ore bodies were formed in a discontinuous manner during the Tertiary period, and are most likely related to the so-called 'western fault zone', a N–S elongate fault >2000 km long (Fig. 14.16).

In the field, the ore bodies occur as dense networks of fractures and sulfide-bearing veinlets, sometimes covering an area of several square kilometres, where the sedimentary or crystalline country rocks, and even the andesitic volcanics directly above the

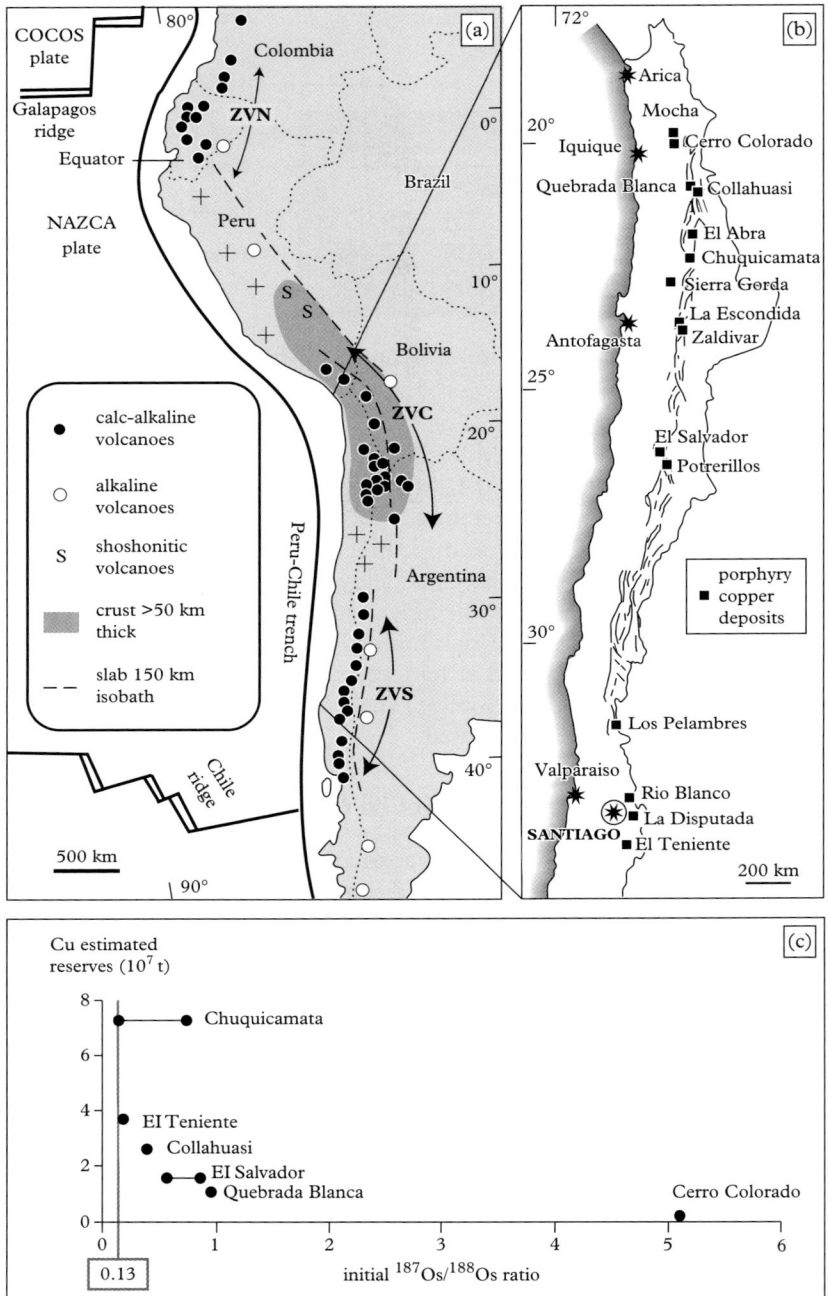

Figure 14.16 *Porphyry copper deposits in Chile. (a) Subduction magmatism in the Andes cordillera; CVZ, NVZ and SVZ: central, northern and southern volcanic zones. (b) Location of copper mines (black squares) and main faults after Sillitoe (2010). (c) Isotopic signature of the major ore bodies. After Mathur et al. (2000).*

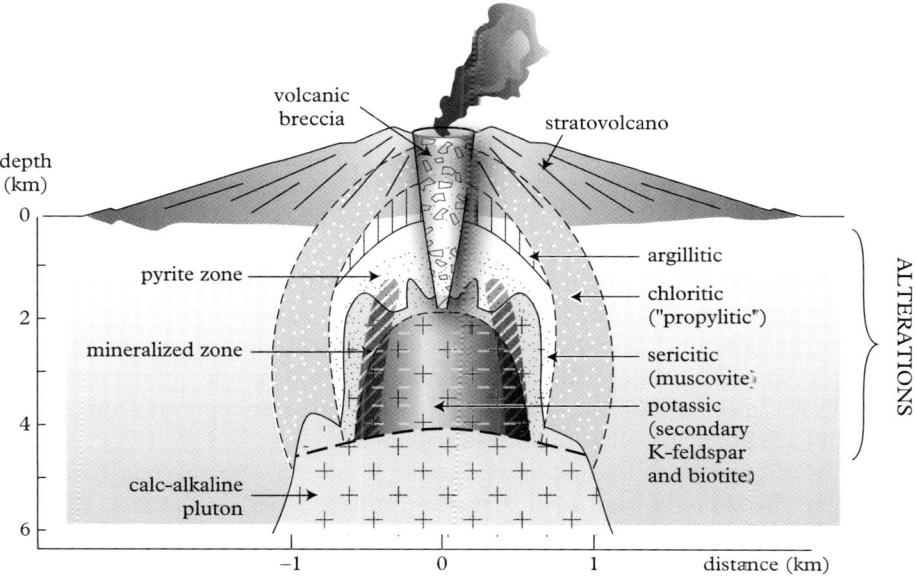

Figure 14.17 *Section of a porphyry copper deposit and its associated alteration zones.*

plutonic intrusions, have been extensively mineralized (Fig. 14.17). All these rocks were strongly modified by potassic alteration responsible for the development of biotite and orthoclase, and then by a phyllitic alteration (development of quartz and sericite). The first generations of sulfides crystallized from the magma. Volatile sulfide complexes also played a role in carrying copper, as mentioned earlier (Section 14.1.4 and Fig. 14.5). Then, meteoric waters progressively invaded the system and promoted the precipitation of another generation of sulphides. Supergene processes were also important: they do not bring any new elements, but they can contribute to the formation of an economic ore deposit, because they may rework and concentrate the minerals of interest.

Finally, the tectonic setting played a major role depending on the variable speed and direction of convergence of the Nazca and South-American plates. Changes of convergence rate were noticed by Pardo-Casas and Molnar (1987). The rate decreased from 20 to 10 cm/yr between 50 and 35 Ma ('incaic' compression phase), and then to less than 50 mm/yr at a time around 30 Ma, hence partly relaxing the compressive stress. In the case of Chuquicamata, the sense of motion along the plate border has been generally dextral since 50 Ma in global reconstructions, but may have changed to sinistral for a short time between 35 and 31 Ma during the slow convergence epoch (Reutter et al., 1996). A transitory trans-tensional regime and the development of fractures in another direction may have favoured voluminous supergene circulation, which also contributed to an additional enrichment stage of the ore body.

14.2.5 Tin and fluorine in Bushveld granites

The giant Bushveld layered complex of South Africa has been described elsewhere in Chapter 12. Mineralization (Cr, V and platinum-group elements) in the lower mafic series is well known, but the upper A-type granite is also related to economic ores of tin (Zaaiplaats mine) and fluorine (Vergenoeg mine); see Fig. 12.2a for the location of these mines.

Tin at Zaaiplaats mine is hosted as cassiterite in the Bobbejaankop granite that is the altered facies of the Nebo granite. The distribution of tin mineralization (Fig. 14.18) is consistent with an early separation of an H_2O–NaCl fluid to produce pipe-like conduits, without significant fracturing of the granites. Circulation of this fluid likely occurred at $T > 600$ °C. Then, the accumulation of later F-rich fluids beneath the downward-advancing crystallization front produced the miarolitic zone at the upper contact of the granite with Rashoop granophyre. Infill phases in the miaroles comprise quartz, albite, fluorite and haematite, and other hydrothermal minerals.

The giant Vergenoeg ore body is located in the Rooiberg rhyolites, but the F-bearing hydrothermal fluids could have been derived from the underlying granite (Fig. 12.2). The Vergenoeg pipe has a reverse cone shape, measuring 650 m high and 300 m in diameter at the outcrop level (Fig. 14.19). Its surficial part is made of a weathered iron cap or gossan, containing haematite and fluorite. The fluorite (CaF_2) quarry has been operating since 1956. About 300,000 tons of ores (containing 40% fluorite) were extracted in 2007; the ore reserves (with a grade of 10 to 40%) amount to 200 million tons, representing 10% of the world's whole global resources (Lagny, 2008). The cone is an intrusive brecciated pipe, where the ore has a pegmatitic texture characterized by centimetre-sized crystals of fluorite and magnetite in a fine-grained matrix of magnetite, fluorite and siderite. It includes several angular fragments of rhyolite. The rhyolitic country rocks are hydrothermally altered. The lower part of the cone is made of a fluorite, ilmenite and fayalite, regarded as the primary mineral assemblage, which has been

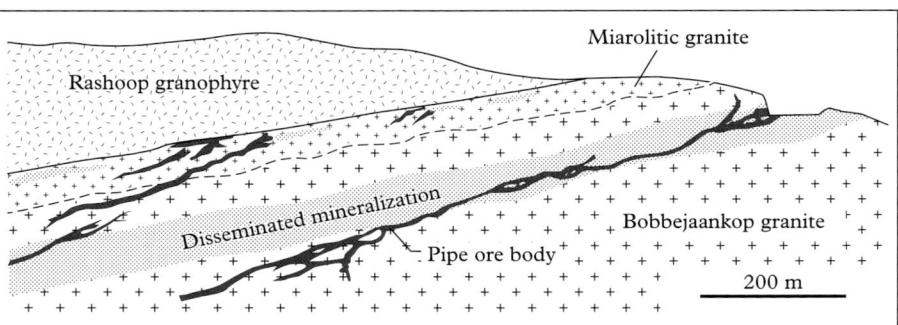

Figure 14.18 *Schematic cross-section of the roof zone of the Bobbejaankop granite showing the different styles of tin mineralization: tabular zones containing disseminated cassiterite and shallow-plunging pipe ore bodies, after Pollard et al. (1991).*

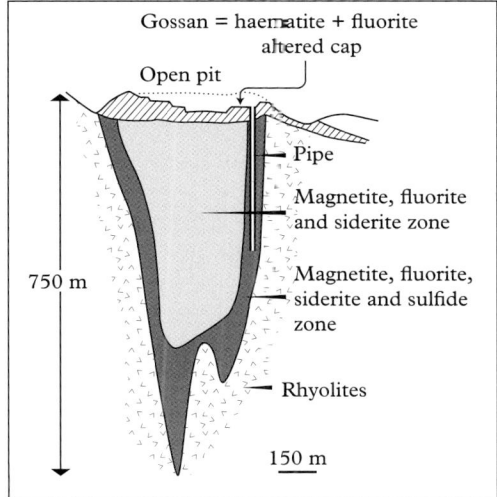

Figure 14.19 *Section of the Vergenoeg pipe, Bushveld complex, South Africa (see location in Fig. 12.2a). After Crocker (1985).*

modified upward to form magnetite and siderite and then oxidized to haematite closer to the surface. Fluorite is present throughout the Vergenoeg ore body.

Abundant fluid inclusions can be observed in fluorite. Primary inclusions found at depth contain water or CO_2, as well as ubiquitous crystals of NaCl and $FeCl_2$. They become homogenized between 300 and 500 °C. At shallower levels, only secondary inclusions are present, which have lower homogenization temperatures (200–500 °C). Methane (CH_4) fluid inclusions may also be observed locally. It is worth noting that CO_2 is generally not an important fluid in granitic magmas (whereas it is in mafic magmas). However, CO_2 is often found in granulitic facies (Touret, 2001). It was said above that H_2O and CO_2 are normally miscible in crustal conditions (see Chapter 7) and specific ligands may occur in this case. Borrok et al. (1998) regarded the aqueous and carbonic fluids identified in the Vergenoeg pipe as exsolved from the granitic magma, in as much as similar fluid inclusions were found in the Zaaiplaats tin deposit formed at deeper levels, namely in the Bushveld granite (and not in the overlying rhyolites). In both cases, fluid inclusions display the same isotopic signature (^{18}O = 9–10 ‰), with a value consistent with a granitic origin (see Chapter 2). Kinnaird et al. (2004) have shown that the Vergenoeg fluorite deposit has rather low initial strontium ratios from 0.716 to 0.722 confirming that the F-source was the Bushveld granite magma.

14.2.6 Uranium at Olympic Dam, Australia

Olympic Dam is a mining centre located some 550 km NNW of Adelaide in South Australia (Fig. 14.20a). It is the largest known single deposit of uranium (U oxides) in

306 Granite metallogeny

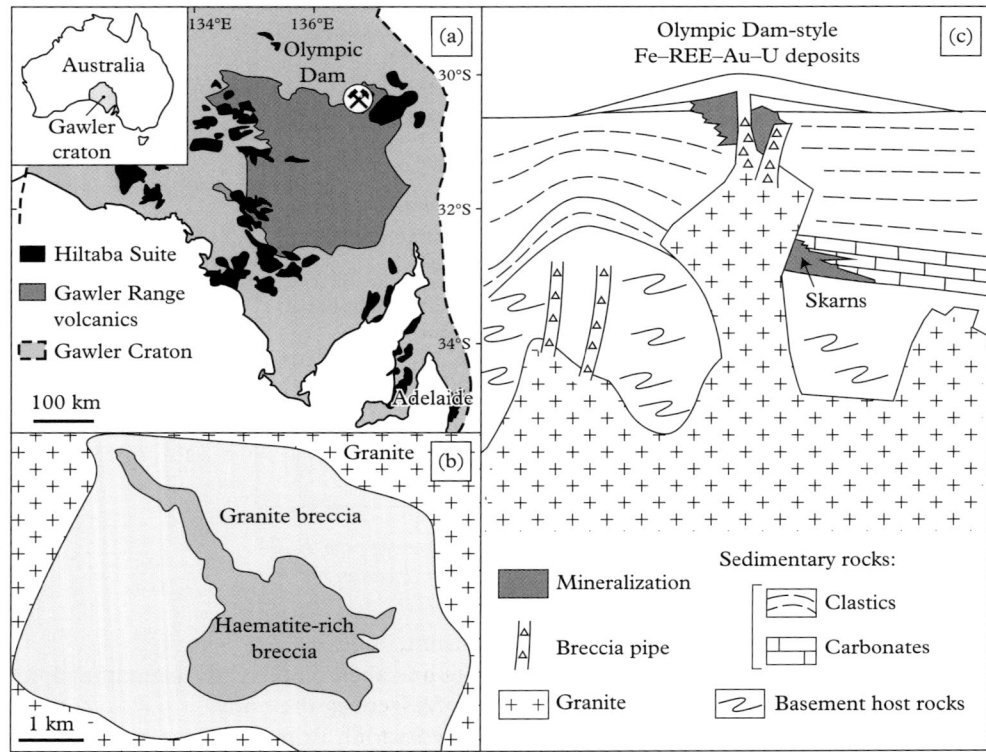

Figure 14.20 *(a) Location of the Olympic Dam mine in the Gawler craton. (b) Detailed map of the U-hosting breccia. (c) Interpretative section of an Olympic Dam-type deposit, after Pirajno (2000).*

the world. An underground mine produces uranium as well as copper, gold and silver. The deposit is hosted by an haematite-rich breccia, within a barren granite-breccia that has gradational contacts with the enclosing Roxby Downs granite. This granite is an A-type granite belonging to the Hiltaba suite (*ca.* 1590 Ma), which is coeval with the silicic Gawler Range volcanics. Granites and volcanics constitute the Gawler silicic large igneous province. A magmatic-hydrothermal origin is favoured for the ore deposit. Reeve et al. (1990) proposed that the magmatic-hydrothermal activity was focussed within maar volcanoes, but, due to later erosion, there is no evidence that the hydrothermal system breached the surface. Oreskes and Einaudi (1992) and Bastrakov and Skirrow (2007) recognized two stages of hydrothermal activity. The first one was due to high-temperature ($T > 400$ °C) fluids and was responsible for a magnetite-rich assemblage. The second hydrothermal stage corresponds to lower temperature fluids (200–300 °C) that induced replacement of magnetite by haematite.

Olympic Dam and other IOGC (iron oxide–copper–gold) deposits have been recognized as belonging to a peculiar mineral deposit class. All of these deposits are linked to continental alkaline magmatism, mainly of Mid-Proterozoic age, and are characterized by enrichment in iron and fluorine. The Vergenoeg deposit from South Africa, described earlier, can be regarded as an example of the same deposit class. The source of fluids and transported elements are still a matter of discussion. Gold and copper may have been leached from the country rocks rather than derived from a granitic magma. Nevertheless, Agangi et al. (2010) note that the Gawler Range rhyolites are characterized by high contents of REE, HFSE and F. These elements behave incompatibly during magma crystallization, as indicated by textural evidence. Indeed, fluorite and REE–F–carbonates are accessory minerals that crystallized late in interstices between mineral phases, or were deposited from a fluid phase in vesicles or miaroles. A similar mineral association and the same high REE and F abundances are also characteristic of the granite-hosted Olympic Dam deposit. Therefore, the mineralizing hydrothermal system is suggested to be genetically related to A-type felsic magmas of the Hiltaba granite suite and coeval with the Gawler Range volcanics, at least at the first stage. Then, mixing with an oxidized meteoric fluid is needed to obtain the haematite-associated uranium deposit at the second stage.

14.3 Conclusions

The economic interest of granitic rocks is significant, but is unequally distributed worldwide. The genesis of mineral deposits often requires a succession of favourable conditions. Among these conditions is the formation of a free aqueous fluid phase, eventually enriched in some elements, that is regarded as essential. The consequence is that granite-related ore deposits mainly formed in the upper crust. Continental areas characterized by outcrops of high grade metamorphic rocks are generally devoid of such ores. Besides the role of the erosion level, the granite-type and the geodynamic setting are also important. To conclude this chapter, the main controls of ore deposit are phase equilibria, conditions of speciation of elements in relation with the oxygen fugacity of granite magmas, and chemical reactions between these species. In addition, structural studies can explain the location of ore deposits. Therefore, granite metallogeny requires the integration of several approaches from petrology to tectonics that have all been examined in this book.

Glossary

Adiabatic taking place without gain or loss of heat (into or out of the system), but not necessarily without any temperature change (example: an adiabatic decompression results in some cooling of the system).

AMS anisotropy of magnetic susceptibility (see also susceptibility).

Anatexis partial melting (more specifically: partial melting of the continental crust). Thus, an anatectic granite is a granite derived from partial melting of the continental crust. Actually, many granites (especially the S-type granites) are indeed anatectic granites; exceptions are granites derived from the differentiation of a parent magma of mafic (mantle-derived) or intermediate composition (although intermediate magmas can themselves be derived from partial melting of the crust!).

Anisotropic having different physical properties in different directions.

Aplite fine-grained granitic rock, containing only quartz and feldspars, hence of light colour. Aplites often occur as veins regarded as fast crystallization products of an evolved granitic magma (close to eutectic composition).

Archaean subdivision of Precambrian times; older than 2.5 Ga (2500 Ma).

ARM anisotropy of remanent magnetization (see also remanence).

Batholith a group of several plutons, sometimes reaching a huge size (example: >1000 km long in the case of the coastal batholith of Peru).

Charnockite hypersthene (orthopyroxene) granite; name given in memory of Job Charnock, who founded the city of Calcutta at the end of seventeenth century and whose tombstone was actually made of such a rock. The name 'charnockite' is sometimes misused for any hypersthene-bearing magmatic rock, including non-granitic ones. Charnockites are often dark greenish in colour, because of many minute (microscopic) ferrous inclusions in their alkali feldspars.

Compatible (of a chemical element) easy to introduce in the crystal lattice of a mineral; the opposite is incompatible.

Conduction heat transfer by propagation of the thermal agitation of atoms or microscopic particles within a body or between two bodies in contact with each other. This mode of heat transfer does not require a material carrier (unlike convection). Conduction is not an efficient heat transfer mode in rocks.

Cotectic line connecting two eutectics in a phase diagram.

Craton continental area that has remained stable since (at least) the end of the Precambrian.

Cumulate igneous rock formed by the accumulation of early crystals (the so-called 'cumulus') during a crystal fractionation process.

Diapir geological structure resulting from the gravity-driven ascent of a light-density material (salt, magma . . .); diapirism is due to the negative buoyancy of a body with respect to its country rocks and requires that the country rocks can be ductilely deformed.

Dislocation linear crystallographic defect, whose ability to move or interact with other dislocations controls the plastic deformation of a crystal.

Distribution coefficient concentration ratio of an element in two different phases.

Dyke a (sub)vertical sheet of magmatic rock. A dyke thickness may range from a few centimetres to several metres for a length up to many kilometres; co-magmatic dykes sometimes appear in swarms.

Eutectic the most fusible composition of a mixture (i.e. with the lowest melting temperature); depending on the phase number (two or three) within the mixture, the eutectic is said to be 'binary' or 'ternary'.

Exsolution separation of an unstable solid solution into two phases of different compositions; exsolution processes are common in alkali feldspars (perthites) and in pyroxenes (lamellae).

Fabric preferred orientation (of shape and/or lattice) of rock-forming minerals.

Foliation planar structure due to the preferred orientation of minerals parallel to a plane (foliation plane), that sometimes corresponds to an easy cleavage (ex: schists and mica schists). A foliation may appear in a rock at the solid state (during metamorphism) or during the magmatic stage (magmatic foliation).

Gabbro plutonic rock of basaltic composition.

Geotherm curve representing the temperature distribution at depth.

Granitoid granitic rock *sensu lato*, i.e. a plutonic rock containing from 20–60% quartz among its light minerals (quartz and feldspars); granitoids include granites, granodiorites, tonalites and trondhjemites.

Haplogranitic of simple granitic composition (quartz plus potassic and sodic alkali feldspars, that is, quartz–orthoclase–albite).

Hercynian of an orogeny beginning during the Late Devonian and ending during the beginning of the Permian, i.e. approximately from 370 to 270 Ma. The word comes from the Latin *hercynia silva*, the Hercynian forest that covered central Germany from the Rhine to the Bohemian massif, through the Black Forest and Harz mountains. Hercynian belts include the Urals or belts in North America. The Hercynian orogeny formed the supercontinent of Pangea. Variscan is sometimes regarded as synonymous with Hercynian in Europe. Former European geologists distinguished the so-called 'Hercynian' strike (NW–SE) from the 'Variscan' strike (NE–SW). The Appalachian orogeny of North America is contemporaneous of the Hercynian orogeny in Europe.

Hornfels dark, hard and fine-grained rock formed by contact metamorphism of dominantly pelitic material. Hornfels often break conchoidally, i.e. forming shell-shaped concave fractures.

Hypersolvus (of a granite) crystallized and cooled in conditions leading to an exsolution process in alkali feldspar crystals: alkali feldspars of original intermediate (potassic-sodic) composition develop perthites (either potassic or sodic domains) during cooling.

Incompatible (of a chemical element) not easily included in the crystalline lattices, hence remaining preferentially dissolved in the melt or in a hydrous fluid (synonymous: 'hygromagmatic'). Compatibility or incompatibility depends on the composition of the magma and on the nature of the present minerals. For instance, zirconium (Zr) may behave incompatibly in a basaltic melt, but compatibly in a more felsic melt, as soon as zircon (a zirconium silicate) begins to crystallize.

Intercept refers to crosscutting of an object (example: crystal) by a line of the reference grid in image analysis or refers to the intersections of Discordia and Concordia curves in geochronology.

Isograd line (on a map) or surface (in a 3D diagram) corresponding to the appearance (or disappearance) of an indicator mineral witnessing peculiar metamorphic conditions.

Laccolith thick (sub)horizontal sheet magmatic intrusion, for instance emplaced between two sedimentary layers and resulting in the doming of the upper layer.

Leucogranite light-coloured granite, nearly devoid of dark (ferromagnesian) minerals (<10%).

Leucosome light-coloured, quartz- and feldspar-rich (granitic) part of a migmatite.

Lineation linear structure determined by the preferential alignment of elongate or stretched minerals; such a mineral lineation results from the ductile strain of the rock and materializes the principal stretching direction (X) of the rock. Therefore, it is a major kinematic indicator. Intersection lineations are also sometimes observed, but only have a geometric significance (intersections of planar surfaces).

Liquidus line (in a binary diagram) or surface (in a ternary diagram) separating the stability field of melt (liquid) from the field where both liquid and crystals are present; liquid crystallization begins as soon as the liquidus conditions are reached.

Lithosphere strong outer shell of the Earth comprising the crust and the uppermost mantle; it is broken into tectonic plates, that are mobile with respect to each other. The lithosphere is about 100 km thick (less in the case of oceanic lithosphere, and more in the case of continental lithosphere, up to 200 km for old cratons). Note that both types of lithosphere (oceanic and continental) may be present in the same plate. The base of the lithosphere corresponds to an isotherm (ca. 1300 °C), below which the mantle is submitted to convective motion. Therefore, the asthenosphere (below the lithosphere) does not differ from the lithosphere in composition, but is mainly hotter (hence rheologically different). Heat is transferred by conduction in the lithosphere and by convection in the asthenosphere; the corresponding geothermal gradients are very different, because convection is a much more efficient heat transfer mode than conduction. In geophysics, reference is often made to the elastic lithosphere, whose thickness is variable and depends (among others) on the duration of the applied stress.

Magma silicate liquid (melt) derived of rock melting, usually containing (inherited and/or newly crystallized) crystals; therefore, it can be described as a suspension.

Major elements chemical elements analysed at the per cent level in magmatic rocks. The rock-forming major elements are Si, Al, Ti, Fe, Mg, Mn, Ca, Na, K, P. Analyses are provided as oxide weight percentages (because O is also a major element, and even the most abundant one, although of less significance in petrology).

Metasomatose solid-state transformation (without melting) due to (mainly) hydrous fluids that modified the rock composition (open-system reaction).

Migmatite heterogeneous rock comprising a gneissic part and a granitic part (or leucosome, crystallized from a magma).

Mode composition of a magmatic rock, expressed as percentages of the observed minerals.

Neosome newly formed part of a migmatite, including leucosome (former melt) and melanosome (dark ferromagnesian crystals in equilibrium with the melt, which are either of restitic origin or produced by the melting reaction).

Norm virtual mineral composition, calculated from the rock chemical composition following a standardized method. Sometimes called CIFW norm, using the initials of the four petrologists, who invented this calculation mode in 1931 (Cross, Iddings, Pirsson and Washington).

Pegmatite magmatic rock characterized by a coarse- to very coarse-grained texture (crystal sizes are typically >2 cm in diameter, and may reach up to 1 m); pegmatites generally display a leucogranitic composition (although other compositions may also occur; example: syenitic pegmatite). Pegmatites are often enriched in incompatible elements and may contain gem-quality minerals (example: beryl).

Perthites association of domains of different compositions (potassic and sodic) within the same alkali feldspar grain; perthites result from a solid-state exsolution process during cooling.

Petrogenesis rock formation mode.

Petrology rock study.

Pluton igneous massif made of mainly cogenetic rocks, with typical size of one or a few tens of kilometres.

Proterozoic subdivision of Precambrian times, corresponding to ages <2.5 Ga (2500 Ma).
Rapakivi word of Finnish origin used for a peculiar granitic texture, where plagioclase crystallized as rims around alkali feldspar grains.
Remanence (or remanent magnetization) magnetization left behind after removal of an external magnetic field.
Rheology etymologically, it is the study of the flow of a matter. More generally, it is the study of rock deformation modes, either in the brittle or in the ductile domains, including the viscous deformation of magmas.
Rift (or graben) down-faulted geological structure formed by extensional tectonics.
Sanukitoid Mg-rich dioritic to monzodioritic plutonic rock, common at the end of Archaean times.
Sill relatively thin (sub)horizontal igneous intrusion.
Skarn calcareous, dolomitic or calc-silicate rock, produced by contact metamorphism, that has been modified by metasomatic processes with respect to its original composition.
Solidus line (in a binary diagram) or surface (in a ternary diagram) separating the field where both liquid and crystals are present from the solid stability field. Partial melting of a rock begins at its solidus.
Solvus line corresponding to the beginning of an exsolution process during cooling of a mineral phase.
Subsidence slow downward motion of tectonic or thermal origin.
Subsolvus (of a granite) crystallized in conditions, that will not involve any exsolution process in alkali feldspars, namely a granite comprising two different feldspars (potassic and sodic) crystallized originally from the magma.
Susceptibility (magnetic susceptibility) property describing the degree of temporary magnetization (M) of a material submitted to an applied magnetic field (H). The magnetic susceptibility is defined as the dimensionless proportionality constant of the M/H ratio. In an anisotropic body, the susceptibility is not a constant but a tensor. The SI indication after a susceptibility value indicates that this value was calculated using the international system of units.
Trace elements chemical elements present in very small quantities in magmatic rocks, hence analysed at the ppm level, i.e. in mg/kg.
Transcurrent (of a fault or shear zone) mainly corresponding to a horizontal relative displacement (or heave) of blocks on either sides.
Trondhjemite leucotonalite or plagiogranite.
TTG initials of an association of plutonic rocks (tonalites, trondhjemites and granodiorites) that are typical of the Archaean continental crust.
Variscan synonym of Hercynian (see above); first coined by the Austrian geologist Suess in 1880 and derived from the Latin name (*Variscia*) of a district in Germany.
Viscosity (dynamic viscosity) resistance to deformation by shear stress (characterizing fluids, magmas and rocks). The kinematic viscosity is the ratio of the dynamic viscosity to the density of the fluid. The dynamic viscosity is measured in pascal seconds (Pa s); this unit is equivalent to 10 poises (the former unit, named after the French physicist J. Poiseuille). A fluid with a viscosity of 1 Pa s placed between two plates will move of a distance equivalent to the layer thickness in one second if the upper plate is pushed sideways with a shear stress of one pascal.
Wavelet mathematical oscillatory function used as a tool signal processing (for instance in exploration geophysics) or image analysis. Wavelets offer more possibilities than Fourier transforms.
Xenolith enclave or rock fragment included in a magmatic rock.

References

Acocella V. (2000). Space accommodation by roof lifting during pluton emplacement at Amiata (Italy). *Terra Nova*, 12, 149–155.

Agangi A., Kamenetsky S.V. & McPhie J. (2010). The role of fluorine in the concentration and transport of lithophile trace elements in felsic magmas: insights from the Gawler Range volcanics, South Australia. *Chem. Geol.*, 273, 314–325.

Albarède F., Dupuis C. & Taylor H.P. Jr. (1980). $^{18}O/^{16}O$ evidence for non-cogenetic magmas associated in a 300 Ma old concentric pluton at Ploumanac'h. *J. Geol. Soc.*, 137, 641–647.

Allègre C.J. (2005). *Géologie isotopique*. Paris: Belin, 495 pp.

Améglio L., Vigneresse J.L. & Bouchez J.L. (1997). Granite pluton geometry and emplacement mode inferred from combined fabric and gravity data. In: Bouchez J.L., et al. (eds.) *Granite: from segregation of melt to emplacement fabrics*. Dordrecht: Kluwer Academic Publishers, pp.199–214.

Améglio L., Vigneresse J.L., Darrozes J. & Bouchez J.L. (1994). Forme du massif granitique du Sidobre (Montagne Noire, France): sensibilité de l'inversion des données gravimétriques au contraste de densité. *C. R. Acad. Sci. Paris*, 319, 1183–1190.

Ames D.E. (2002). *Sudbury targeted initiative (TGI): overview and update. Ontario Geol. Surv. Open File Rep.*, 6100, 17, 1–10.

Anderson J.L. (1996). Status of thermobarometry in granitic batholiths. *Trans. Earth Sci.*, 87, 125–138.

Archanjo C.J., Launeau P. & Bouchez J.L. (1995). Magnetic fabrics *vs* magnetite and biotite shape fabrics of the magnetite-bearing granite pluton of Gameleiras (Northeast Brazil). *Phys. Earth Planet. Interiors*, 89, 63–75.

Armstrong R.A., Compston W., de Wit M.J. & Williams I.S. (1990). The stratigraphy of the 3.5–3.2 Ga Barberton Greenstone Belt revisited: a single zircon ion microprobe study. *Earth Planet. Sci. Lett.*, 101, 90–106.

Arth J.G. (1976). Behavior of trace element during magmatic processes. a summary of theoretical models and their applications. *J. Res. US Geol. Surv.*, 4, 41–47.

Audétat A. & Pettke T. (2003). The magmatic–hydrothermal evolution of two barren granites: a melt and fluid inclusion study of the Rito del Medio and Cañada Pinabete plutons in northern New Mexico (USA). *Geochim. Cosmochim. Acta*, 67, 97–121.

Audrain J., Amice M., Vigneresse J.L. & Bouchez J.L. (1989). Gravimétrie et géométrie tri-dimentionnelle du pluton granitique de Cabeza de Araya (Estrémadure, Espagne). *C. R. Acad. Sc. Paris*, 309, 1757–1764.

Auréjac J.B., Gleizes G., Diot H. & Bouchez J.L. (2004). The Quérigut Complex (Pyrenees, France) revisited by the AMS technique: a syntectonic pluton of the Variscan transpression. *Bull. Soc. Géol. France*, 175, 157–174.

Avrami M. (1939). Kinetics of phase change. I. General theory. *J. Chem. Phys.*, 7, 1103–1112.

Ayres M., Harris N. & Vance D. (1997). Possible constraints on anatectic melt residence times from accessory mineral dissolution rates: an example from Himalayan leucogranites. *Min. Mag.*, 61, 29–36.

Bachl C.A., Miller C.F., Miller J.S. & Faulds J.E. (2001). Construction of a pluton: evidence from an exposed cross section of the Searchlight pluton, Eldorado Mountains, Nevada. *Geol. Soc. Am. Bull.*, 113, 1213–1228.

Bachmann O. & Bergantz G.W. (2004). On the origin of crystal-poor rhyolites: extracted from batholithic crystal mushes. *J. Petrol.*, 45, 1565–1582.

Bagdassarov N., Dingwell D.B. & Webb S.L (1993). Effect of boron, phosphorus and fluorine on shear stress relaxation in haplogranite melts. *Eur. J. Mineral.*, 5, 409–425.

Bailey R.A., Dalrymple G.B. & Lanphere M.A. (1976). Volcanism, structure and geochronology of Long Valley caldera, Mono County, California. *J. Geophys. Res.*, 81, 725–744.

Bailey E.H. & Stevens R.E. (1960). Selection staining of K-feldspar and plagioclase on rock slabs and thin sections. *Am. Mineral.*, 45, 1020–1025.

Baïlon J.P. & Dorlot J.M. (2000). *Des matériaux*. Montréal: Presses International Polytechnique, 768 pp.

Balk R. (1937). Primary structure of granite massifs. *Geol. Soc. Am. Bull.*, 36, 679–696.

Barbarin B. (1988). Field evidence for successive mixing and mingling between the Piolard diorite and the Saint-Julien-la-Vêtre monzogranite (Nord-Forez, Massif Central, France). *Can. J. Earth Sci.*, 25, 49–59.

Barbarin B. (1999). A review of the relationships between granitoid types, their origins and their geodynamic environments. *Lithos*, 46, 605–626.

Barbey P. (2009). Layering and schlieren granitoids: a record of interactions between magma emplacement, crystallization and deformation in growing plutons. *Geol. Belg.*, 12, 109–133.

Bard J.P. (1980). *Microtextures des roches magmatiques et métamorphiques*. Paris: Masson, 192 pp.

Bard J.P. (1983). Metamorphism of an obducted island arc: example of the Kohistan sequence (Pakistan) in the Himalayan collided range. *Earth Planet. Sci. Lett.*, 65, 133–144.

Baronnet A. (1988). *Minéralogie*. Paris: Bordas-Dunod, 184 pp.

Barrière M. (1977). Deformation associated with the Ploumanac'h intrusive complex, Brittany. *J. Geol. Soc.*, 134, 311–324.

Bastrakov E.N. & Skirrow R.G. (2007). Fluid evolution and origins of iron oxide-Cu-Au prospects of the Olympic Dam district, Gawler craton, South Australia. *Econ. Geol.*, 102, 1415–1440.

Bau M. (1996). Controls on the fractionation of isovalent trace elements in magmatic and aqueous systems: evidence from Y/Ho, Zr/Hf, and lanthanide tetrad effect. *Contrib. Mineral. Petrol.*, 123, 323–333.

Bédard J.H. (2006). A catalytic delamination-driven model for coupled genesis of Archaean crust and sub-continental lithospheric mantle. *Geochim. Cosmochim. Acta*, 70, 1188–1214.

Begg G.C., Griffin W.L., Nataopov L.M., O'Reilly S.Y., Grand S.P., O'Neill C.J., Hronsky J.M.A., Poudjom Djomani Y., Swain C.J., Deen T. & Bowden P. (2009). The lithospheric architecture of Africa: seismic tomography, mantle petrology and tectonic evolution. Magmatic fabrics in batholiths as markers of regional strains and plate kinematics: example of the Cretaceous Mt. Stuart Batholith. *Geosphere*, 5, 23–50.

Belousova E.A., Kostistyn Y.A., Griffin W.L., Begg G.C., O'Reilly S.Y. & Pearson N.J. (2010). The growth of the continental crust: constraints from zircon Hf-isotope data. *Lithos*, 119, 457–466.

Benn K., Paterson S.R., Lund S.P., Pignotta G.S. & Kruse S. (2001). Magmatic fabrics in batholiths as markers of regional strains and plate kinematics: example of the Cretaceous Mt. Stuart Batholith. *Phys. Chem. Earth A*, 26, 343–354.

Berthè D., Choukroune P. & Jègouzo P. (1979). Orthogneiss, mylonite and non-coaxial deformation of granites: the example of the South Armorican Shear Zone. *J. Struct. Geol.*, 1, 31–42.

Blumenfeld P. & Bouchez J.L. (1988). Shear criteria in granite and migmatite deformed in the magmatic and solid states. *J. Struct. Geol.*, 10, 361–372.

Blumenfeld P., Mainprice D. & Bouchez J.L. (1986). C-slip in quartz from subsolidus deformed granites. *Tectonophysics*, 127, 95–115.

Bodnar R.J. & Student J.J. (2006). Melt inclusions in plutonic rocks: petrography and microthermometry. *Mineral. Assoc. Can. Short Course Ser.* 36, 1–25.

Bolle O., Diot H. & Duchesne J.C. (2000). Magnetic fabric and deformation in charnockitic igneous rocks of the Bjerkreim–Sokndal layered intrusion (Rogaland, Southwest Norway). *J. Struct. Geol.*, 22, 647–667.

Bolle O., Trindade R.I.F., Bouchez J.L. & Duchesne J.C. (2002). Imaging downward granitic magma transport in the Rogaland Igneous Complex, SW Norway. *Terra Nova*, 14, 87–92.

Bonin B. (1982). Les granites des complexes annulaires. In: *Manuels et méthodes*, Vol. 4. Paris: Bureau de Recherches Géologiques et Minières.

Bonin B. (1996). A-type granite ring complexes: mantle origin through crustal filters and the anorthosite–rapakivi magmatism connection. In: Demaiffe D. (ed.) *Petrology and geochemistry of magmatic suites of rocks in continental and oceanic crusts: a volume dedicated to Professor Jean Michot.* Tervuren: Royal Museum for Central Africa, pp.201–218.

Bonin B., Bowden P. & Vialette Y. (1979). Behaviour of Rb and Sr during mineralization phases: Ririwai (Liruei) granite, Nigerai. *C. R. Acad. Sci. Paris*, 289, 707–710.

Borrel A. (1978). *Le Massif granitique du Sidobre : pétrographie, structure, relation, mise en place, cristallisation*. PhD Thesis, University of Toulouse.

Borrok D.M., Kessler S.K., Boer R.H. & Essene E.J. (1998). The Vergenoeg magnetite–fluorite deposit, South Africa: support for a hydrothermal model for massive iron oxide deposit. *Econ. Geol.*, 93, 564–586.

Bottinga Y. & Weill D. (1972). The viscosity of magmatic silicate liquids: a model for calculation. *Am. J. Sci.*, 272, 438–475.

Bouchez J.L. (1997). Granite is never isotropic: an introduction to AMS studies of granitic rocks. In: Bouchez J.L., et al. (eds.) *Granite: from segregation of melt to emplacement fabrics*. Dordrecht: Kluwer Academic Publishers, pp.95–112.

Bouchez J.L. (2000). Magnetic susceptibility anisotropy and fabrics in granites, *C. R. Acad. Sci.: Earth Planet. Sci.*, 330, 1–14.

Bouchez J.L., Delas C., Gleizes G., Nédélec A. & Cuney M. (1992). Submagmatic microfractures in granites. *Geology*, 20, 35–38.

Bouchez J.L., Guillet P. & Chevalier F. (1981). Structures d'écoulement et mise en place du granite de Guérande (Loire Atlantique, France). *Bull. Soc. Géol. France*, 387–399.

Bouchez J.L., Mi Nguema T., Esteban L., Sicueira R. & Scrivener R. (2006). The tourmaline-bearing granite pluton of Bodmin (Cornwall, UK): magnetic fabric study and regional inference. *J. Geol. Soc.*, 163, 607–616.

Bouhallier H., Choukroune P. & Ballèvre M. (1993). Diapirism, bulk homogeneous shortening and transcurrent shearing in the Archaean Dharwar craton: the Holenarsipur area, southern India. *Precambrian Res.*, 113, 87–120.

Bowden P. & Kinnaird J.A. (1984). The petrology and geochemistry of alkaline granites from Nigeria. *Phys. Earth Planet. Int.*, 35, 199–211.

Bowen N.L. (1928). *The evolution of the igneous rocks.* New York: Princeton University Press.

Bowers T.S. & Helgeson H.C. (1983). Calculation of the thermodynamic and geochemical consequences of non-ideal mixing in the system $H_2O–CO_2–NaCl$ on phase relations in geologic systems: metamorphic equilibria at high pressures and temperatures. *Am. Mineral*, 68, 1059–1075.

Bowring S.A. & Williams I.S. (1999). Priscoan (4.00–4.03 Ga) orthogneisses from northwestern Canada. *Contrib. Mineral. Petrol.*, 134, 3–16.

Breiter K., Förster H.J. & Seltmann E.W. (1999). Variscan silicic magmatism and related tin–tungsten mineralization in the Erzgebirge–Slavkovsky metallogenic province. *Miner. Deposita*, 34, 505–521.

Brown M., Averkin Y.A., McLellan E.L. & Sawyer E.W. (1995). Melt segregation in migmatites. *J. Geophys. Res.*, 100, B8, 15655–15679.

Brown G.C. & Fyfe W.S. (1970). The production of granitic melts during ultra-metamorphism. *Contrib. Mineral. Petrol.*, 28, 310–318.

Brun J.P., Gapais D., Cogné J.P., Ledru P. & Vigneresse J.L. (1990). The Flamanville granite (NW France): an unequivocal example of an expanding pluton. *Geol. J.*, 25, 271–286.

Buck W.R. (1991). Modes of continental lithospheric extension. *J. Geophys. Res.*, 96, 20161–20178.

Buffon G.L. (1783). *Histoire naturelle des minéraux*. Amsterdam: Schneider.

Burg J.P. & Vanderhaeghe O. (1993). Structures and way-up criteria in migmatites, with applications to the Velay dome (French Massif Central). *J. Struct. Geol.*, 15, 1293–1301.

Bussell M.A. & Pitcher W.S. (1985). The structural controls of batholith emplacement. In: Pitcher W.S., et al. (eds.) *Magmatism at a plate edge: the Peruvian Andes*. Blackie, pp.167–176.

Butler R.F. (1992). *Paleomagnetism: magnetic domains to geologic terranes*. Boston: Blackwell Scientific Publications, 319 pp.

Byerlee J.D. (1978). Friction of rocks. *Pure Appl. Geophys.*, 116, 615–626.

Campbell I.H. & Turner J.S. (1986). The influence of viscosity on fountains in magma chambers. *J. Petrol.*, 27, 1–30.

Castro A. (1985). The Central Extremadura batholith: geotectonic implications (European Hercynian Belt): an outline. *Tectonophysics*, 120, 57–68.

Cavosie A.J., Valley J.W., Wilde S.A. & Edinburgh Ion Microprobe Facility. (2005). Magmatic $\Delta^{18}O$ in 4400–3900 Ma detrital zircons: a record of the alteration and recycling of crust in the Early Archean. *Earth Planet. Sci. Lett.*, 235, 663–681.

Cerny P. (1991). Rare-element granitic pegmatites. Part I: anatomy and internal evolution of pegmatite deposits. *Geosci. Can.*, 18, 49–67.

Cerny P. (2005). The Tanco rare-element pegmatite deposit, Manitoba: regional context, internal anatomy and global comparisons. *Geol. Assoc. Can. Short Course*, 17, 127–158.

Chappell B.W. (1999). Aluminium saturation in I- and S-type granites and the characterization of fractionated haplogranites. *Lithos*, 46, 535–551.

Chappell B.W. & White A.J.R. (1974). Two contrasting granite types. *Pac. Geol.*, 8, 173–174.

Chappell B.W. & White A.J.R. (1992). I- and S-type granites in the Lachlan Fold Belt. *Trans. Earth Sci.*, 83, 1–26.

Charoy B. (1979). *Définition et importance des phénomènes deutériques et des fluides associés dans les granites. Conséquences métallogéniques*. PhD Thesis, Ann. Ecole Natl. Sup. Géologie Nancy, 364 pp.

Chopra P.N. & Paterson M.S. (1981). The experimental deformation of dunite. *Tectonophysics*, 78, 453–473.

Clarke D.B. (1981). The mineralogy of peraluminous granites: a review. *Can. Mineral.*, 19, 3–17.

Clemens J.D. (1998). Observations on the origins and ascent mechanisms of granitic magmas. *J. Geol. Soc.*, 155, 843–851.

Clemens J.D. & Droop G.T.R. (1998). Fluids, P-T paths and the fates of anatectic melts in the Earth's crust. *Lithos*, 44, 21–36.

Clemens J.D., Helps P.A. & Stevens G. (2009). Chemical structure in granitic magmas – a signal from the source. *Trans. Earth Sci.*, 100, 159–172.

Clemens J.D. & Vielzeuf D. (1987). Constraints on melting and magma production in the crust. *Earth Planet. Sci. Lett.*, 86, 287–306.

Clemens J.D. & Wall V.J. (1981). Origin and crystallization of some peraluminous (S-type) granitic magmas. *Can. Mineral.*, 19, 111–131.

Clemens J.D., Yearron L.M. & Stevens G. (2006). Barberton (South Africa) TTG magmas: geochemical and experimental constraints on source-rock petrology, pressure of formation and tectonic setting. *Precambrian Res.*, 151, 53–78.

Cloos H. (1925). Bau and Bewegung der Gebirge in Nordamerica, Skandinavien and Mitteleuropa. *Fortsch. Geol. Pol., Berlin*, 7, no 21.

Cocherie A. (1985). *Interaction manteau–croûte, son rôle dans la genèse d'associations plutoniques calco-alcalines, contraintes géochimiques.* Orléans: Bureau de Recherches Géologiques et Minières, BRGM document 90.

Cocherie A. (1986). Systematic use of trace element distribution patterns in log-log diagrams for plutonic suites. *Geochim. Cosmochim. Acta*, 50, 2517–2522.

Cocirta C. (1986). Les enclaves microgrenues sombres du massif de Bono (Sardaigne septentrionale). Signification pétrogénétique des plagioclases complexes et de leurs inclusions. *C. R. Acad. Sci. Paris*, 302, 441–446.

Coleman R.G. & Peterman Z. (1975). Oceanic plagiogranite. *J. Geophys. Res.*, 80B, 1099–1108.

Collins W.J., Beams S.D., White A.J.R. & Chappell B.W. (1982). Nature and origin of A-type granites with particular reference to southeastern Australia. *Contrib. Mineral. Petrol.*, 80, 189–200.

Compston W. & Pidgeon R.T. (1986). Jack Hills, evidence of more very old detrital zircons in Western Australia. *Nature*, 321, 766–769.

Condie K.C. & Aster R.C. (2010). Episodic zircon age spectra of orogenic granitoids: the supercontinent connection and continental growth. *Precambrian Res.*, 180, 227–236.

Cook S.J. & Bowman J.R. (2000). Contact metamorphism surrounding the Alta stock: thermal constraints and evidence of advective heat transport from calcite + dolomite geothermometry. *Am. Mineral.* 79, 513–525.

Coster M. & Chermant J.L. (1989). *Précis d'analyse d'images.* Paris: Presses du CNRS, 560 pp.

Courrioux G. (1987). Oblique diapirism: the Criffel granodiorite/granite pluton (southwest Scotland). *J. Struct. Geol.*, 9, 313–330.

Cox K.G., Bell J.D. & Pankhurst R.J. (1979). *The interpretation of igneous rocks.* London: Allen & Unwin, 450 pp.

Creaser R., Price R. & Wormald R. (1991). A-type granites revisited: assessment of residual-source model. *Geology*, 19, 163–166.

Crocker I.T. (1985). Volcanogenic fluorite–hematite deposits and associated pyroclastic rock suite at Vergenoeg, Bushveld complex. *Econ. Geol.*, 80, 1181–1200.

Cuney M. & Barbey P. (1982). Mise en évidence de phénomènes de cristallisation fractionnée dans les migmatites. *C. R. Acad. Sci. Paris*, 295, 37–42.

Dahlquist J.A., Galindo C., Pankhurst R.J., Rapela C.W., Alasino P.H., Saavedra J. & Fanning C.M. (2007). Magmatic evolution of the Peñón Rosado granite: petrogenesis of garnet-bearing granitoids. *Lithos*, 95, 177–207.

Danishwar S., Stern R.J. & Khan M.A. (2001). Field relationships and structural constraints for the Teru volcanic formation, northern Kohistan terrane, Pakistani Himalayas. *J. Asian Earth Sci.*, 19, 683–695.

Darrozes J., Gaillot P., Saint-Blanquat M. (de) & Bouchez J.L. (1997). Software for multiscale image analysis: the normalized optimized anisotropic wavelet. *Comput. Geosci.*, 23, 889–895.

Darrozes J., Moisy M., Olivier P., Améglio L. & Bouchez J.L. (1994). Structure magmatique du granite du Sidobre (Tarn, France): de l'échelle du massif à celle de l'échantillon. *C. R. Acad. Sci. Paris*, 318, 243–250.

Day R., Fuller M. & Schmidt V.A. (1977). Hysteresis properties of titanomagnetites: grain size and compositional dependence. *Earth Planet. Sci. Lett.*, 13, 260–267.

Debon F. (1972). Notice explicative pour la carte géologique au 1/50 000ème des massifs granitiques de Cauterets et Panticosa (Pyrénées occidentales). Orléans: Bureau de Recherches Géologiques et Minières, 37 pp.

Debon F. & Lemmet M. (1999). Evolution of Mg/Fe ratios in Late Variscan plutonic rocks from the External Crystalline Massifs of the Alps (France, Italy, Switzerland). *J. Petrol.*, 40, 1151–1185.

Dehls J.D., Cruden A.R. & Vigneresse J.L. (1998). Fracture control of late-Archean pluton emplacement in the Northern Slave Province, Canada. *J. Struct. Geol.*, 20, 1145–1154.

Dehoff R.T. (1991). A geometrical general theory of diffusion controlled coarsening. *Acta Metall. Mater.*, 39, 2349–2360.

Déléris J., Nédélec A., Ferré E., Gleizes G., Ménot R.P., Obasi C.K. & Bouchez J.L. (1996). The Pan-African Toro complex (northern Nigeria): magmatic interactions and structures in a bimodal intrusion. *Geol. Mag.*, 133, 535–552.

Denele Y., Olivier P., Gleizes G. & Barbey P. (2007). The Hospitalet gneiss dome (Pyrenees) revisited: lateral flow during Variscan transpression in the middle crust. *Terra Nova*, 19, 445–453.

DePaolo D.J. (1981). Trace-element and isotopic effects of combined wall-rock assimilation and fractional crystallization. *Earth Planet. Sci. Lett.*, 53, 189–202.

DePaolo D.J. & Wasserburg G.J. (1976). Nd isotope variations and petrogenetic models. *Geophys. Res. Lett.*, 4, 465–468.

Dingwell D.B. (1999). Granitic melt viscosities. *Geol. Soc. Lond. Spec. Publ.*, 168, 27–38.

Dioh E., Béziat D., Debat P., Grégoire M. & Ngom P.M. (2006). Diversity of the Palaeoproterozoic granitoids of the Kédougou inlier (eastern Sénégal): petrographical and geochemical constraints. *J. Afr. Earth Sci.*, 44, 351–371.

Djouadi T., Gleizes G., Ferré E., Bouchez J.L., Caby R. & Lesquer A. (1997). Oblique magmatic structures of two epizonal granite plutons, Hoggar, Algeria: late-orogenic emplacement in a transcurrent orogen. *Tectonophysics*, 279, 351–374.

Dubois J. & Diament M. (1997). *Géophysique*. Paris: Masson, 205 pp.

Duchesne J.C. & Wilmart E. (1997). Igneous charnockites and related rocks from the Bjerkreim–Sokndal layered intrusion (southwest Norway): a jotunite (hypersthene monzodiorite)-derived A-type granitoid suite. *J. Petrol.*, 38, 337–369.

Durocher J. (1857). Recherches sur les roches ignées, sur les phénomènes de leur émission et sur leur classification. *C. R. Acad. Sci. Paris*, 44, 325–330 and 459–465.

Eales H.V. & Cawthorn R.G. (1996). The Bushveld Complex. In: Cawthorn R.G. (ed.) *Layered intrusions*. Amsterdam: Elsevier, pp.181–230.

Ebadi A. & Johannes W. (1991). Beginning of melting and composition of first melts in the system Qz–Ab–Or–H_2O–CO_2. *Contrib. Mineral. Petrol.*, 106, 286–295.

Echtler H. & Malavieille J. (1990). Extensional tectonics, basement uplift and Stephano-Permian collapse basin in a Late Variscan metamorphic core complex (Montagne Noire, Southern Massif Central). *Tectonophysics*, 177, 125–138.

Einstein A. (1906). Eine neue Bestimmung der Molekuldimensionnen. *Ann. Phys.*, 19, 289–306.

England P.C. & Thompson A.B. (1984). Pressure–temperature–time paths of regional metamorphism. I. Heat transfer during the evolution of regions of thickened continental crust. *J. Petrol.*, 25, 894–928.

Esmaeily D., Bouchez J.L. & Siqueira R. (2007). Magnetic fabrics and microstructures of the Shah Kuh Jurassic granite pluton (Lut block, Eastern Iran) and geodynamic inference. *Tectonophysics*, 439, 149–170.

Esmaeily D., Nédélec A., Valizadeh M.V., Moore F. & Cotten J. (2005). Petrology of the Jurassic Shah-Kuh granite (eastern Iran), with reference to tin mineralization. *J. Asian Earth Sciences*, 25, 961–980.

Evans N.G., Gleizes G., Leblanc D. & Bouchez J.L. (1997). Hercynian tectonics in the Pyrenees: a new view based on structural observations around the Bassies granite pluton. *J. Struct. Geol.*, 19, 195–208.

Evans N.G., Gleizes G., Leblanc D. & Bouchez J.L. (1998). Syntectonic emplacement of the Maladeta granite (Pyrenees) deduced from relationships between Hercynian deformation and contact metamorphism. *J. Geol. Soc.*, 155, 209–216.

Evensen J.M. & London D. (2002). Experimental silicate mineral/melt partition coefficients for beryllium and the crustal Be cycle from migmatite to pegmatite. *Geochim. Cosmochim. Acta*, 66, 2239–2265.

Faure M. (1995). Late orogenic carboniferous extensions in the Variscan French Massif Central. *Tectonics*, 14, 132–153.

Faure G. (2001). *Origin of igneous rocks: the isotopic evidence*. Berlin: Springer. 496 pp.

Fenn P.M. (1977). The nucleation and growth of alkali feldspars from hydrous melts. *Can. Mineral.*, 15, 135–161.

Fernandez A., Feybesse J.L. & Mezure J.F. (1983). Theoretical and experimental study of fabrics developed by different shaped markers in two-dimensional simple shear. *Bull. Soc. Géol. Fr.*, 25, 319–326.

Fernandez A. & Gasquet D. (1994). Relative rheological evolution of chemically constrained coeval magmas: example of the Tichka plutonic complex (Morocco). *Contrib. Mineral. Petrol.*, 116, 316–326.

Fischer M., Röller K., Küster M., Stöckhert B. & McConnell V.S. (2003). Open fissure mineralization at 2600 m depth in Long Valley Exploratory Well (California): insight into the history of the hydrothermal system. *J. Volc. Geotherm. Res.*, 127, 347–363.

Floyd P.A., Yaliniz M.K. & Goncuoglu M.C. (1998). Geochemistry and petrogenesis of intrusive and extrusive ophiolitic plagiogranites, Central Anatolian Crystalline Complex, Turkey. *Lithos*, 42, 225–241.

Fourcade S. (1998). Les isotopes: effets isotopiques, bases de radio-géochimie. In: Hagemann G. & Treuil M. (eds.) *Introduction à la géochimie et ses applications*. Paris: CEA, pp.195–265.

Fourcade S. & Allègre C.J. (1981). Trace-element behavior in granite genesis: a case study the calc-alkaline plutonic association from the Quérigut complex (Pyrenees, France). *Contrib. Mineral. Petrol.*, 76, 177–195.

France-Lanord C., Sheppard S.M.F. & Le Fort P. (1988). Hydrogen and oxygen isotope variations in the High Himalayas peraluminous Manaslu leucogranite: evidence for heterogeneous sedimentary source. *Geochim. Cosmochim. Acta*, 52, 513–526.

Frost T.P. & Mahood G.A. (1987). Field, chemical and physical constraints on mafic-felsic magma interaction in the Lamarck granodiorite, Sierra Nevada, California. *Geol. Soc. Am. Bull.*, 99, 272–291.

Gaillot P., Darrozes J. & Bouchez J.L. (1999). Wavelet transform: a future of rock fabric analysis? *J. Struct. Geol.*, 21, 1615–1621.

Gardien V., Thompson A.B., Grujic D. & Ulmer P. (1995). Experimental melting of biotite + plagioclase + quartz ± muscovite assemblages and implications for crustal melting. *J. Geophys. Res.*, 100, 15581–15591.

Gardien V., Thompson A.B. & Ulmer, P. (2000). Melting of biotite + plagioclase + quartz gneisses: the role of H_2O in the stability of amphibole. *J. Petrol.*, 41, 651–666.

Glazner A.F., Bartley J.M., Coleman D.S., Gray W. & Taylor R.Z. (2004) Are plutons assembled over millions of years by amalgamation from small magma chambers? *GSA Today*, 14, 4–11.

Gleizes G. (1992). *Structures des granites hercyniens des Pyrénées de Mont-Louis-Andorre à la Maladeta*. PhD Thesis, University of Toulouse, 259 pp.

Gleizes G., Leblanc D., Santana V., Olivier P. & Bouchez J.L. (1998). Sigmoidal structures featuring dextral shear during emplacement of the Hercynian granite complex of Cauterets–Panticosa (Pyrenees). *J. Struct. Geol.*, 20, 1229–1245.

Graessner T. & Schenk V. (1999). Low-pressure metamorphism of Palaeozoic pelites in the Aspromonte, southern Calabria: constraints for the thermal evolution in the Calabrian crustal section during the Hercynian orogeny. *J. Metamorph. Geol.*, 17, 157–172.

Graham J.W. (1954). Magnetic susceptibility anisotropy: an unexploited petrofabric element. *Geol. Soc. Am. Abstr. Program*, 65, 1257–1258.

Grégoire V., Darrozes J., Gaillot P., Nédélec A. & Launeau P. (1998). Magnetite grain shape fabric and distribution anisotropy vs. rock magnetic fabric: a three-dimensional case study. *J. Struct. Geol.*, 20, 937–944.

Grégoire V., Nédélec A., Moniè P., Montel J.M., Ganne J. & Ralison B. (2009). Structural reworking and heat transfer related to the late-Pan-African Angavo shear of Madagascar. *Tectonophysics*, 477, 197–216.

Guillot S. (1999). An overview of the metamorphic evolution in Central Nepal. *J. Asian Earth Sci.*, 17, 713–725.

Guillot S., Le Fort P., Pêcher A., Roy Barman M. & Aprahamian J. (1995). Contact metamorphism and depth of emplacement of the Manaslu granite (Central Nepal): implications for Himalayan orogenesis. *Tectonophysics*, 241, 99–119.

Guineberteau B., Bouchez J.L. & Vigneresse J.L. (1987). The Mortagne granite pluton (France) emplaced by pull-apart along a shear zone: structural and gravimetric arguments and regional implications. *Geol. Soc. Am. Bull.*, 99, 763–770.

Halter W.E. & Webster J.D. (2004). The magmatic to hydrothermal transition and its bearing on ore-forming systems. *Chem. Geol.*, 210, 1–6.

Hammarstron J.M. & Zen E. (1986). Aluminum in hornblende: an empirical igneous geobarometer. *Am. Mineral.*, 71, 1297–1313.

Hansen F.D. & Carter N.L. (1983). Semibrittle creep of dry and wet Westerly granite at 1000 MPa. In: *24th symposium on rock mechanics*.

Hargraves R.D., Johnson D. & Chan C. (1991). Distribution anisotropy, the cause of AMS in igneous rocks. *Geophys. Res. Lett.*, 18, 2193–2196.

Harker A. (1909). *The natural history of igneous rocks*. London: Methuen.

Harrison T.M. & Clarke G.K.C. (1979). A model of the thermal effects of igneous intrusion and uplift as applied to Quotoon pluton, British Columbia. *Can. J. Earth Sci.*, 16, 411–420.

Harrison T.M., Grove M., McKeegan K.D., Coath C.D., Lovera O.M. & Le Fort P. (1999). Origin and episodic emplacement of the Manaslu intrusive complex, Central Himalaya. *J. Petrol.*, 40, 3–19.

Harrison T.M., Schmidt A.K., McCulloch M.T. & Lovera O.M. (2008). Early ($\geq$ 4.5 Ga) formation of terrestrial crust: Lu–Hf, $\Delta^{18}O$ and Ti thermometry results for Hadean zircons. *Earth. Planet. Sci. Lett.*, 268, 476–486.

Harry A.P. & Richey J.E. (1963). Magmatic pulses in the emplacement of plutons. *J. Geol.*, 3, 254–268.

Heinonen W.T., Andersen T. & Rämö O.T. (2010). Re-evaluation of rapakivi petrogenesis: source constraints from the Hf isotope composition of zircon in the rapakivi granites and associated mafic rocks of southern Finland. *J. Petrol.*, 51, 1687–1709.

Hess K.U. & Dingwell D.B. (1996). Viscosities of hydrous leucogranitic melts: a non-Arrhenian model. *Am. Mineral.*, 81, 1297–1300.

Hibbard M.J. & Watters R.J. (1985). Fracturing and diking in incompletely crystallized granitic plutons. *Lithos*, 18, 1–12.

Higgins M.D. (1999). Origin of megacrysts in granitoids by textural coarsening: a crystal size distribution (CSD) study of microcline in the Cathedral Peak Granodiorite, Sierra Nevada, California. *Geol. Soc. Spec. Publ.*, 168, 207–219.

Hildreth W. (1979). The Bishop Tuff: evidence for the origin of compositional zonation in silicic magma chambers. *Geol. Soc. Am. Spec. Pap.*, 180, 43–75.

Hogan J.P., Gilbert M.C. & Price J.D. (1998). Magma traps and driving pressure: consequences for pluton shape and emplacement in an extensional regime. *J. Struct. Geol.*, 20, 1155–1168.

Holtz F., Behrens H., Dingwell D.B. & Johannes W. (1995). Water solubility in haplogranitic melts: compositional, pressure and temperature dependence. *Am. Mineral.*, 80, 94–108.

Holtz F. & Johannes W. (1994). Maximum and minimum water contents of granitic melts: implications for chemical and physical properties of ascending magmas. *Lithos*, 32, 149–159.

Huppert H.E., Parks R.S., Whitehead J.A. & Hallworth M.A. (1986). Replenishment of magma chambers by light inputs. *J. Geophys. Res.*, 91, 6113–6122.

Hutton J. (1794). Observations on granite. *Trans. Roy. Soc. Edinb.*, III, 77–85.

Ildefonse B., Arbaret L. & Diot H. (1997). Rigid particles in simple shear flow: is their preferred orientation periodic or steady-state? In: Bouchez J.L., et al. (eds.) *Granite: from segregation of melt to emplacement fabrics*, Dordrecht: Kluwer Academic Publishers, pp. 177–188.

Ildefonse B. & Fernandez A. (1988). Influence of the concentration of rigid markers in a viscous medium on the production of preferred orientations: an experimental contribution: I. Non-coaxial strain. *Bull. Geol. Inst. Univ. Uppsala*, 14, 55–60.

Ishihara S. (1977). The magnetite-series and ilmenite-series granitic rocks. *Min. Geol.*, 27, 293–305.

Jackson M. (1991). The anisotropy of remanence: a brief review of mineralogical sources, physical origins, geological applications and comparison with susceptibility anisotropy. *Pure Appl. Geophys.*, 136, 1–28.

Jagoutz O., Schmidt M.W., Enggist A., Burg J.P., Hamid D. & Hussein S. (2013). TTG-type plutonic rocks formed in a modern arc batholith by hydrous fractionation in the lower arc crust. *Contrib. Mineral. Petrol.*, 166, 1099–1118.

Jahn B.M., Wu F., Capdevila R., Martineau F., Zhao Z. & Wang Y. (2001). Highly evolved juvenile granites with tetrad REE patterns: the Woduhe and Baerzhe granites from the Great Xing'an Mountains in NE China. *Lithos*, 59, 171–198.

Jahns R.H. & Burnham C.W. (1969). Experimental studies of pegmatites genesis: I. A model for the derivation and the crystallisation of granitic pegmatites. *Econ. Geol.*, 64, 843–864.

Jégouzo P. (1980). The South Armorican Shear Zone. *J. Struct. Geol.*, 39–48.

Jelinek V. (1981). Characterization of the magnetic fabrics of rocks, *Tectonophysics*, 79, 63–67.

Johan Z. & Johan V. (2001). Les micas de la coupole granitique de Cinovec (Zinnwald), République tchèque: un nouvel aperçu sur la métallogenèse de l'étain et du tungstène. *C. R. Acad. Sci. Paris.*, 332, 307–313.

Johannes W. (1983). On the origin of layered migmatites. In: Atherton M.P. & Gribble C.D. (eds.) *Migmatites, melting and metamorphism*. Nantwich: Shiva, pp. 142–162.

Johannes W. & Holtz F. (1996). *Petrogenesis and experimental petrology of granitic rocks*. Berlin: Springer, 335 pp.

Johnson M.R.W. (2003). Insight into the nature of Indian crust underthrusting High Himalaya. *Terra Nova*, 15, 46–51.

Jover O., Rochette P., Lorand J.P., Maeder M. & Bouchez J.L. (1989). Magnetic mineralogy of some granites from the French Massif Central: origin of their low-field susceptibility. *Phys. Earth Planet. In.*, 55, 79–92.

Kemp A.I.S., Hawkesworth C.J., Foster G.L., Paterson B.A., Woodhead J.D., Hergt J.M., Gray C.M. & Whitehouse M.J. (2007). Magmatic and crustal differentiation history of granitic rocks from Hf–O isotopes in zircon. *Science*, 315, 980–983.

Kemp A.I.S., Wormald R.J., Whitehouse M.J. & Price R.C. (2005). Hf isotopes in zircon reveal contrasting sources and crystallization histories for alkaline to peralkaline granites of Temora, southeastern Australia. *Geology*, 33, 797–800.

Khan M.A., Stern R.J., Gribble R.F. & Windley B.F. (1997). Geochemical and isotopic constraints on subduction polarity, magma sources, and palaeogeography of the Kohistan intra-oceanic arc, northern Pakistan Himalaya. *J. Geol. Soc.*, 154, 935–946.

Kinnaird J.A., Kruger F.J. & Cawthorn F.G. (2004). Rb–Sr and Nd–Sm isotopes in fluorite related to the granites of the Bushveld complex. *South Afr. J. Geol.*, 107, 413–430.

Kinny P.D. & Maas R. (2003). Lu–Hf and Sm–Nd isotope systems in zircon. *Rev. Mineral. Geochem.*, 53, 327–341.

Kleemann G.J. & Twist D. (1989). The compositionally-zoned sheet-like granite pluton of the Bushveld complex: evidence bearing on the nature of A-type magmatism. *J. Petrol.*, 30, 1383–1414.

Konopelko D., Seltmann R., Biske G., Lepekhina E. & Sergeev S. (2009). Possible source dichotomy of contemporaneous post-collisional barren I-type versus tin-bearing A-type granites lying on opposite sides of the South Tien Shan suture. *Ore Geol. Rev.*, 35, 206–216.

Koyaguchi T. (1987). Magma mixing in a squeezed conduit. *Earth Planet. Sci. Lett.*, 84, 339–344.

Krogh T.E. (1973). A low-contamination method for hydrothermal decomposition of zircon and extraction of U and Pb for isotopic age determination. *Geochim. Cosmochim. Acta*, 48, 505–511.

Kuznir N.J. & Park R.G. (1986). Continental lithosphere strength: the critical role of lower crustal deformation. *Geol. Soc. Lond. Spec. Publ.*, 24, 79–93.

Laduron D. (1966). Sur les procédés de coloration sélective des feldspaths en lame mince. *Ann. Soc. Géol. Belgique*, 89, B281–B294.

Lagny P. (2008). Le gisement de fluorine de Vergnenoeg. In: *La fluorine*. Géochronique, Document no. 106.

Laporte D., Rapaille C. & Provost A. (1997). Wetting angles, equilibrium melt geometry, and the permeability threshold of partially molten crustal protoliths. In: Bouchez J.L., et al. (eds.) *Granite: from segregation of melt to emplacement fabrics*. Dordrecht: Kluwer Academic Publishers, pp. 31–54.

La Roche H. (de), Leterrier J., Grandclaude P. & Marchal M. (1980). A classification of volcanic and plutonic rocks using R1R2-diagram and major element analyses: its relationships with current nomenclature. *Chem. Geol.*, 29, 183–210.

Lasaga A.C. (1998). *Kinetic theory in the earth sciences*. Princeton University Press, 811 pp.

Launeau P. (1990). *Analyse numérique des images et orientations préférentielles de forme des agrégats polyphasés*. PhD Thesis, University of Toulouse, 180 pp.

Launeau P., Archanjo C.J., Picard D., Arbaret L. & Robin P.Y. (2010). Two- and three-dimensional shape fabric analysis by the intercept method in grey levels. *Tectonophysics*, 492, 230–239.

Launeau P. & Bouchez J.L. (1992). Mode et orientation préférentielle de forme des granites par analyse d'images numériques. *Bull. Soc. Géol. France*, 163, 721–732.

Launeau P. & Cruden A. (1998). Magmatic fabric acquisition mechanisms in a syenite: results of a combined anisotropy of magnetic susceptibility and image analysis study. *J. Geophys. Res.*, 103, 5067–5089.

Launeau P., Cruden A.R. & Bouchez J.L. (1994). Mineral recognition in digital images of rocks: a new approach using multichannel classification. *Can. Mineral.*, 32, 919–933.

Launeau P. & Robin P.Y.F. (1996). Fabric analysis using the intercept method. *Tectonophysics*, 267, 91–119.

Laznicka P. (1999). Quantitative relationships among giant deposits of metal. *Econ. Geol.*, 94, 455–473.

Le Fort P. (1975). Himalayas: the collided range. Present knowledge of the continental arc. *Am. J. Sci.*, 275A, 1–44.

Leloup P.H. & Kienast J.R. (1993). High temperature metamorphism in a major Tertiary ductile continental strike-slip shear zone: the Ailao Shan-Red River. *Earth Planet. Sci. Lett.*, 118, 213–234.

Leloup P.H., Ricard Y., Battaglia J. & Lacassin R. (1999). Shear heating in continental strike-slip shear zones: models and field examples. *Geophys. J. Int.*, 136, 19–40.

Lespinasse M. & Pêcher A. (1986). Microfracturing and regional stress field: a study of the preferred orientations of fluid-inclusion planes in a granite from the Massif Central, France. *J. Struct. Geol.*, 8, 169–180.

Loiselle M.C. & Wones D.R. (1979). Characteristics and origin of anorogenic granites. *Geol. Soc. Am. Abstr.*, 11, 468.

London D. (1986). The magmatic–hydrothermal transition in the Tanco rare-element pegmatite: evidence from fluid inclusions and phase equilibrium experiments. *Am. Mineral.*, 71, 376–395.

London D. (2008). Pegmatites. *Mineral. Assoc. Can. Spec. Publ.*, 10, 1–368.

London D. (2009). The origin of primary textures in granitic pegmatites. *Can. Mineral.*, 47, 697–724.

Louis L., Robion P. & David C. (2004). A single method for the inversion of anisotropic data sets with application to structural studies, *J. Struct. Geol.*, 26, 2065–2072.

Lowenstern J.B., Smith R.S. & Hill D.P. (2006). Monitoring super-volcanoes: geophysical and geochemical signals at Yellowstone and other large caldera systems. *Phil. Trans. Roy. Soc. A*, 364, 2055–2072.

Lowrie W. (2007). *Fundamentals of geophysics*. 2nd ed. Cambridge University Press.

Luth W.C., Jahns R.H. & Tuttle O.F. (1964). The granite system at pressures of 4 to 10 kb. *J. Geophys. Res.*, 69, 759–773.

Lyell C. (1830–1833). *Principles of geology*. London: John Murray.

MacDonald G.A. & Katsura T. (1964). Chemical composition of Hawaiian lavas. *J. Petrol.*, 5, 82–133.

Mamtani M.A., Piazolo S., Greiling R.O., Kontny A. & Hrouda F. (2011). Process of magnetic fabric development during granite deformation. *Earth Planet. Sci. Lett.*, 308, 77–89.

Manning D.A.C. (1981). The effect of fluorine on liquidus phase relationships in the system Qz–Ab–Or with excess water at 1 kb. *Contrib. Mineral. Petrol.*, 76, 206–215.

Manning D.A.C. (1998). Granites and associated igneous activities. In: Selwood E.B., et al. (eds.) *The geology of Cornwall*. University of Exeter Press, pp. 120–135.

Marignac C. & Cuney M. (1999). Ore deposits of the French Massif Central: insight into the metallogenesis of the Variscan collision belt. *Mineral. Deposita*, 34, 472–504.

Marsh B.D. (1988). Crystal size distribution (CSD) in rocks and the kinetics and dynamics of crystallization. I. Theory. *Contrib. Mineral. Petrol.*, 99, 277–291.

Marsh B.D. (1989). Magma chambers. *Ann. Rev. Earth Planet. Sci.*, 17, 439–474.

Marsh B.D. (1998). On the interpretation of crystal size distributions in magmatic systems *J. Petrol.*, 39, 553–599.

Marsh B.D. (2000). Magma chambers. In *Encyclopedia of volcanoes*. Academic Press.

Martin H. (1986). Effect of steeper Archaean geothermal gradient on geochemistry of subduction-zone magmas. *Geology*, 14, 753–756.

Martin H. (1987). Petrogenesis of Archaean trondhjemites, tonalites and granodiorites from eastern Finland: major and trace element geochemistry. *J. Petrol.*, 28, 921–953.

Martin H. (1993). The mechanism of petrogenesis of the Archaean continental crust: comparison with modern processes. *Lithos*, 30, 373–388.

Martin H. (1999). Adakitic magmas: modern analogues of Archaean granitoids. *Lithos*, 46, 411–429.

Martin R. & Bowden P. (1981). Peraluminous granites produced by rock–fluid interaction in the Ririwai non-orogenic ring complex, Nigeria: mineralogical evidence. *Can. Mineral.*, 19, 65–82.

Martin H. & Moyen J.F. (2002). Secular changes in TTG composition as markers of the progressive cooling of the Earth. *Geology*, 30, 319–322.

Martin H., Moyen J.F. & Rapp R. (2009). The sanukitoid series: magmatism at the Archaean–Proterozoic transition. *Trans. Earth Sci.*, 100, 15–33.

Martin H., Smithies R.H., Rapp R., Moyen J.F. & Champion D. (2005). An overview of adakite, tonalite–trondhjemite–granodiorite (TTG) and sanukitoid: relationships and some implications for crustal evolution. *Lithos*, 79, 1–24.

Mathur R., Ruiz J. & Munizaga F. (2000). Relationship between copper tonnage of Chilean base-metal porphyry deposits and Os isotope ratios. *Geology*, 28, 555–558.

McKenzie D.P. (1985). The generation and compaction of partially molten rocks. *J. Petrol.*, 25, 713–765.

McQuarrie N., Horton B.K., Zandt G., Beck S. & DeCelles P.G. (2005). Lithospheric evolution of the Andean fold-thrust belt, Bolivia, and the origin of the central Andean plateau. *Tectonophysics*, 399, 15–37.

Mehnert K.R. (1968). *Migmatites and the origin of granitic rocks*. Amsterdam: Elsevier, 393 pp.

Michel J., Baumgartner L., Putlitz B., Schaltegger U. & Ovtcharova M. (2008). Incremental growth of the Patagonian Torres del Paine laccolith over 90 k.yr. *Geology*, 36, 459–462.

Miller C.F. & Miller J.S. (2002). Contrasting stratified plutons exposed in tilt blocks, Eldorado Mountains, Colorado River Rift, NV, USA. *Lithos*, 61, 209–224.

Mintsa mi Nguema T., Trindade R.I.F., Bouchez J.L. & Launeau P. (2002). Selective thermal enhancement of magnetic fabrics from the Carnmenellis granite (British Cornwall). *Phys. Chem. Earth*, 27, 1281–1287.

Mollier B. (1984). *Le Granite de Brâme-Saint-Sylvestre-Saint-Goussaud: ses structures magmatiques et une étude de la distribution de l'uranium à l'échelle du grain*. PhD Thesis, University of Nantes, Nancy, CREGU, Mém. no5, 150 pp.

Mollier B. & Bouchez J.L. (1982). Magmatic structure of the Brame–St Sylvestre–St Goussaud (Limousin, Massif Central, France) granitic complex. *C.R. Acad. Sci. Paris*, 294, 1329–1334.

Monecke T., Dulski P. & Kempe U. (2007). Origin of convex tetrads in rare earth element patterns of hydrothermally altered siliceous igneous rocks from the Zinnwald Sn–W deposit, Germany. *Geochim. Cosmochim. Acta*, 71, 335–353.

Moyen J.F. (2011). The composite Archaean grey gneisses: petrological significance and evidence for a non-unique tectonic setting for Archaean crustal growth. *Lithos*, 123, 21–36.

Moyen J.F. & Martin H. (2012). Forty year of TTG research. *Lithos*, 148, 312–336.

Moyen J.F., Martin H. & Jayananda M. (2001). Multi-element geochemical modelling of crust-mantle interactions during late-Archaean crustal growth: the Closepet granite (south India). *Precambrian Res.*, 112, 87–105.

Moyen J.F., Nédélec A., Martin H. & Jayananda M. (2003). Syntectonic granite emplacement at different structural levels: the Closepet granite, south India. *J. Struct. Geol.*, 25, 611–631.

Nabelek P.I. (2002). Calc-silicate reactions and bedding-controlled isotopic exchange in the Notch Peak aureole, Utah: implications with differential fluid fluxes with metamorphic grade. *J. Metamoph. Geol.*, 20, 429–440.

Naney M.T. & Swanson S.E. (1980). The effect of Fe and Mg on crystallization in granitic systems. *Am. Mineral.*, 65, 639–653.

Nédélec A., Chevrel M.O., Moyen J.F., Ganne J. & Fabre S. (2012). TTGs in the making: natural evidence from Inyoni shear zone (Barberton, South Africa). *Lithos*, 153, 25–38.

Nédélec A., Minyem D. & Barbey P. (1993). High-P–high-T anatexis of Archean tonalitic gray gneisses: the Eseka migmatites, Cameroon. *Precambrian Res.*, 62, 191–205.

Nédélec A., Paquette J.L., Bouchez J.L., Olivier P. & Ralison B. (1994). Stratoid granites of Madagascar: structure and position in the Pan-African orogeny. *Geodinamica Acta*, 7, 48–56.

Nédélec A., Ralison B., Bouchez J.L. & Grégoire V. (2000). Structure and metamorphism of the granitic basement around Antananarivo: a key to the Pan-African history of central Madagascar and its Gondwana connections. *Tectonics*, 19, 997–1020.

Nédélec A., Stephens W.E. & Fallick A.E. (1995). The Pan-African stratoid granites of Madagascar: alkaline magmatism in a post-collisional extensional setting. *J. Petrol.*, 36, 1367–1391.

Nesen G. (1981). *Le modèle exogranite–endogranite à Stockscheider et la métallogénèse Sn–W*. PhD Thesis, University of Nancy, 360 pp.

Neumann E.R., Olsen K.H. & Baldridge W.S. (1995). The Oslo rift. *Dev. Geotectonics*, 25, 345–374.

Nicolas A. (1989). *Principes de tectonique*. Masson, 223 pp.

Nicolas A. (1992). Kinematics in magmatic rocks with special reference to gabbros. *J. Petrol.*, 33, 891–915.

Nielsen F.M., Campbell I.H., McCulloch M. & Wilson J.R. (1996). A strontium isotopic investigation of the Bjerkreim–Sokndal layered intrusion, southwest Norway. *J. Petrol.*, 37, 171–193.

Njanko T., Nédélec A. & Affaton P. (2006). Synkinematic high-K calc-alkaline plutons associated with the Pan-African Central Cameroon shear zone (W-Tibati area): petrology and geodynamic significance. *J. Afr. Earth Sci.*, 44, 494–510.

Norton D. & Knight J. (1977). Transport phenomena in hydrothermal systems: cooling plutons. *Am. J. Sci.*, 277, 937–981.

Ohmoto H. & Goldhaber M.B. (1997). Sulfur and carbon isotopes. In: Barnes H.L. (ed.) *Geochemistry of hydrothermal ore deposits*. 3rd ed. Wiley, pp. 517–611.

Olivier P., Saint-Blanquat M. (de), Gleizes G. & Leblanc D. (1997). Homogeneity of granite fabrics at the metre and dekametre scales. In: Bouchez J.L., et al. (eds.) *Granite: from segregation of melt to emplacement fabrics*. Dordrecht: Kluwer Academic Publishers, pp. 113–128.

Oreskes N. & Einaudi M.T. (1992). Origin of hydrothermal fluids at Olympic Dam: preliminary results from fluid inclusions and stable isotopes. *Econ. Geol.*, 87, 64–90.

Ouillon G., Castaing C. & Sornette D. (1996). Hierarchical geometry of faulting. *J. Geophys. Res.*, 101, 5477–5487.

Panozzo-Heilbronner R. (1992). The autocorrelation function: an image processing tool for fabric analysis. *Tectonophysics*, 212, 351–370.

Papanastassiou D.A. & Wasserburg G.J. (1969). Initial strontium isotopic abundances and the resolution of small time differences in the formation of planetary objects. *Earth Planet. Sci. Lett.*, 5, 361–376.

Pardo-Casas F. & Molnar P. (1987). Relative motion of the Nazca (Farallon) and South American plates since Late Cretaceous times. *Tectonics*, 6, 233–248.

Paterson S.R., Fowler T.K., Schmidt K.L., Yoshinobu A.S., Yuan E.S. & Miller R.B. (1998). Interpreting magmatic fabrics patterns in plutons. *Lithos*, 44, 53–82.

Patiño Douce A.E. (1995). Experimental generation of hybrid silicic melts by reaction of high-Al basalt with metamorphic rocks. *J. Geophys. Res.*, 100, 15623–15639.

Patiño Douce A.E. & Beard J.S. (1996). Effects of P, f(O2) and Mg/Fe ratio on dehydration melting of model metagreywackes. *J. Petrol.*, 37, 999–1024.

Pattison D.R.M. & Tracy R.J. (1991). Phase equilibria and thermobarometry of metapelites. In: Kerrick D.M. (ed.) *Contact metamorphism*. Mineralogical Society of America, pp. 105–206.

Pearce J.A., Harris N.B.W. & Tindle A.G. (1984). Trace element discrimination diagrams for the tectonic interpretation of granitic rocks. *J. Petrol.*, 25, 956–983.
Petford N. (1996). Dykes or diapirs? *Trans. Earth Sci.*, 87, 105–114.
Petford N., Lister J.R. & Kerr R.C. (1994). The ascent of felsic magmas in dykes. *Lithos*, 32, 161–168.
Petford N., Paterson B., McCaffrey K. & Pugliese S. (1996). Melt infiltration and advection in microdioritic enclaves. *Eur. J. Mineral.*, 8, 405–412.
Pichavant M. (1987). Effects of B and H_2O on liquidus phase-relations in the haplogranite system at 1 kb. *Am. Mineral.*, 72, 1056–1070.
Pili E., Ricard Y., Lardeaux J.M. & Sheppard S.M.F. (1997). Lithospheric shear zones and mantle–crust connections. *Tectonophysics*, 280, 15–29.
Pirajno F. (2000). *Ore deposits and mantle plumes*. Dordrecht: Kluwer Academic Publishers, 556 pp.
Pitcher W.S. (1993). *The nature and origin of granite*. London: Blackie, 322 pp.
Pitcher W.S., Atherton M.P., Cobbing E.J. & Beckinsale R.D. (1985). *Magmatism at a plate edge: the Peruvian Andes*. Glasgow: Blackie.
Pognante U. & Benna P. (1993). Metamorphic zonation, migmatization and leucogranites along the Everest transect of eastern Nepal and Tibet: record of an exhumation history. *Geol. Soc. Spec. Publ.*, 74, 323–340.
Poitrasson F., Duthou J.L. & Pin C. (1995). The relationship between petrology and Nd isotopes as evidence for contrasting anorogenic granites: examples of the Corsican province (SE France). *J. Petrol.*, 36, 1251–1274.
Pokrovski G.S., Borisova A.Y. & Harrichoury J.C. (2008). The effect of sulfur on vapor–liquid fractionation of metals in hydrothermal systems. *Earth Planet. Sci. Lett.*, 266, 345–362.
Pollard P.J., Andrews A.S. & Taylor R.G. (1991). Fluid inclusion and stable isotope evidence for interaction between granites and magmatic hydrothermal fluids during the formation of disseminated and pipe-style mineralisation at the Zaaiplaats tin mine. *Econ. Geol.*, 86, 121–141.
Prouteau G. & Scaillet B. (2003). Experimental constraints on the origin of the 1991 Pinatubo dacite. *J. Petrol.*, 44, 2203–2241.
Pupier E., Barbey P., Toplis M.J. & Bussy F. (2008). Igneous layering, fractional crystallization and growth of granitic plutons: the Dolbel batholith in SW Niger. *J. Petrol.*, 49, 1043–1068.
Ramberg H. (1981). *Gravity, deformation and the Earth crust*. London: Academic Press.
Rämö O.T. & Haapala I. (1996). Rapakivi granite magmatism: a global review with emphasis on petrogenesis. In: Demaiffe D. (ed.) *Petrology and geochemistry of magmatic suites of rocks in the continental and oceanic crusts*. ULB-MRAC, pp.177–200.
Razanatseheno M.O.M., Nédélec A., Rakotondrazafy M., Ralison B. & Meert J. (2009). Four stage building of the Cambrian Carion pluton (Madagascar) revealed by rock magnetic properties. *Trans. Earth Sci.*, 100, 133–145.
Read H.H. (1948). Granites and granites. *Mem. Geol. Soc. Am.*, 28, 1–19.
Reeve J.S., Cross K.C., Smith R.N. & Oreskes N. (1990). The Olympic Dam copper–uranium–gold–silver deposit, South Australia. *Australas. Inst. Min. Metall. Mon.*, 14, 1009–1035.
Reutter K.J., Scheuber E. & Chong G. (1996). The Precordilleran fault system of Chuquicamata, Northern Chile: evidence for reversals along arc-parallel strike-slip faults. *Tectonophysics*, 259, 213–228.
Rickwood P.C. (1989). Boundary lines within petrologic diagrams which use oxides of major and minor elements. *Lithos*, 22, 247–263.
Robie R.A. & Hemingway B.S. (1984). Entropies of kyanite, andalusite and sillimanite: additional constraints on the pressure and temperature of the Al_2SiO_5 triple point. *Am. Mineral.*, 69, 298–306.

Robin P.Y.F. (1979). Theory of metamorphic segregation and related processes. *Geochim. Cosmochim. Acta*, 43, 1587–1600.
Robin P.Y.F. (2002). Determination of fabric and strain ellipsoids from measured sectional ellipses: theory. *J. Struct. Geol.*, 24, 531–544.
Rochette P. (1987). Magnetic susceptibility of the rock matrix related to magnetic fabric studies. *J. Struct. Geol.*, 9, 1015–1020.
Rochette P., Jackson M. & Aubourg C. (1992). Rock magnetism and the interpretation of anisotropy of magnetic susceptibility. *Rev. Geophys.*, 30, 209–226.
Roedder E. (1984). Fluid inclusions. *Rev. Mineral. Min. Soc. Am.*, 12.
Roig J.Y., Faure M. & Truffert C. (1998). Folding and granite emplacement inferred from structural, strain, TEM and gravimetric analyses: the case study of the Tulle antiform, SW French Massif Central. *J. Struct. Geol.*, 20, 1169–1189.
Roscoe R. (1952). The viscosity of suspensions of rigid spheres. *Br. J. Appl. Phys.*, 3, 267–269.
Rushmer T. (1995). An experimental deformation study of partially molten amphibolite: application to low-melt fraction segregation. *J. Geophys. Res.*, 100, B8, 15681–15695.
Rutter E.H. & Neumann D.H.J. (1995). Experimental deformation of partially molten Westerly granite under fluid-absent conditions with applications for the extraction of granitic magmas. *J. Geophys. Res.*, 100, B8, 15697–15715.
Sadeghian M., Bouchez J.L., Nédélec A. Siqueira R. & Valizadeh M.V. (2005). The granite pluton of Zahedan (SE-Iran): a petrological and magnetic fabric study of a syntectonic sill emplaced in a transtensional setting. *J. Asian Earth Sci.*, 25, 301–325.
Saint-Blanquat M. (de) & Tikoff B. (1997). Development of magmatic to solid-state fabrics during syntectonic emplacement of the Mono Creek granite, Sierra Nevada batholith. In: Bouchez J.L., et al. (eds.) *Granite: from segregation of melt to emplacement fabrics*. Dordrecht: Kluwer Academic Publishers, pp. 231–252.
Saltykov, S.A. (1958). *Stereometric metallography*. 2nd ed. Moscow: Metallurgizdat.
Sawyer E.W. (1991). Disequilibrium melting and the rate of melt-residuum separation during migmatization of mafic rocks in the Grenville front, Quebec. *J. Petrol.*, 32, 701–738.
Sawyer E.W. (2001). Melt segregation in the continental crust: distribution and movement of melt in anatectic rocks. *J. Metamorph. Geol.*, 19, 291–309.
Scaillet S., Cheilletz A., Cuney M., Farrar E. & Archibald D.A. (1996a). Cooling pattern and mineralization history of the St Sylvestre and western Marche leucogranite pluton, French Massif Central: I. $^{39}Ar/^{40}Ar$ isotopic constraints. *Geochim. Cosmochim. Acta*, 60, 4653–4671.
Scaillet S., Cuney M., Le Carlier de Veslud C., Cheilletz A. & Royer J.J. (1996b). Cooling pattern and mineralization history of the St Sylvestre and western Marche leucogranite pluton, French Massif Central: II. Thermal modelling and implications for the mechanisms of uranium mineralization. *Geochim. Cosmochim. Acta*, 60, 23, 4673–4681.
Scaillet B., Holtz F. & Pichavant M. (1998). Phase equilibrium constraints on the viscosity of silicic magmas. 1. Volcanic–plutonic comparison. *J. Geophys. Res.*, 103B, 27257–27266.
Schmidt, M.W. (1992). Amphibole composition in tonalite as a function of pressure: an experimental calibration of the Al-in-hornblende barometer. *Contrib. Mineral. Petrol.*, 110, 304–310.
Schmidt, M.W. & Thompson A.B. (1996). Epidote in calcalkaline magmas: an experimental study of stability, phase relationships and the role of epidote in magmatic evolution. *Am. Mineral.*, 81, 462–474.
Schweitzer J.K., Hatton C.J. & De Waal S.A. (1997). Link between the granitic and volcanic rocks of the Bushveld complex, South Africa *J. Afr. Earth Sci.*, 24, 95–104.
Searle M.P., Metcalfe R.P., Rex A.J. & Norry M.J. (1993). Field relations, petrogenesis and emplacement of the Bhagirathi leucogranite, Garhwal Himalaya. *Geol. Soc. Lond. Spec. Publ.*, 74, 429–444.

Sederholm J.J. (1891). Über die finnländischen Rapakivigesteine. *Tschermaks Miner. Petrogr. Mitt.*, 12, 1–31.
Sederholm J.J. (1907). Om granit och gneiss. *Bull. Comm. Géol. Finlande*, 4, no. 23.
Shand S.J. (1943). *Eruptive rocks: their genesis, composition, classification, and their relations to ore-deposits.* New York: Wiley, 444 pp.
Shaw H.R. (1980). The fracture mechanism of magma transport from the mantle to the surface. In: Hargraves R.B. (ed.) *Physics of magmatic processes.* Princeton University Press, pp.201–264.
Shaw S.E. & Flood R.H. (1981). The New England Batholith, Eastern Australia: geochemical variations in space and time. *J. Geophys. Res.*, 86, 530–544.
Shepherd T., Rankin A.H. & Alderton D.H.M. (1985). *A practical guide to fluid inclusion studies.* Glasgow: Blackie and Son, 239 pp.
Sheppard S.M.F. (1977). The Cornubian batholith, SW England: D/H and $^{18}O/^{16}O$ studies of kaolin and other alteration minerals. *J. Geol. Soc.*, 133, 573–591.
Shimura T., Fraser G.L., Tsuchiya N. & Kagami H. (1998). Genesis of the migmatites of Breidvagnipa, East Antarctica. *Mem. Natl. Inst. Polar Res. Spec. Publ.*, 53, 109–136.
Shirey S.B. & Richardson S.H. (2011). Start of the Wilson cycle at 3 Ga shown by diamonds from subcontinental mantle. *Science*, 333, 434–436.
Sillitoe R.H. (2010). Porphyry copper systems. *Econ. Geol.*, 105, 3–41.
Simon A.C., Franck M.R., Pettke T., Candela P.A., Piccoli P.M. & Heinrich C.A. (2005). Gold partitioning in melt–vapor–brine systems. *Geochim. Cosmochim. Acta*, 69, 3321–3335.
Simon J.I. & Reid M.R. (2005). The pace of rhyolite differentiation and storage in an 'archetypical' silicic magma system, Long Valley, California. *Earth Planet. Sci. Lett.*, 235, 123–140.
Skjerlie K.P. & Johnston A.D. (1993). Fluid-absent melting behavior of an F-rich tonalite gneiss at mid-crustal pressures: implications for the generation of anorogenic granites. *J. Petrol.*, 34, 785–815.
Sorby H.C. (1858). On the microscopical structure of crystals. *Q. J. Geol. Soc. Lond.*, 14, 453–500.
Sowerby J.R. & Keppler H. (2002). The effect of fluorine, boron and excess sodium on the critical curve in the albite–H_2O system. *Contrib. Mineral. Petrol.*, 143, 32–37.
Spear F. (1993). *Metamorphic phase equilibria and pressure–temperature–time paths.* Mineral. Soc. Am. Monogr., 799 pp.
Spray J.G., Thompson L.M., Biren M.B. & O'Connell-Cooper C. (2010). The Manicouagan impact structure as a terrestrial analogue site for lunar and Martian planetary science. *Planet. Space Sci.*, 58, 538–551.
Stacey F.D. & Banerjee S.K. (1974). *The physical principles of rock magnetism.* Amsterdam: Elsevier, 195 pp.
Stephens W.E. (1992). Spatial, compositional and rheological constraints on the origin of zoning in the Criffell pluton, Scotland. *Trans. Earth Sci.*, 83, 191–199.
Stephens W.E. & Halliday A.N. (1979). Compositional variation in the Galloway plutons. In: Atherton M.P. & Tarney J. (eds.) *Origin of granite batholiths: geochemical evidence.* Nantwich: Shiva Publications.
Stephenson A. (1994). Distribution anisotropy: two simple models for magnetic lineation and foliation. *Phys. Earth Planet. Int.*, 82, 49–53.
Stern R.A., Hanson G.H. & Shirey S.B. (1989). Petrogenesis of mantle-derived, LILE-enriched Archean monzodiorites and trachyandesites (sanukitoids) in southwestern Superior Province. *Can. J. Earth Sci.*, 26, 1688–1712.
Stilling A., Cerny P. & Vanstone P.J. (2006). The Tanco pegmatite at Bernic Lake, Manitoba: zonal and bulk compositions and their petrogenetic significance. *Can. Mineral.*, 44, 599–623.

Stormer J.C. & Nicholls J. (1978). XLFRAC: a program for the interactive testing of magmatic differentiation models. *Comput. Geosci.*, 4, 143–159.

Streckeisen A. (1976). To each plutonic rock its proper name. *Earth Sci. Rev.*, 12, 1–33.

Streckeisen A. & Le Maître R.W. (1979). A chemical approximation to the modal QAPF classification of the igneous rocks. *Neues Jahrbuch Mineral.*, 136, 169–206.

Takagi T. (2004). Origin of magnetite- and ilmenite-series granitic rocks in the Japan arc. *Am. J. Sci.*, 304, 169–202.

Takahashi Y., Mikoshiba M.U., Takahashi Y., Kausar A.B., Khan T. & Kubo K. (2007). Geochemical modelling of the Chilas Complex in the Kohistan terrane, northern Pakistan. *J. Asian Earth Sci.*, 29, 336–349.

Talbot J.Y., Faure M., Chen Y. & Martelet G. (2005). Pull-apart emplacement of the Margeride granitic complex (French Massif Central): implications for the late evolution of the Variscan orogen. *J. Struct. Geol.*, 27, 1610–1629.

Taylor S.R. & McLellan S.M. (1985). *The continental crust: its composition and evolution.* Oxford: Blackwell, 312 pp.

Taylor H.P. Jr & Sheppard S.M.F. (1986). Igneous rocks. I. Processes of isotopic fractionation and isotope systematics. *Rev. Mineral.*, 16, 227–271.

Therriault A.M., Reimold W.U. & Reid A.M. (2004). Geochemistry and impact origin of the Vredefort granophyre. *South Afr. J. Geol.*, 100, 115–122.

Thomas R. & Davidson P. (2012). Water in granite and pegmatite-forming melts. *Ore Geol. Rev.*, 46, 32–46.

Thomas R., Webster J.D. & Heinrich W. (2000). Melt inclusions in pegmatite quartz: complete miscibility between silicate melts and hydrous fluids at low pressure. *Contrib. Mineral. Petrol.*, 139, 394–401.

Thomas R., Webster J.D., Rhede D., Seifert S., Rickers K., Förster H.J., Heinrich W. & Davidson P. (2006). The transition from peraluminous to peralkaline granitic melts: evidence from melt inclusions and accessory minerals. *Lithos*, 91, 137–149.

Thompson A.B. & Connolly J.A.D. (1995). Melting of the continental crust: some thermal and petrologic constraints on anatexis in continental collision zones. *J Geophys. Res.*, 100, 15, 565–579.

Thornton C.P. & Tuttle O.F. (1960). Chemistry of igneous rocks. I: Differentiation index. *Am. J. Sci.*, 258, 664–684.

Tizzani P., Battaglia M., Zeni G., Atzori S., Berardino P. & Lanari R. (2009). Uplift and magma intrusion at Long Valley caldera from InSAR and gravity measurements. *Geology*, 37, 63–66.

Touret J. (2001). Fluids in metamorphic rocks. *Lithos*, 55, 1–25.

Trémolet de Lacheisserie E. (du) (1999). *Magnétisme (Tome I) Fondements.* Grenoble Sciences, 496 pp.

Trindade R.I.F., Bouchez J.L., Bolle O., Nédélec A., Peschler A. & Poitrasson F. (2001). Secondary fabrics revealed by remanence anisotropy: methodological study and examples from plutonic rocks. *Geophys. J. Int.*, 147, 310–318.

Trindade R.I.F., Raposo M.I.B., Ernesto M. & Siqueira R. (1999). Magnetic susceptibility and partial anhysteretic remanence anisotropies in the magnetite-bearing granite pluton of Tourao, NE Brazil. *Tectonophysics*, 314, 443–468.

Tronnes R.G. & Brandon A.D. (1992). Mildly peraluminous high-silica granites in a continental rift: the Drammen and Finnemarka batholiths, Oslo rift, Norway. *Contrib. Mineral. Petrol.*, 109, 275–294.

Turner D.C. & Bowden P. (1979). The Ningi–Burra complex, Nigeria: dissected calderas and migrating magmatic centres. *J. Geol. Soc.*, 136, 105–119.

Tuttle O.F. & Bowen N.L. (1958). Origin of granite in the light of experimental studies in the system NaAlSi$_3$O$_8$–KAlSi$_3$O$_8$–SiO$_2$–H$_2$O. *Mem. Geol. Soc. Am.*, 74.

Underwood E.E. (1970). *Quantitative stereology*. Addison-Wesley Publishing, 274 pp.

van Breemen O., Hutchinson J. & Bowden P. (1975). Age and origin of the Nigerian Mesozoic granites: a Rb–Sr isotopic study. *Contrib. Mineral. Petrol.*, 50, 157–172.

van den Driessche J. & Brun J.P. (1989). Un modèle cinématique d'extension paléozoïque supérieur dans le sud du Massif Central. *C. R. Acad. Sci. Paris*, 309, 1607–1613.

Vanderhaeghe O. & Teyssier C. (2001). Partial melting and flow of orogens. *Tectonophysics*, 342, 451–472.

van Kranendonk M.J., Collins W.J., Hickman A. & Pawley M.J. (2004). Critical tests of vertical *vs.* horizontal tectonic models for the East Pilbara granite-greenstone terrane, Pilbara Craton, Western Australia. *Precambrian Res.*, 131, 173–211.

van Lichtervelde M. (2006). *Métallogénie du tantale: applications aux différents styles de minéralisations en tantale dans la pegmatite de Tanco, Manitoba, Canada*. PhD Thesis, University of Toulouse III.

Vegas N., Aranguren A. & Tubia J.M. (2001). Granites built by sheeting in a fault stepover (the Sanabria massifs, Variscan orogen, NW Spain). *Terra Nova*, 13, 180–187.

Vellutini P. (1977). *Le Magmatisme permien du nord-ouest de la Corse. Son extension en Méditerranée occidentale*. PhD Thesis, University of Marseille, 276 pp.

Vernon R.H. & Paterson S.R. (2008). How late are K-feldspar megacrysts in granites? *Lithos*, 104, 327–336.

Vidal P., Bernard-Griffiths J., Cocherie A., Le Fort P., Peucat J.J. & Sheppard S.M.F. (1984). Geochemical comparison between Himalayan and Hercynian leucogranites. *Phys. Earth Planet. Int.*, 35, 179–190.

Vielzeuf D. & Montel J.M. (1994). Partial melting of metagreywackes: fluid-absent experiments and phase relationships. *Contrib. Mineral. Petrol.*, 117, 375–393.

Vielzeuf D. & Schmidt M.W. (2001). Melting reactions in hydrous systems revisited: application to metapelites, metagreywackes and metabasalts. *Contrib. Mineral. Petrol.*, 141, 251–267.

Vigneresse J.L. (2005). The specific case of the Mid-Proterozoic rapakivi granites and associated suite within the context of the Columbia supercontinent. *Precambrian Res.*, 137, 1–34.

Vigneresse J.L., Barbey P. & Cuney M. (1996). Rheological transitions during partial melting and crystallization with application to felsic magma segregation and transfer. *J. Petrol.*, 37, 1579–1600.

Vigneresse J.L. & Burg J.P. (2000). Continuous vs. discontinuous melt segregation in migmatites: insights from a cellular automat model. *Terra Nova*, 12, 188–192.

Vigneresse J.L. & Tickoff B. (1999). Strain partitioning during partial melting and crystallizing felsic magmas. *Tectonophysics*, 312, 117–132.

von Platen H. (1965). Experimental anatexis and the genesis of migmatites. In: Pitcher W.S. & Flinn G.W. (eds.) *Controls of metamorphism*. Edinburgh: Oliver and Boyd, pp. 203–218.

Wager L.R. & Deer W.A. (1939). Geological investigations in East Greenland: the petrology of the Skaergaard intrusion, Kangerdlugssuaq, East Greenland. *Meddelser Øm Gronland*, 105, 1–352.

Watson E.B. & Harrison T.M. (1983). Zircon saturation revisited: temperature and composition effects in a variety of crustal magma types. *Earth Planet. Sci. Lett.*, 64, 295–304.

Weber C. & Barbey P. (1986). The role of water, mixing process and metamorphic fabric in the genesis of the Baume migmatites (Ardèche, France). *Contrib. Mineral. Petrol.*, 92, 481–491.

Wedepohl K.H. (1991). Chemical composition and fractionation of the continental crust. *Geol. Rundsch.*, 80, 207–223.

Wedepohl K.H. (1995). The composition of the continental crust. *Geochim. Cosmochim. Acta*, 59, 1217–1232.
Weinberg R.F. & Podlachikov Y. (1994). Diapiric ascent of magmas through power-law crust and mantle. *J. Geophys. Res.*, 99, 9543–9559.
Weiss S. & Troll G. (1988). The Ballachulish igneous complex, Scotland: petrography, mineral chemistry and order of crystallization in the monzodiorite–quartz diorite suite and in the granite. *J. Petrol.*, 30, 1069–1115.
Westerman D.S., Dini A., Innocenti F. & Rocchi S. (2004). Rise and fall of a Christmas-tree laccolith complex, Elba Island, Italy. *Geol. Soc. Spec. Publ.*, 234, 195–213.
Whalen J.B., Currie K.L. & Chappell B.W. (1987). A-types granites: geochemical characteristics, discrimination and petrogenesis. *Contrib. Mineral. Petrol.*, 95, 407–419.
Whitney J.A. (1975). The effects of pressure, temperature and XH_2O on phase assemblage in four synthetic rock compositions. *J. Geol.*, 83, 1–31.
Whitney D.L., Tessier C. & Fayon A.K. (2004). Isothermal decompression, partial melting and exhumation of deep continental crust. *Geol. Soc. Spec. Publ.*, 227, 313–326.
Wiebe R.A. (1993). The Pleasant Bay layered gabbro-diorite, coastal Maine: ponding and crystallization of basaltic injections into a silicic magma chamber. *J. Petrol.*, 34, 461–489.
Wiebe R.A., Frey H. & Hawkins D.P. (2001). Basaltic pillow mounds in the Vinalhaven intrusion, Maine. *J. Volcanol. Geotherm. Res.*, 107, 171–184.
Wilde S.A., Valley J.W., Peck W.H. & Graham C.M. (2001). Evidence from detrital zircons for the existence of continental crust and oceans on the Earth 4.4 Gyr ago. *Nature*, 409, 175–178.
Winkler H.G.F. & von Platen H. (1961). Experimentelle Gesteinsmetamorphose: Bildung anatektischer Schmelzen aus ultrametamorphisierten Grauwacken. *Geochim. Cosmochim. Acta*, 24, 48–69.
Wolf M.B. & Wyllie P.J. (1995). Liquid segregation parameters from amphibolite dehydration melting experiments. *J. Geophys. Res. Solid Earth*, 100, 15611–15621.
Wolfram S. (1986). *Theory and applications of cellular automata*. Singapore: World Scientific, 560 pp.
Wyart J. & Sabatier G. (1959). Transformation des sédiments pélitiques à 800°C sous une pression d'eau de 1800 bars et granitisation. *Bull. Soc. Fr. Minéral. Cristallogr.*, 82, 201–210.
Wyllie P.J. (1984). Sources of granitoid magmas at convergent plate boundaries. *Phys. Earth Planet. Int.*, 35, 12–18.
Yoshinobu A.S., Fowler T.K., Paterson S.R., Llambias E., Tickyj H. & Sato A.M. (2003). A view from the roof: magmatic stoping in the shallow crust, Chita pluton, Argentina. *J. Struct. Geol.*, 25, 1037–1048.
Zeh A., Gerdes A. & Jackson M.B. Jr. (2009). Archean accretion and crustal evolution of the Kalahari craton: the zircon age and Hf isotope record of granitic rocks from Barberton/Swaziland to the Francistown arc. *J. Petrol.*, 50, 933–966.
Zen E. & Hammarstrom J.M. (1984). Magmatic epidote and its petrological significance. *Geology*, 12, 515–518.
Zieg M.J. & Marsh B.D. (2005). The Sudbury igneous complex: viscous emulsion differentiation of a superheated impact melt sheet. *Geol. Soc. Am. Bull.*, 117, 1427–1450.

Index

A
A (-type), 4, 5, 11, 37–39, 157, 230, 232, 248, 250, 271, 304, 306, 307
AAR, 201–204
adakite, 264, 265
Africa (West Africa), 231, 272
Alps, 8, 9, 246
AMCG, 271–275
Amiata (Monte, Italy), 93, 94
amphibole, 2, 7, 14, 21, 25, 44, 101, 137, 177, 195, 218, 236, 261–263
AMS, 190, 191, 194
andalusite, 7, 101, 102, 115, 118
Andes, 233, 238
annular (complex), 91, 231
anorogenic, 4, 91, 92, 232
Antilles, 235
aplite, 6, 155, 156, 300, 301
Archaean, 252, 256, 267, 268
Archimedes, 74, 77
Armorican (massif), 95, 243, 245
Armorican (shear zone), 95, 98, 245
assimilation, 150, 151
Australia, 32, 216, 256, 259, 269, 305, 306

B
Barberton (South Africa), 257, 262, 263
Bassiès (Pyrenees), 116, 210
batholith, 36, 84, 85, 234, 237, 238
biotite, 2, 21, 22, 137, 140, 162, 177, 188
Bouguer (anomaly), 102–104
Brazil, 204, 271
Brittany (France), 222, 245, 295
Bushveld (South Africa), 229, 230, 304, 305

C
Calabria (Italy), 122, 123, 248
calc-alkaline, 10, 25, 235, 239, 246, 250
California, 55, 135, 193, 208, 221
Canada, 86, 135, 255, 268, 279
Carion (Madagascar), 215, 216, 225
Cauterets (Pyrenees), 116, 192, 193
Central (massif, France), 103, 178, 243, 245, 295–297
charnockite, 22, 271, 273, 276
Chile, 264, 265, 301, 302
China, 125, 126, 234
Closepet (India), 84, 85, 98, 269
coltan, 300
collision, 238–243, 250
compatible (incompatible), 145–149
Concordia (Discordia), 252, 253
conduction (conductive heat transfer), 109–112, 218
contact (metamorphism), 114–113, 120, 121, 294
contamination, 68
convection (convective heat transfer), 113, 114
cooling, 128, 130, 253, 254
copper, 281, 283, 301–303, 306, 307
cordierite, 7, 22, 140, 241
core complex (migmatite dome), 100
Cornwall, 243, 293
Corsica, 39, 91, 92, 243, 248
CSD (crystal size distribution), 131–135
cumulate, 143, 149, 150, 263
Curie, 187, 188

D
Day (diagram), 189
diamagnetic (diamagnetism), 186–187
diapirism, 73–77, 257
differentiation, 141, 226
dislocation, 161, 162, 164, 165
dyke, 77, 81, 84, 85, 93, 110, 111, 117, 136, 274, 278

E
epidote, 7, 101, 140, 263
eutectic, 15–17, 20, 273
experimental melting, 16–22, 265

F
fabrics, 159, 169–171, 179, 185, 195–198
ferromagnetic (ferromagnetism), 187–191, 193, 196
filter-press, 52
Finland, 271, 274
Flamanville (France), 116, 117, 246
fluid (phase), 18, 153–155, 158, 284, 287
fluorine (F), 152, 230, 232, 284–287, 289, 291, 304, 307
fractional crystallization, 141–144, 146, 148, 150, 264, 268, 273

G
garnet, 7, 14, 22, 24, 25, 140, 241, 262, 263, 265
geotherm, 82, 122, 123, 125, 240, 247
granophyre, 6, 131, 229, 278
granulites, 10, 36, 123

H
Hadean, 255, 256
haplogranitic, 17, 19, 20

Index

Harker (diagram), 60, 63, 64, 142
Hercynian, 179, 222, 241, 243, 244, 248, 295
HFSE (high field strength elements), 5, 145, 307
Himalayas, 35, 55, 95, 118, 236
Hoggar (Algeria), 206, 207
hornblende, (see amphibole)
hot spot, 227, 250
hybrid, 58, 64, 69, 70–72
hydrothermal, 114, 156, 294, 295, 297, 299, 306, 307

I

I (-type), 4, 7, 11, 25, 37, 248, 250, 282, 283
ilmenite, 8, 25, 190, 191, 234
India, 85, 236, 258, 269
island arc, 235
Iran (Shah Kuh pluton), 290, 291
Italy, 63, 94, 248
isotopes (Hf), 32–34, 37, 38, 255, 272, 273, 275
isotopes (Nd), 28–31, 34–36, 39, 249, 275
isotopes (O), 27, 31–33, 35–38, 121, 222, 275, 305
isotopes (Sr), 26–28, 30, 31, 34–36, 231, 238, 249
Ivrea (zone, Italy), 8, 248

J

Japan, 191, 234

K

kaolin, 293–295, 297
Kerguelen (archipelago), 228
Kohistan (arc), 235–237

L

layering (layered), 14, 54, 76, 217, 218
leucogranite, 7, 34, 84, 95, 192, 242, 296
leucosome, 13, 14
LILE (large ion lithophile elements), 145
Limousin (France), 244, 298
liquidus, 16, 19, 20, 138–140, 157, 264

M

Madagascar, 38, 98–100, 157, 197, 215, 216
magnetic susceptibility, 185, 186, 190, 191, 193–195, 213
magnetite, 8, 25, 134, 188, 189, 191, 196–198, 304–306
Maine (USA), 118, 224, 225
Maladeta (Pyrenees), 115, 116
Manaslu (granite, Himalayas), 34, 35, 242
Margeride (granite, France), 243, 245
MCT (thrust), 35, 241, 242
migmatite, 12–15, 53–56, 84, 99, 126
mingling, 58, 59, 66, 68, 70–72
mixing, 58, 59, 61, 62, 64–66, 68, 70–72
Mohr (diagram), 53
Mont-Louis (pluton, Pyrenees), 116, 162

N

Neptunism, 12
New Mexico (USA), 154
Niger, 217, 218
Nigerian Jurassic granites, 216, 217
Norway, 249, 273, 276

O

Oslo (rift), 249
oxygen fugacity, 8, 24, 25, 282

P

P-T-t (pressure-temperature-time) path, 239–241, 247
paramagnetic, 186–188, 191, 192, 195
pegmatite, 6, 155, 156, 286, 287, 290, 300, 301
Peru, 93, 237, 238
plagiogranite, 227, 228
Ploumanac'h (pluton, France), 222, 225
Pyrenees, 115, 116, 121, 150, 162, 192, 195, 209, 210

Q

quadrants (diagram), 30, 35, 36

R

rapakivi, 6, 271–275
Rare earth elements (REE), 37, 38, 145, 228, 232, 260–262, 272, 284, 287, 289, 307
Rayleigh (number), 113
remanent magnetization, 188, 201, 202
Reynolds (number), 68–69, 77, 78
rheology, 114
root, 84, 85, 106

S

S (type), 4–8, 31, 32, 228, 231, 232, 241, 244–246, 280–282
sanukitoid, 268, 269
Sardinia (Italy), 63
saturation (water-), 18, 20, 153, 155
Scotland, 131, 225
Searchlight (pluton, Nevada), 212, 214
series (magmatic series), 10, 11, 235, 250
shear, 46, 55, 96, 97, 126, 167, 172–176, 245
SHRIMP, 253
Sidobre (pluton, France), 103, 105, 106, 177, 181–183
Sierra Nevada (USA), 36, 59, 133, 135, 191, 196
sill, 98, 105, 123, 244, 247, 279
skarn, 119, 121
solidus, 18–21, 23, 138–140, 157, 263
Spain, 96, 97, 106
shape fabric, 171, 173–175, 182, 184
Streckeisen classification, 2, 273
stress, 45, 47, 52
subduction, 232–234, 250, 264, 265
Sudbury (Canada), 276–279
syenite, 3, 11, 38, 98, 99, 134, 197, 231, 249

T

texture, 6, 131, 137, 271
thermal (numerical) modelling, 110–112, 114, 240, 247
tin (Sn), 232, 281, 285, 287, 289–293, 304

tonalite, 2, 3, 5, 11, 64, 123, 235, 237, 243, 244, 256, 258, 264
Toro (pluton, Nigeria), 64, 212, 213
TTG, 256–271
Turkey, 228
trondhjemite, 2, 256, 258, 267, 272
tourmaline, 7, 241, 293, 295, 300

U
uranium (U), 154, 282, 295, 298, 300, 305, 306
Utah (USA), 120, 121

V
Velay (dome, France), 54, 76, 77, 248
viscosity, 40–42, 59–62, 77, 148
Vosges (massif, France), 176, 243

W
wavelets, 181–183
wetting, 41, 43

Y
Younger Granites (Nigeria), 216, 217

Z
zircon, 32–34, 252–256, 270, 275